New Arctic Cinemas

The publisher and the University of California Press Foundation gratefully acknowledge the generous support of the Eric Papenfuse and Catherine Lawrence Endowment Fund in Film and Media Studies.

New Arctic Cinemas

MEDIA SOVEREIGNTY AND
THE CLIMATE CRISIS

*Scott MacKenzie and
Anna Westerstahl Stenport*

UNIVERSITY OF CALIFORNIA PRESS

University of California Press
Oakland, California

Library of Congress Cataloging-in-Publication Data

Names: MacKenzie, Scott, author. | Stenport, Anna Westerståhl, author.
Title: New Arctic cinemas : media sovereignty and the climate crisis /
    Scott MacKenzie and Anna Westerstahl Stenport.
Description: Oakland, California : University of California Press, [2023] |
    Includes bibliographical references and index.
Identifiers: LCCN 2022028213 (print) | LCCN 2022028214 (ebook) |
    ISBN 9780520390546 (cloth) | ISBN 9780520390553 (paperback) |
    ISBN 9780520390560 (ebook)
Subjects: LCSH: Motion pictures—Arctic regions—21st century.
Classification: LCC PN1995.9.A6955 M33 2023 (print) | LCC PN1995.9.A6955
    (ebook) | DDC 791.430911/30905—dc23/eng/20220816
LC record available at https://lccn.loc.gov/2022028213
LC ebook record available at https://lccn.loc.gov/2022028214

32   31   30   29   28   27   26   25   24   23
10   9   8   7   6   5   4   3   2   1

CONTENTS

FIGURES

ACKNOWLEDGMENTS

Over the years as we researched and wrote this book, many people, groups, festivals, archives, museums, and funding agencies have provided support and encouragement, for which we will always be grateful. The Social Sciences and Research Council of Canada (Insight and Connection Grants), the US Department of Education Title VI program, and the European Union through the Jean Monnet Center of Excellence Program awarded funding to undertake research at archives and film festivals, meet and interview artists and filmmakers, organize conferences and screening programs, and present work-in-progress to an array of audiences. We also thank for funding and support the following: Queen's University (Funds for Scholarly Research and Creative Work and Professional Development; Department of Film and Media, Faculty of Arts and Science); University of Illinois (Conrad Humanities Professorial Fund); Georgia Institute of Technology (Ivan Allen College of Liberal Arts); and Rochester Institute of Technology (College of Liberal Arts).

At museums, archives, and research centers where we conducted onsite research and presented findings, much gratitude and appreciation go to those who made it possible by granting access and discussing their holdings and their importance. At the Danish Film Institute, Copenhagen: Birgit Granhøj, Lars-Martin Sørensen, and Thomas C. Christensen; at the Danish National Museum, Copenhagen: Martin Appelt and Anne Mette Jørgensen; at the National Audiovisual Institute (KAVI), Helsinki: Timo Kinnunen, Mari Kiiski, Tiina Junttila, and Tommi Partanen; at the Arctic Studies Center, National Museum of Natural History, Smithsonian Institution, Washington, DC: Bill Fitzhugh, Aron Cromwell, and Igor Krupnik; at the Royal Library, Stockholm: Anna-Karin Lundgren; at the Russian State Museum of the Arctic and Antarctic, Saint Petersburg: Mikhail Lamakin; at the Norwegian Film Archive, Oslo: Åse Meyer; at Library and Archives Canada, Ottawa: Caroline Forcier-Holloway; at the Swedish Film Institute, Stockholm: Ola Törjas and Ingela Utterström; at the Queen's University Archive (Arnait Video Productions fonds), Kingston: Heather Home; at the American Museum of Natural History, New York City: Greg Raml; at the Scott Polar Institute, University of Cambridge: Richard Powell; at the Arctic Research Center (CER-ARCTIC), Universitat Autónoma de Barcelona: Eduard Ariza; at the Byrd Polar and Climate Research Center, Ohio State University, Columbus: Laura J. Kissel; at the Nord-Europa Institut, Humboldt University, Berlin: Stefanie von Schnurbein; at Groenlandica, National Library of Greenland, Nuuk: Charlotte D. Andersen; and at the National Science Foundation's Polar Geospatial Center, University of Minnesota, Minneapolis: Paul Morin.

Many Indigenous and Arctic filmmakers and programmers and their collaborators generously shared their time and entrusted us with their work. We offer sincere thanks to Ivalo Frank, Lene Borch Hansen, Inuk Silis Høegh, and Emile Hertling Péronard, Madeline Ivalu and Marie-Hélène Cousineau, Britt Kramvig, Anastasia Lapsui and Markku Lehmuskallio, Jorma Lehtola, Sardana Savvina, Greta Stocklassová, Elle-Máijá Tailfeathers, Anne Lajla Utsi, and Liselotte Wajstedt.

We were fortunate to be asked to curate programs at festivals, which allowed us to make new connections and bring little-known works to larger audiences. We'd especially like to thank Stefan Kindberg and Milbry Polk, the Explorers' Club, NYC, who for several years have invited us to program special screenings at the Polar Film Festival; Oksana Sarkisova and Eniko Gyuresko, Verzio International Human Rights Documentary Film Festival, Budapest, for asking us to program a series of Arctic works; Brenda Longfellow for inviting us to program Arctic documentaries at Visible Evidence XXII in Toronto; Barbara Scharres at the Gene Siskel Film Center in partnership with the Chicago European Union Film Festival, which allowed us to bring Amanda Kernell's *Sami Blood* to North America; and Beth Watkins at the Spurlock Museum in Urbana, Illinois, for inviting us to pursue a film program in conjunction with the exhibition *North of the Northern Lights: Exploring the Crocker Land Arctic Expedition 1913–1917*. Our research has been greatly enhanced by the opportunity to organize conferences and programs on Arctic research, including the "Arctic Stream" at the Society for the Advancement of Scandinavian Studies (2013–16); the conference Arctic Cinemas and the Documentary Ethos (University of

Illinois at Urbana-Champaign, August 2015); and the workshops Visualizing Climate Change: Connecting Arts, Media, Science, Technology, and Activism I (Georgia Institute of Technology, November 2017) and II (Queen's University, April 2018).

We are lucky to have had an amazing array of research assistants from the University of Illinois at Urbana-Champaign, the Georgia Institute of Technology, and Queen's University. Our heartfelt thanks to Caitie Annear, Jared Aronoff, Anu Babuji, Noelle Belanger, Victoria Chai, Eunji (Emily) Kim, Carina Magazzeni, Alejandra Pires, Deirdre Ruscitti Harshman, Daniel Simpson, and Garrett Traylor. We also thank Eric Feinberg and Richard Earles for their detailed and judicious copyediting efforts.

Many colleagues offered us unwavering support, friendship, and intellectual feedback and insight over the years, including Nancy Abelmann, Dag Avango, Henry Bacon, Karine Bertrand, Sean Cubitt, Chris Cuomo, Jan Anders Diesen, Klaus Dodds, Monika Kin Gagnon, Harald Gaski, Faye Ginsburg, Rebecca Hearne, Mette Hjort, Helene Hokland, Anna V. Hudson, Gunnar Iversen, Lilya Kaganovsky, Gary Kibbins, Lill-Ann Körber, Mariah Larsson, Brenda Longfellow, Susan Lord, Clarke Mackey, Janine Marchessault, Toby Miller, Andy K. Nestingen, Dorit Naaman, Ryan Randall, Eva Novrup Redvall, Michael Renov, Johannes Riquet, Linda Rugg, Mark Safstrom, Mark Sandberg, Scott Schnur, Meryl Shriver-Rice, Sverker Sörlin, Troy Storfjell, Kirsten Thisted, Pegi Vail, Hunter Vaughn, Jerry White, Nina Wormbs, and Gregory Zinman.

Thanks to the external reviewers of *New Arctic Cinemas,* who offered us invaluable feedback, and to the team at UC Press, particularly Raina Polivka, who has enthusiastically shepherded the project forward.

Finally, thanks to our families for their support, love, and encouragement over the years of this project.

ONE

Twenty-First-Century Arctic Cinemas and Global Media Studies, Media Sovereignty, the Anthropocene, and Interventionist Historiography

FOR CENTURIES, THE ARCTIC was visualized as unchanging, stable, and rigidly alien. These well-known visual tropes are both inadequate and inaccurate. This recognition does not obscure the enduring powers of long-standing Western, white, imperialist, patriarchal, and outsider views that sought to construct it as such, or the significance those views have had in situating the Arctic as both embedded within global contexts (when colonized, for instance) yet outside of twenty-first-century globalization (for example, prohibiting Inuit from commercial trades in seal skin). The impacts of the climate crisis, expanding resource extraction and massive infrastructure and industry projects that exemplify Anthropocene priorities, and Indigenous political mobilization are constituent parts of the global present. This book takes these complementary yet contradictory claims of the Arctic region's simultaneous connection to and absence from globalization discourse as its starting point. We articulate a new history of media representation and production in the circumpolar North, leveraging juxtapositions and comparisons to situate the Arctic in a context it rightfully deserves to occupy: as central to global environmental concerns and to a global media system of intertwined production contexts, circulation opportunities, and imaginaries.

The increasing reach to global audiences also demonstrates that films and media made in and about the Arctic are not a fringe category addressing marginal issues. Instead, Arctic cinemas engage the most challenging issues of our time, from the climate crisis to cultural, social, gendered, ethnic, racial, and social justice. These works also speak to the need for artistic and linguistic autonomy—the right to produce and consume media in one's own language being one of many aspects of Indigenous self-determination.

Recent films engaging implicitly and explicitly with the Arctic have garnered worldwide attention and span a wide range of genres, forms, and modes
of circulation. A few key examples illustrate this diversity. Set in Nunavut and
retelling a centuries-old legend, Zacharias Kunuk's *Atanarjuat: The Fast
Runner* (Canada, 2001) won the Caméra d'Or prize at Cannes and was effectively, for the first time since Robert Flaherty's ethnofiction *Nanook of the
North* (USA, 1922), a story of Arctic Indigenous life that reached a global film
audience. *Atanarjuat* remains the best-known work of art cinema from the
circumpolar North to date. Jennifer Lee's and Chris Buck's massively popular
Disney productions, *Frozen* (USA, 2013) and *Frozen II* (USA, 2019) (see
chapter 11), shed light on colonialism's negative impact on Sámi Indigenous
land and its fragile environment, with shapeshifting ice and the trauma of
ecological destruction as key themes. Other works by or about Arctic
Indigenous women activists address global Indigenous mobilization and transnational solidarity, such as Alethea Arnaquq-Baril's *Angry Inuk* (Canada,
2014), Jan van den Berg's *Silent Snow* (Netherlands, 2010), and Elle-Máijá
Tailfeathers's *Rebel* (*Bihttoš*, Canada/Norway, 2014). By contrast, Davis
Guggenheim's *An Inconvenient Truth* (USA, 2006) and Jeff Orlowski's
Chasing Ice (USA, 2013) feature didactic male talking heads addressing the
global repercussions of a diminishing polar ice cap and thinning glaciers,
expanding the reach of documentary to multiple audiences. This reach is also
an aspect of Roland Emmerich's *The Day after Tomorrow* (USA, 2004) and
Christopher Nolan's *Interstellar* (USA, 2014), which popularized cli-fi and
the climate crisis dystopia. Experimental documentaries—such as Bill
Morrison's remediated found-footage work, *Dawson City: Frozen Time*
(USA, 2016), about Yukon settler cultures and the excavation of nitrate film
from the Arctic tundra—have also gained audiences in both experimental
and documentary festivals. Andrey Zvyagintsev's bleak Russian marital social
critique drama, *Leviathan* (*Leviafan,* Russia, 2014), has broadened the representation of the Arctic in global cinemas. This diversity of work reflects the
global output of media products that take the Arctic—or metaphorical versions thereof—as their central concern. Because of this pattern of globalized
production and circulation, one requires an approach such as global media
studies to fully examine and comprehend the ramifications of these works.

A global media studies approach provides the foundational paradigm for
this book. Global media studies, in going beyond histories of (trans)national
relationships and methodologies of close reading, address how (1) infrastructure and technologies, (2) industries and institutions, and (3) cultures and

content shape both the production and societal implications of media in relation to and outside of the work itself (see Flew 2018). Similarly, circulation across, beyond, and within territorial borders, linguistic contexts, and political agendas cannot be disassociated from a work's significance. This is an important perspective for the study of twenty-first-century Arctic cinemas, since global media studies assume geographical complexities where local, regional, cross-border, and virtual connections coexist. Indeed, Arctic cinemas—and the significance they have played in advancing debates and policies on climate change, environmental protection, self-governance, linguistic sustainability, and social justice—must be examined within this expansive and interconnected framework, rather than as isolated and dispersed works, or as subsets of national film histories. One of the few areas of nonmainstream media production that is truly global is media associated with and produced in and about the Arctic, not least because of its linguistic and cultural diversity and the small and heterogeneous populations of the circumpolar North. For these reasons, moving images remain central for global knowledge and awareness of the region and its peoples.

Global media studies scholars emphasize the significance of technology—from the role of radio and television in the 1950s and 1960s, to the internet, mobile devices, and social media in the current moment—as inseparable from a normative modernity of linguistically unified and integrated nation-states. Toby Miller and Marwan Kraidy highlight the "migratory, linguistic, and politico-economic changes" that make it "imperative that we study the media in their global context, and slough off monolingual, disciplinary, parthenogenesis," because "the complexity of processes subsumed in the media makes linguistically, analytically, and geographically narrow approaches to the topic simply untenable" (Miller and Kraidy 2016, 22). Building on Miller and Kraidy, we note that global media and therefore Arctic cinema "disobeys frontiers again and again. And mass migration and the spread of ideas make extrapolating from English-language work almost facetious" (2016, 22). While the Arctic is often overlooked in global media studies, it is critical that the region—including a wide range of Indigenous multilingual and environmentally oriented production—receives due attention.

To this end, we turn to media theorist Marshall McLuhan's analysis of what he describes as the "global village" and trace the concept's intrinsic connections to the Arctic. McLuhan first develops the concept in *The Gutenberg Galaxy: The Making of Typographic Man* (1962), emphasizing that the term *global village* does not equate to homogenization; instead, by bringing a wide range of communities and societies into contact with each other, the global village creates greater diversity through the juxtaposition of ideas and beliefs. As McLuhan argued in a 1966 television roundtable about his work, "it doesn't necessarily mean harmony and peace and quiet, but it does mean huge involvement in everybody else's affairs. And so, the global village is as big as a planet and as small as the village post office" (McLuhan et al. 1966). What McLuhan proposes here is that the global village is cacophonous and dependent on specific connective mechanisms that relate what's geographically separate through communicative missives. There is a history to this.

McLuhan's communication theories were developed in tandem with his colleague Edmund Carpenter, an anthropologist whose work was based in the Canadian Arctic. Carpenter's first study, *Aivilik Eskimo: Time-Space Concepts* (1955), connected, like a letter from an isolated village, Inuit expression and worldviews with McLuhan's work. For instance, in their article "Acoustic Space," McLuhan and Carpenter argue: "To [Inuit], truth is given through oral tradition, mysticism, intuition, all cognition, not simply by observation and measurement of physical phenomena. To them, the ocularly visible apparition is not nearly as common as the purely auditory one; *hearer* would be a better term than *seer* for their holy men" (Carpenter and McLuhan 1960, 66). McLuhan and Carpenter demonstrate that the primacy of the visual is cultural, not biological. This has ramifications for the development of global media studies and the fact that the field has continued to downplay the aural—especially when expressed in Indigenous or "small" languages—in favor of a seemingly more transparently traveling visual. By contrast, the linguistic component of the audiovisual in many of the works we address equally privileges the *audio*—or, rather, the linguistic, the voiced, the uttered, and the nonspoken. Moreover, the concept of the global village is a multimedial form of intercultural and transcultural communication, involving the full sensorium, while also integrating into its very essence recognition of technology hubs—a remote post office, for example, in Nunavut—for connecting different parts of the world. For the works we examine herein, the aural and linguistic are often as important as the visual.

The connections between global media studies, the global village, and the Arctic intersect with other systems that redefine and reframe the circumpolar North and its moving images in the twenty-first century. In contrast to the concept of the global village, the field of "global change science" is not used extensively in global media studies, though it is contemporaneous with its development. Emerging during the 1980s as an outcome of the World Climate

Research Programme (1980), the International Geosphere-Biosphere Program (1987), and the United Nations' Brundtland Commission (1987), "global change science" focuses on large and interrelated planetary natural systems, such as the atmosphere and oceans, where the polar regions are especially critical (Robin et al. 2013). Dependent on massive computational processing and technological interconnectivity, *global change science* provides yet another connection point between global media technology, media circulation, and their impacts on the Arctic, since study of the polar regions—and especially glaciology—has been foundational for the emergence of the field. The Brundtland Commission's report, *Our Common Future,* is also resolutely global, emphasizing multilateral collaboration and the international community's role in addressing the concerns of global change scientists and promoting "sustainable development which meets the needs of current generations without compromising the ability of future generations to meet their own needs" (Brundtland Commission 1987). The connections between global media, the Arctic, sustainability, and environmental discourse are thus inextricable.

ARCTIC IMAGINARIES, GLOBAL CHANGE SCIENCE, AND THE GREAT ACCELERATION

The history of the emergence of the Anthropocene as a concept and the subsequent challenges to the determinism upon which it relies are well known (see Demos 2017; Haraway 2016; Yusoff 2018). Yet the Anthropocene and climate crisis aesthetics are closely intertwined with representations of the Arctic in twenty-first-century film and media cultures. The lone emaciated polar bear on a shrinking ice floe, the deep and monumental landscape scars caused by invasive mining infrastructures, and the melting Greenland ice cap are all well-known and well-worn climate crisis images. At the same time, the growth of lens-based and moving-image representations of the Arctic since the 1950s is a product of globalization, industrialization, and the movement north of settler cultures. With documentary and news media's voracious appetites for ethnographic images, a parallel development of Arctic visual representation continues to configure the global circumpolar North as empty and open for resource exploration and extraction, while strategically deemed inhabited if the nation-state defense needs of Cold War geopolitics so dictate.

The increased circulation of global media also relates to what is now known as the "Great Acceleration" (figure 1.1). Starting in the 1950s, the key indicators

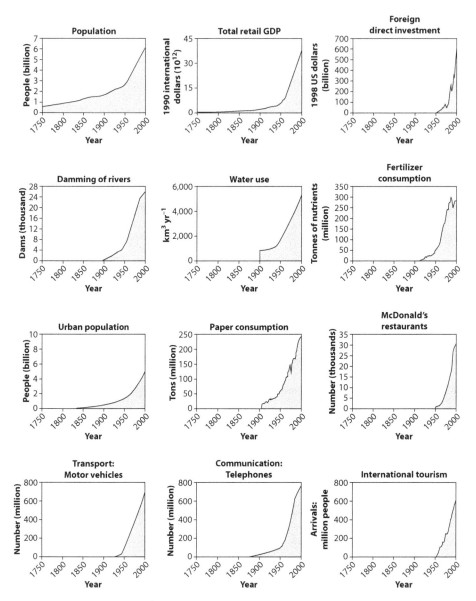

FIGURE 1.1. "The Great Acceleration, 1950–present." Reprinted by permission of Springer Nature from Will Steffen et al., "The Anthropocene Era: How Humans are Changing the Earth System," in *Global Change and the Earth System: A Planet under Pressure* (Berlin: Springer Verlag, 2004).

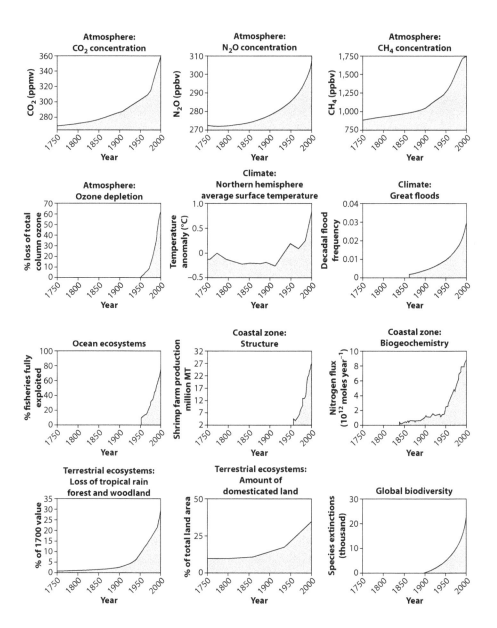

of the Anthropocene "we can measure and graph—carbon levels in the atmosphere, global temperatures, pollution of the air and seas, population, species loss, Gross Domestic Product, and even globalization itself ... could be graphed in the succeeding time period" (Robin et al. 2013, 9). The quantification of planetary data upon which the "Great Acceleration" depends, conveyed through communication and computational technology, is also mediated.

Without mediation, this narrative could not be told (see also McNeill 2016). As media historians Birgit Schneider and Thomas Nocke emphasize, "most phenomena studied in climate science are invisible. Climate change as a long-term process cannot be seen. It needs to be constructed on the basis of physics, chemistry, and big data: measurements, simulations, and statistics. Because climate is a scientifically constructed object, there is no way to learn about it other than through media devices. We *need* media to learn about climate change" (Schneider and Nocke 2014, 12). We emphasize that the effects of the climate crisis are not empirically visible in real time; moving images, especially ones about the Arctic that circulate globally, make them so (though this has recently changed somewhat with the incremental rise of forest fires, and how they are seen as more of a threat to Anglo-America than what happens in the distant Arctic).

Yet, for many of the filmmakers and directors whose work we discuss, the effects of the climate crisis and the Anthropocene are lived, conscious experiences and very much part of everyday existence. The kinds of films we examine demonstrate this in ways that scientific data cannot. Though the rapid climatological and environmental changes in the Arctic lend themselves to the visualization of melting ice, there is another aspect to note. As Nocke and Schneider argue, "we need to differentiate between questions of visualization and imagination. Some findings are so vast and huge—they contradict all experiences we have about the world today; in fact they are *unimaginable*. We can look at the IPCC [Intergovernmental Panel on Climate Change] curves as long as we want—the meaning of climate futures stays unimaginable although the colored lines follow a clear rationality" (2014, 13). In global media coverage, the Arctic has become the most important location for representing the climate crisis, at least partly because the accelerated melting of ice can be effectively represented by moving images. At the same time, film and media represent how melting ice impacts the world at large, as sea levels rise and the atmosphere warms (with extreme climatological effects, regardless of latitude), just as these changes have profound impact on local communities in the circumpolar North.

Cultural and ideological constructions of ice connect global media representations, global change science, and the Arctic. On the one hand, ice, for
many scientists, is understood as a repository of knowledge (see Frank and
Jakobsen 2019). Several millennia of climate data can be accessed through ice
cores drilled from the Antarctic and Greenlandic ice sheets, for instance. At
the same time, geophysicist Henry Pollack argues: "Ice asks no questions,
presents no arguments, reads no newspapers, listens to no debates. It is not
burdened by ideology and carries no political baggage as it crosses the threshold from solid to liquid. It just melts" (Pollack 2012, A26). The study of ice in
the climate crisis is a complex construction, filling the role, in part, of a passive
ethnographic, feminized, and infantilized subject to be studied by a polar
explorer or the Arctic anthropologist of yore, connecting global-scale issues of
colonialism and ecosystem destruction to inequitable politics. Ice melt impacts
Arctic populations directly in terms of increased opportunities for resource
extraction, shipping, and diminished possibilities for the maintaining of ice-
and snow-based Indigenous practices and sources of livelihood (such as winter
sealing and reindeer herding or transport over frozen ground and water).
Furthermore, what is referred to as the "Arctic paradox" is a key symptom of
the Anthropocene's effects: Arctic environments are among the most toxin-
polluted on earth, yet the most remote from the sources of rampant consumption that drive industrial and agricultural pollution, while the environmental
and social changes that follow in the wake of global warming impact Arctic
populations disproportionately (see Gabrielsen 2005; Palosaari 2019).

Though the Anthropocene is not synonymous with the climate crisis, early
on, global warming, exemplified by melting polar ice and rising sea levels,
became the most significant example of the Anthropocene, as Johan
Rockström and others have argued (Rockström et al. 2009; see also Robin et
al. 2013, 491–501). Moreover, the visualization of melting ice, disappearing as
if into thin air, became an unequivocal and media exemplification of the
impending crisis. Moving images about Arctic climate change and the
Anthropocene are not only aesthetic or cultural artifacts, but also images that
circulate through the discourses of different publics, including artists, scientists, media producers, popular audiences, and politicians on a global scale.

Many of the films we discuss reveal a fundamentally different relationship
to ice and its relevance to the climate crisis—a relationship further explored in
recent Indigenous studies of ice-integral cultures (see Cruikshank 2005;
Wright 2014; Callison 2015). The cinematic is a significant, and perhaps even
inseparable, part of the Anthropocene. Jennifer Fay proposes in her analysis

of the Anthropocene in European, Hollywood, and other Western cinemas that it is one of fabrication, spectacle, and scale. She argues that "filmmaking occasions the creation of artificial worlds, unnatural and inclement weather, and deadly environments produced as much for the sake of entertainment as for scientific study and military strategy" (Fay 2018, 4). While not explicitly addressing the representation of Arctic environments and populations, Fay argues that there is a "philosophical relationship between the histories, temporalities, and aesthetics of human-driven climate change and the politics, environmentalism, and ethics of cinema" (2018, 5; see also Kara 2016; Bloom 2022). Indeed, this tendency demonstrates how varying aesthetics of the Anthropocene have emerged, with the range and breadth of the conversations matching the complexity of the concept itself. Throughout *New Arctic Cinemas,* various film examples address these complementary and competing notions of the climate crisis, the Anthropocene, and the Arctic as globally mediated. Recent Indigenous works are especially significant in this context.

INDIGENOUS MEDIA SOVEREIGNTY

Within an Arctic global media framework, the great increase in Indigenous media production both counters the "othering" of film cameras in the hands of ethnographers and explorers throughout the twentieth century (see Rony 1996) and posits film and media as political and activist tools of self-determination and autonomy. Indigeneity emerged as a global solidarity movement from the 1970s onward, with Indigenous media proliferating from Nunavut to New Zealand and throughout the global Arctic. In recent decades, questions of Indigenous media practices within a global frame have come to the forefront. In response to this globalization, anthropologist Faye Ginsburg coined the term *media sovereignty* to address the burgeoning role of Indigenous media production, representation and circulation:

> As Indigenous media has grown more robust over the past two decades— in part because of the increasing convergence of media forms that blur the boundaries delineating television from film, web-based work or phone made media. The remarkably diverse array of works suggest that this emergence of media sovereignty—the synthesis of command over media technology with new and ongoing forms of collective self-production and the control over circulation—has much to offer Indigenous communities as they redefine their lives to themselves, the world and future generations. (Ginsburg 2016, 593)

Media sovereignty is akin to what Indigenous legal scholar Jessica Shadian affirms as the importance of the control and autonomy over ideas and knowledge for Indigenous peoples: while "indigenous groups often seek political control, it is not necessarily with secessionist aims. Rather, their goal is to attain political control over various aspects related to place, as opposed to simply seeking ownership of physical territory (or territorial integrity). These alternative political aims often relate to land and resource rights and control over indigenous knowledge and ideas" (Shadian 2015, 11). The UN's *Declaration of the Rights of Indigenous Peoples* (UNDRIP) closely links self-determination with media production and cultural representation. The declaration asserts that Indigenous peoples have "the right to maintain, protect and develop the past, present and future manifestations of their cultures, such as archaeological and historical sites, artefacts, designs, ceremonies, technologies and visual and performing arts and literature," and to "maintain, control, protect and develop their cultural heritage, traditional knowledge and traditional cultural expressions" (United Nations 2007, Art. 11, 31). To maintain and enhance these rights, media sovereignty is significant in that it encompasses access to production and circulation as well as the representation and stories of Indigenous peoples and cultures.

UNDRIP also calls attention to the significance of Indigenous languages and their central role in maintaining and promoting Indigenous rights. Film and media works of the global circumpolar Arctic, and the funding and production schemes that support their making and circulation, have increasingly emphasized language as a vehicle consciously mobilized for political purposes. While not the main focus of Ginsburg's notion of media sovereignty, and largely unaddressed in Michelle Raheja's well-known concept of "visual sovereignty" (Raheja 2007), the activist power of creating and producing narrative in one's own language should not be underestimated. That impetus goes far beyond preserving or even giving new life to languages labeled as "dying" or facing "extinction." Placing emphasis on Indigenous production in Indigenous languages counteracts the pervasive globalization of "world English" as the lingua franca, disregarding Indigenous languages as the political tools they are.

Indeed, the colonialist enterprise of settler-language Christianity, commerce, governmentality, and nomenclature removed Indigenous agency by abolishing language sovereignty. The agency embedded in Indigenous language used in many twenty-first-century Arctic films goes well beyond language and cultural revitalization to encompass the imagining of new vistas, opportunities, and potentials (see Cocq and Dubois 2019). Yet the language politics of media sovereignty are complex. As Julian Brave NoiseCat argues, the work of cinema in Indigenous languages fosters new communities of Indigenous artists "who have begun to share questions about form, audience, and access. Among these questions about the politics of applying intellectual-property regimes to indigenous art and cultures, the accuracy of labelling legends as nonfiction, the appropriateness of dubbing animated characters who are also spiritual beings, and the implications of nominating indigenous movies for foreign-language-film awards" (NoiseCat 2019). This complex process, then, is one of negotiation between the tenets of media art and its presuppositions about aesthetic categories, terminology, practices, and Indigenous knowledge. These priorities also reflect what filmmakers and activists such as Alethea Arnaquq-Baril and Zacharias Kunuk have expressed: when Indigenous peoples craft their own histories, they do so less in the form of the written and printed word, and more through oral storytelling, video, and performance (Judell 2002; Arnaquq-Baril 2018).

Global connectivity and the networks of exchange among Arctic Indigenous media producers have been foundational to the development of media sovereignty. This has engendered a shared notion of film and media works as among the most significant political tools for all aspects of self-determination priorities. Pamela Wilson and Michelle Stewart argue: "In recent years, Indigenous media—which we loosely define as forms of media expression conceptualized, produced, and/or created by Indigenous peoples across the globe—have emerged from geographically scattered, locally based production centers to become part of globally linked media networks with increased effectiveness and reach. Simultaneously, Indigenous media have begun to receive greater attention from scholars, critics, and global activists" (Wilson and Stewart 2008, 3).

Emphasizing connectivity, juxtaposition, and coalitional politics under-lies the observations articulated by Chadwick Allen: "My goal in staging purposeful Indigenous juxtapositions is to develop a version of Indigenous literary [and by extension, media] studies that locates itself firmly in the spe-cificity of the Indigenous local while remaining always cognizant of the complexity of the relevant Indigenous global. . . . What can we see or under-stand differently by juxtaposing distinct and diverse Indigenous texts, con-texts, and traditions?" (Allen 2012, xix). Digital media plays a key role in this development. For instance, in Nunavut, as Nancy Wachowich and Willow Scobie suggest, "through the act of uploading clips and inviting dialogue,

Inuit assert their presence in the world and forge new online and offline (transnational and local) social networks" (Wachowich and Scobie 2010, 83–84). In light of this development, we deploy the term *media sovereignty* to examine global Arctic Indigenous media. Yet, given the cultural, linguistic, political, and geographical diversity of the global circumpolar Arctic, we are also mindful of the central role played by "purposeful Indigenous juxtaposition"—echoing McLuhan's formulation of the global village, with which we engage throughout. We also note that the shared language for many (trans)Indigenous collaborations and settler-Indigenous collaborations is often English, a partial riposte to totalizing claims about globalization and "world English" being only a detrimental development.

INTERVENTIONIST HISTORIOGRAPHY:
A CRITICAL CINEMATIC PRACTICE

Several twenty-first-century Arctic Indigenous films—such as the anticolonialist historical epic *The Kautokeino Rebellion* (*Kautokeino-opprøret,* Nils Gaup, Norway, 2008); the family reunification and Truth and Reconciliation drama *Uvanga* (Marie-Hélène Cousineau and Madeline Piujuq Ivalu, Canada, 2013); the European art cinema, coming-of-age narrative *Sami Blood* (*Sameblod,* Amanda Kernell, Sweden, 2016); and the "Arctic Western" *Searchers* (*Maliglutit,* Zacharias Kunuk, Canada, 2016)—engage in actions of media sovereignty that recognize Indigenous realities as always already in contact with settler cultures. These films do so by mobilizing dominant cinematic forms while positioning Indigenous experiences in a relational manner to settler European cultures of colonialism. As a part of this emergent practice, these films provide a political and public intervention in relation to colonialist, imperialist, and nation-state histories.

The concept of interventionist historiography builds on Walter Benjamin's arguments in "Theses on the Philosophy of History" (Benjamin 1969) and Gayatri Spivak's development of the principles in "Subaltern Studies: Deconstructing Historiography" (1988). In recent years, the term *interventionist historiography* has been applied to postcolonial and Indigenous media works that destabilize realist, positivist narratives of progress, erupting these histories through the reclamation of silenced histories, and placing them in dialectical juxtaposition with their traces in contemporary culture and politics (see Huang 2015), along with new feminist film historiographies (see Cobb and Williams 2020). We argue for the significance of the term *interventionist historiography* as especially apt to examine the contributions of Arctic circumpolar media.

Indigenous women activists have played a key role in developing a politics of interventionist historiography in Arctic media and politics. For instance, Sheila Watt-Cloutier, then chair of the Inuit Circumpolar Council, filed a petition in 2005 to "The Inter-American Commission on Human Rights Seeking Relief from Violations Resulting from Global Warming Caused by Acts and Omissions of the United States" (Watt-Cloutier 2005). Though denied, this was one of the first Inuit human rights and climate justice actions. In her activism and writings, Watt-Cloutier works to reshape and reimagine Inuit agency and representation, deconstructing colonial paradigms to bring to light suppressed histories, such as her experience in residential schools. Her goal is to foster the possibility of mobilizing history to postulate a new future for Inuit with agency on the global stage: "We also need to remember that we Inuit have been fighting for our *right to be cold*. We have been trying to draw the world's attention to the devastation caused by human-produced or accelerated climate change. And we have been trying to educate the world that the vast majority of greenhouse gases that affect the Arctic are produced well outside it, by largely the United States and China" (Watt-Cloutier 2015, 294). Because of this disproportionate effect, "the right to be cold" needs to be mobilized as a means of global education and as an interventionist historiography of the contemporary climate crisis: "To be the most effective global citizens we can be, to be sentinels of climate change and the models of sustainable development, we Inuit still have much work to do to educate the world about our Arctic communities" (2015, 307). Watt-Cloutier's role in political mobilization operates on multiple levels.

Culture and media—Watt-Cloutier references films, music, printmaking, and many other forms of cultural production—are central to interventionist historiography as tools to reimagine Arctic diversity and circumpolar connections. Describing a global tour with Adrienne Clarkson (then governor general of Canada), Watt-Cloutier notes that the "emphasis was not on politics and business, but on culture and connection with our circumpolar neighbours. It was about meeting the people of the Arctic and acknowledging the Arctic as a homeland, one connecting many nations, not just as a frozen, resource-rich landscape. I know that when people learn more about the people of the Arctic they will be moved to work with us" (2015, 313). What Watt-Cloutier describes is a coalitional politics with the Canadian nation-state to build allyship. This example of allyship speaks to the manner in which interventionist historiography is not solely the providence of the disenfranchised; indeed, in the face of the climate crisis, both settlers and Indigenous peoples need to intervene and counteract dominant modes of history making and master narratives about the climate crisis.

Sámi Indigenous studies scholar Rauna Kuokkanen offers another frame in terms of interventionist historiography. Kuokkanen is critical of the ways in which self-determination is conceptualized and deployed in Sápmi. In her view, individual Sámi parliaments seek self-determination internally while paralleling the political models of Nordic nation-state parliaments, leading the Sámi to lose the "basic premise that we are one people," and "little, if any, thought has been given to Sámi models of organization" (Kuokkanen 2011, 51–52). Kuokkanen shows how the dominant narratives of self-determination replicate structures of Anglo-European models of governance. Yet, if Sámi self-governance mimics the Scandinavian model, gender equity does not. Within Sápmi, Kuokkanen notes that gender inequality remains a systemic issue, though not immediately apparent. She points to the long-standing myth of the strong Sámi woman as part of the "unease among many Sámi leaders and laypeople alike toward adopting mainstream Nordic gender equality policies. The resistance stems in part from a desire to distinguish the Sámi society from the dominant Nordic society" (Kuokkanen 2019, 150; see also Kuokkanen 2007). She also cautions against interpreting "the high representation of Sámi women in political and other Sámi institutions, [as if] gender equality has been achieved and hence, gender discrimination no longer exists" (2019, 149). Kuokkanen's overall project, then, is to establish an interventionist historiography that challenges normative, interconnected accounts of gender, state politics, and Indigenous self-determination. This is equally true about many of the Sámi and Nunavut film and media works discussed in subsequent chapters. Watt-Cloutier and Kuokkanen differ as to what they see as the relationship between the settler nation-state and Indigenous peoples, but both seek, through interventionist historiography, to destabilize dominant narratives of Indigenous life and rights.

Many of the book's chapters address recent work by women filmmakers and artists. There are a number of reasons for this. The first ties into the concept of interventionist historiography itself, as a critical and aesthetic practice to bring to light suppressed, marginalized, and/or neglected narratives, including those of women, to imagine new futures and new histories. Second, the Arctic works made by these practitioners are in no way limited to addressing women's stories about women's lives. Instead, following the principles of Indigenous knowledge and intersectionality, these works are a constitutive part of discourses of Indigeneity, climate change, geopolitics, resource extraction, and transnational politics of affiliation. Third, there has been a great expansion of works by Arctic Indigenous women and of Indigenous women as decision makers in film production in the twenty-first century, which parallels the rising levels of access women are attaining in media overall. Fourth, these films both build on and challenge the history of earlier Arctic Indigenous works made by men, functioning as an intervention to challenge emergent heteronormative narratives.

In the twenty-first century, Arctic works by both women and men, settler and Indigenous, have demonstrated a great formal range, from fiction features and activist documentaries to experimental film and expanded cinema. Moreover, paralleling the way in which Bill Nichols categorizes the hybridity of documentary modes and models (Nichols 1991, 1994), and subsequent challenges and revisions by Toni de Bromhead (1996, 22–24), these modes are often porous and interstitial, both in style and modes of address, bridging fiction, documentary, experimental media, and animation. These works break with the long history of visual representations of the Arctic (empty, pristine, rigid) and the dominant mode of representation (realism) and, in so doing, create new, hybrid, diverse, and polyvocal taxonomies of Arctic moving images.

ACTIVIST SCHOLARSHIP AND ALLY METHODOLOGIES

In *New Arctic Cinemas,* we engage with Indigenous works as forms of knowledge that are not, and cannot be, simply assimilated into paradigms of Western research. Avoiding hermeneutical approaches, and as settler scholars, we understand Indigenous knowledge, art, and cultural production as autonomous. We strive to dialogue with the works, at times placing them in juxtaposition, both to destabilize the settler colonial idea that Indigenous media is a homogeneous singularity and to use our position as Western scholars to amplify the generative dialogue between these works and the theoretical paradigms that have shaped our practice. Some of the directors we engage with highlight, specifically, how they themselves have been trained in those paradigms and are seeking to make works that bridge and supersede them.

Moreover, we are as concerned with modes of production, circulation, and counter-publics as we are with the works as stand-alone texts. Our discussions with Indigenous directors, artists, producers, and practitioners were oriented toward what kinds of generative dialogue within and between communities they thought our scholarship could offer to connect global Arctic Indigenous communities and global audiences. *New Arctic Cinemas,* for both ethical considerations and necessity, is only a partial account of the meanings and possibilities of global Arctic Indigenous media. Building on our own curatorial practices, developed in tandem with the expressed priorities of Indigenous artists and producers, we have selected works throughout the book that strive to tell stories, offer counter-histories, and engage in decolonization. Through juxtaposition, they offer a different history of the contemporary Arctic, outside of those found in dominant Anglo-American, Nordic, or Russian media, however politically committed those works may be. At the book's outset and conclusion, we engage with forms of mainstream media but decenter these works by placing them at the margins, and not at center stage where they are so often placed without any self-reflection.

The ethos of activist scholarship we have sought to pursue has followed many of the principles outlined by Linda Tuhiwai Smith in *Decolonizing Methodologies* (1999), engaging with communities of Indigenous and ally filmmakers at multiple points, who shared their artistic and professional practices. It has also meant organizing a number of screening programs to further a broadened public engagement with Arctic Indigenous cinemas, especially in contexts where such works might not otherwise have circulated, including at The Explorer's Club in New York City, at the European Union Film Festival in Chicago, at the Human Rights Film Festival in Budapest, at academic conferences such as Visible Evidence, or in conjunction with polar explorer exhibitions at museums such as the Spurlock in Urbana, Illinois. Similarly, our work has contributed to ensuring the longevity of vulnerable media, such as working with the Arnait collective's deposit at the Queen's University archive. Our research has also involved conversation and dialogue with filmmakers at their locus of practice, from FILM.GL in Nuuk to the International Sámi Film Institute in Guovdageaidnu (Kautokeino). These actions have been a deliberate component of the *New Arctic Cinemas* methodology: to amplify a diverse set of Arctic Indigenous voices; to learn in the context and communities where filmmakers, artists, and producers are professionally active; to support enhanced funding for Indigenous film production; and to share the work with broader audiences.

A unifying set of principles in *New Arctic Cinemas* align with the notion of "refusing research" in a traditional, Western, anthropological, ethnographic, or social science sense, as articulated by Mohawk scholar Audra Simpson, Unangax̂ scholar Eve Tuck, and ethnic studies scholar K. Wayne Yang. As Simpson argues, the concept of the "Indigenous" subject only emerges out of "colonial contact" and, as she notes, "in those moments, people left their own spaces of self-definition and *became* 'Indigenous'" (Simpson 2007, 69). In using the term *Indigenous media sovereignty,* and discussing principles of *Indigenous self-determination,* we call attention to a large number of examples of different kinds of film and media works, not aiming for totality, but rather to foreground the theoretical, political, and interventionist historiographic work of films and media that are, as Simpson calls them, "ethnographies of the familiar, ethnographies of refusal" (2007, 70). Similarly, we have sought to follow Tuck's and Yang's guiding principles to "resist the draw to traffic theories that cast communities as in need of salvation" (Tuck and Yang 2013, 245). This approach includes avoiding "damage-centered" modes of inquiry as "empirically substantiating the oppression and pain of Native communities . . . within a theory of change in which harm must be recorded or proven in order to convince an outside adjudicator that reparations are deserved." As they note, such approaches are "both colonial and flawed" (2013, 226–27). While of course it is true that some works of Arctic Indigenous cinema and media address systemic oppression and pain, the works considered in this book are not solely about this colonial history of violence and therefore the need for white saviors, as "outside adjudicators" (2013, 227), to engage in reparations.

Instead, the works from across the global Indigenous Arctic addressed in *New Arctic Cinemas,* even when considering pain and oppression, are also about many other things: a vibrant present, a counter-history, celebration, the inversion of dominant Western modes of expression, humor, and global forms of activism, with other Indigenous peoples and allies alike. Therefore, as activist ally scholars we are also interested in amplifying the theoretical components of the knowledge conveyed through the art forms addressed in the book. Thus, we follow Tuck and Wang in seeking to uncover how "literature and art as theory—especially decolonial literature and art—intervenes upon modes of theorizing . . . , setting limits to . . . research and also making those limits permeable to other forms of inquiry. The relationship between research and art can be one of epistemological respect and reciprocity rather than epistemological assimilation or colonization" (Tuck and Yang 2013,

237). As such, the majority of Indigenous and ally works addressed in *New Arctic Cinemas* are self-reflective and self-reflexive about the connections between aesthetic form, voice and language, political purpose, mode and method of production and circulation, and the community of practitioners involved in the work's arc of life, from creative inception to circulation and reception by a range of audiences—Indigenous, settler, and neither. This methodological choice is based in recognition of a multitude of knowledge formations and in our awareness that "theory" (in the Western academic sense) insufficiently conveys the richness of the interventionist historiography and media sovereignty the works themselves—and the practitioners involved in making them—constitute and advocate.

PRODUCTION AND CIRCULATION IN THE CIRCUMPOLAR NORTH

While mainstream features and documentaries about the Arctic are often produced and distributed by major production companies, the funding and production of Arctic Indigenous media takes on several forms. In Canada and the Nordic countries, various public funding schemes have emerged for both film and television production. For instance, since the 1940s, the National Film Board of Canada/Office national du film (NFB/ONF)—an arm's-length, state-run nonprofit—has produced films in the Canadian Arctic. These works range from ethnographic documentaries to recent Inuit Indigenous media. The NFB/ONF, from the perspective of Arctic cinemas, is a globally unique public initiative. Its catalog includes over two hundred films and videos documenting life in the Arctic; indeed, the NFB/ONF's interests in the Canadian Far North extend back to its inception in 1939. There are no other organizations within Arctic nation-states that have supported such a consistently diverse body of cinematic work. The NFB/ONF's mandate as a public entity is to represent all aspects of Canada and the lives of Canadians (though the exact wording of these goals has changed over time since the passing of the National Film Act in 1950). Currently, the mandate of the NFB/ONF

is to provide new perspectives on Canada and the world from Canadian points of view, perspectives that are not provided by anyone else and that serve Canadian and global audiences by an imaginative exploration of who we are and what we may be. We will do this by creating, distributing and engaging audiences with innovative and distinctive audiovisual works and immersive experiences that will find their place in classrooms, communities, and cinemas, and on all the platforms where audiences watch, exchange and network around creative content. (National Film Board of Canada 2014)

This mandate reflects how Canada imagines itself as operating as a "cultural mosaic," versus the American "melting pot" of cultural assimilation. Canada celebrates this aspect of its imagined identity, though many have challenged the egalitarianism of the concept, with John Porter (1965), for instance, arguing that the country is actually a "vertical mosaic," with British descendants at the upper echelon, and Inuit and First Nations peoples at the bottom. Nevertheless, the NFB/ONF does self-identify as having diverse constituencies and a global mission. Productions supported by the NFB/ONF in the Northwest Territories, the Yukon, Nunavut, and more recently Nunavik set the stage for the wave of independent Inuit film and video productions in twenty-first-century Canada, with Isuma (founded by Zacharias Kunuk, Norman Cohn, and Paul Apak Angilirq in 1990) and Arnait Video Productions (founded by Marie-Hélène Cousineau, Madeline Ivalu, Susan Avingaq, Carol Kunuk, and Atuat Akkitirq in 1991) being key examples. As such, the Canadian Arctic is presented, through the decentralization of the NFB/ONF studio system, as heterogeneous, challenging the often homogeneous image projected from outside the Arctic (see MacKenzie and Stenport 2019a). The NFB/ONF's Indigenous Action Plan furthers these goals through its response to the recommendations of Canada's Truth and Reconciliation Commission, including "supporting the imagineNATIVE Film + Media Arts Festival research initiative to develop proposed screen protocols for the Canadian industry" and "developing and adopting (in tandem with the imagineNATIVE process) internal NFB protocols for the production and distribution of documentary, animation and interactive/immersive works" (National Film Board of Canada 2017a, 9), along with achieving representational employment parity by 2025 (2017a, 8).

Similar undertakings have developed in Sámi media production, though outcomes of these initiatives are less well known internationally. Support and infrastructure for facilitating Nordic Indigenous film and TV production has increased during the past decade. The International Sámi Film Center (ISF)—now the International Sámi Film Institute (ISFI)—was launched in Norway in 2007 and has ambitiously fostered local filmmaking culture by coproducing shorts, features, TV series, and documentary films. Owned and operated by the municipality of Guovdageaidnu, a location rich in Sámi cultural heritage and home of one of the first Sámi theater ensembles, the ISFI is funded by the Norwegian government with contributions from the Sámi Parliament. While the primary aim of the ISFI is to encourage filmmaking in the Sámi languages for Sámi audiences, it also seeks to foster connections to Indigenous film production globally and has developed guidelines for the film industry when engaging with Sámi communities and cultures.

On its website, the ISFI also markets itself as located in a pristine and dramatic landscape, seemingly inviting "runaway" productions and location shooting. The establishment of the ISFI reflects Europe-wide developments in the promotion of regional film production and film funding centers in the interest of employment and tourism. The Finnish Film Foundation has, since 2012, allocated funding targeted for Sámi film production, supporting Skábma, the Indigenous Peoples' Film Center in Aanaar (Inari). This is a regional resource center for film and audiovisual production, operated in conjunction with the Sámi Parliament of Finland. The center's mandate includes furthering Sámi language and culture and providing ways for Sámi and other Indigenous peoples to actively participate in the Nordic and global film and media industry.

Television has also played a key role in production in these countries. In Canada, the Inuit Broadcasting Corporation (IBC) is an example of a local, culturally specific, and resistant form of broadcasting. IBC was formed in 1980 to counteract the pervasive images of the South coming to Nunavut (then part of the Northwest Territories) through satellite technology (see chapter 3). In a similar vein, the public broadcasting companies of Norway, Sweden, and Finland have developed and supported radio, television, and internet programming in different Sámi languages across Sápmi for many decades, NRK (Norwegian Broadcasting Corporation) Sápmi being the largest, with approximately one hundred employees.

Along with broadcast media, Indigenous populations throughout the Arctic have deployed a plethora of moving-image technologies to produce and circulate their own images of themselves. For both economic and aesthetic reasons, Super 8, home video technologies, consumer digital technologies, and smartphone video incorporate the technical limitations of these cameras into the aesthetic choices made by practitioners. Arctic Indigenous media, then, works both outside and inside of dominant modes of image-making— whether through state-funded initiatives, public broadcasting, DIY works, or digital media—to engage in forms of cultural counterprogramming.

In contrast to the relatively robust structures in place to support production and circulation of Indigenous works in Canada and the Nordic countries, media ecosystems look different in Alaska, Russia, and Greenland. We
address the example of Alaska here in some detail, before turning to Russia,
as we do not cover Alaska extensively elsewhere. Unlike Canada and the
Nordic countries, twenty-first-century locally produced feature films, especially fiction features, are few and far between in Alaska. Television, though,
has played a role. The native-owned KYUK-TV is a local Alaskan PBS affili-
ate, broadcasting since 1972. Native American–operated, the station serves
rural Alaska and Alaska Native populations in the Yukon-Kuskokwim Delta
region. Its stated mission is to "educate, stimulate and inform as well as provide cultural enrichment, entertainment, and opportunity for public access
and language maintenance for cultural survival" (KYUK n.d.). KYUK-TV's
mandate is to bridge communication between settlers and Indigenous inhabitants, creating a broadcast space for dialogue between the inhabitants of the
American Arctic, and to counterprogram against television and film emanat-
ing from the South (see also Fienup-Riordan 1995, 2015). In the Alaskan
context, where local television provided support for film shorts as well, the
emergence of new technology, and of lightweight and inexpensive video cameras in particular, became especially important as both a means of Indigenous
self-expression and a form of counterprogramming.

In contrast to Alaskan television, the Alaska Film Office was founded in
2008, and made part of the Alaska Department of Taxation in 2013, its main
function being to provide generous tax credits for runaway productions that
brought in revenue and used local crews. The program ended in 2015 because
of falling oil revenues. Alaska's only locally produced Indigenous fiction feature to date, Andrew Okpeaha MacLean's crime thriller *On the Ice* (USA,
2011), was made with a $1 million budget funded by Sundance and other
private entities in the continental United States, garnering no public support
from Alaskan sources. Made in Iñupiat and shot in Utqiaġvik (Barrow) with
local actors, *On the Ice* won awards at Berlin and many other festivals around
the world. As Joanna Hearne argues, *On the Ice* is a New Wave Indigenous
coming-of-age story within popular global Hollywood genres (the crime
story, the road movie) with a focus on articulating a "new and powerful youth
self-definition in the context of both Arctic specificity and the 'placelessness'
of the internet, for if sea ice was [once] the condition of isolation for northern
peoples, that isolation is now an artifact of the predigital age, and the film
reaches out to Indigenous youth by recognizing that this audience is as

immersed in mainstream global popular culture as an audience in New Jersey" (Hearne 2017, 187). The film's plot also resonates with how the production team pursued an international Kickstarter campaign to get *On the Ice* a limited theatrical release (Renninger 2011). This example demonstrates that Indigenous films made in and about the Arctic reach global audiences in widely divergent ways, and that festivals such as Sundance continue to play an important role in supporting Indigenous Arctic film. A related example is Chinonye Chukwu's *alaskaLand* (USA, 2012), about cross-cultural tension between an Alaska-raised teenager and his Nigerian-born parents, which was funded on a shoestring budget through support provided by the University of Alaska, and shot with student actors. The film thematizes Indigenous self-determination and the global African diaspora through the form of a high school road movie in which the distinctive Alaskan environment becomes part of the main character's identity exploration (Riquet and Stenport 2023). Yet, without public film funding schemes, Alaska's independent filmmakers, especially Indigenous ones, face a series of constraints.

While there is little to no state or arm's-length government funding for film production in Alaska, allyship works of interventionist historiography have been made, which have become significant activist documentaries for both Indigenous cultures and climate justice. *The Last Days of Shishmaref* (Jan Louter, The Netherlands, 2008) was made by a fly-in/fly-out documentarian, and independently from the Alaskan or US government or any Indigenous funding support. Yet, because the documentary gives voice exclusively to the experiences of the climate crisis by the six-hundred-member Indigenous community of Shishmaref, on Sarichef Island, it appears very much as the perspective of an insider familiar with the impact of rising sea levels and melting permafrost on local Arctic communities. Inhabited for millennia, Shishmaref is part of a rich historical cultural contact zone across the Bering Strait and up and down the coast, where Iñupiat subsistence practices have flourished. During the past fifteen years, Shishmaref has been consistently portrayed by global media as "the end of the earth," with both the environment and the community on the verge of "disappearance," as the sandbank island faces increased erosion from a changing climate, stronger storms, and loss of permafrost. By 2015, hundreds of journalists from major media outlets like the *New York Times,* CNN, and *Time* had visited Shishmaref to report about a "disappearing" community of climate refugees (Marino 2015, 9). For the global media, Shishmaref became a way to visualize, sensationalize, and spectacularize the community and environmental impact of climate change. The premise of *The Last Days of Shishmaref* is that life goes on, with the documentary's aesthetics, themes, and production contexts redressing the discourses of "disappearance" that are part of the history of Arctic visual cultures, as well as those based in more recent spectacularized and media-saturated climate change visualization. Without state or local funding, *The Last Days of Shishmaref* is an example of how an outsider perspective can allow local voices to be heard that are otherwise silenced, either through the lack of media production or through the elision of Indigenous voices in mainstream news reportage. With the production of short works in Alaska increasing over the past few years, and a growing attention to fostering local talent in Anchorage, there may be more independent films from Alaska on the horizon.

Russia poses its own set of problems in terms of the analysis of Arctic film production, especially Indigenous film production. Most films and television made in Russia obtain some kind of public funding, whether through broadcast organizations or through cultural and arts programming. Unlike in Canada and Scandinavia, these productions are not arm's-length; most are either deliberate propaganda for domestic audiences or vehicles of soft-power cultural diplomacy for international circulation, such as *Leviathan*. The role of the Russian state in these productions means that the works reflect—in implicit and explicit ways—the goals of the state and the master narrative it wants to tell of the Arctic region as central to the Russian national imaginary. We discuss these aspects in detail in chapters 9 and 10, contextualizing Arctic film and media culture in relation to Russia's extensive resource extraction in the Far North and the downplaying of climate change effects or self-determination movements on behalf of Russian Arctic Indigenous peoples. Perhaps not surprisingly, support for Indigenous production has been extremely limited. The war in Ukraine has also led to complications in terms of existing collaborations across Sápmi. Indeed, one of Russia's best-known Nenets directors, Anastasia Lapsui, left her home on the Yamal Peninsula for Finland early in her career; her internationally known Nenets-language productions have since been funded, produced, and marketed by Finnish organizations. The Yakutsk region and the regional film production center in Yakutia, affectionately called "Sakhawood," provides a counterexample that is beginning to influence film production by Indigenous practitioners in the Russian interior. Many of these films, such as the award-winning *The Lord Eagle* (*Toyon Kyyl*, Eduard Novkikov, Russia, 2018) and *Black Snow* (*Khara Khaar*, Stephan Burnashev, Russia, 2020), are made on a small budget and foreground a distinctive winter environment or the long days of summer, images that are often constitutive of Arctic cinema (see Haynes and Roache 2020).

Various film commissions in the northernmost regions of the Nordic countries take a different approach from the ISFI to filming in the Sápmi region, promoting the locations as a vast wilderness. In this way, they reinforce the stereotypes of a sublime landscape that many of the works we discuss in the book challenge, while also championing the use of local film crews for runaway productions. For instance, the Swedish Lapland Film Commission, founded in 2005, celebrates the "natural wonder" of the area as a means by which to promote location shooting. While many of the films it promotes are set in the Swedish Far North, others use the area for its landscape, often divorced from any cultural specificity. The Finnish Lapland Film Commission, founded in 2008, provides similar scouting services, as does the North Finnish Film Commission, whose tagline is "Tundra and taiga forests. Frozen harbours and the Sun that never sets. In the middle of nowhere but still close and well connected. Northern Finland offers unique settings for unique stories" (North Finland Film Commission n.d.). Similar marketing strategies underlie film commissions around the world, but most do not celebrate the fact that visually, they seem to be "in the middle of nowhere." The film commissions of northernmost Scandinavia, moreover, do not present themselves as connected to Indigenous cultures and film production, but rather, their web presence conveys a conventional Southern approach to the Arctic North, namely as a blank canvas onto which imagery of a depopulated and supra-locational wondrous sublime can be conjured.

Greenland offers a hybrid model of both local production support and promotion of location shooting in Greenland for runaway productions. The burgeoning film industry is still small. There are roughly fifty active practitioners, according to the nonprofit industry association FILM.GL, which supports developing talent, promoting production, strengthening capacity, and attracting runaway productions interested in Greenland's spectacular landscapes. Founded in 2012, three years after the act of self-government and increasing independence from Denmark, FILM.GL's hybrid model is reflected in the organization's desire "to professionalize the local film industry," expand access to financing, and "raise international awareness on Greenland as a film producing nation." Like the ISFI, FILM.GL is explicitly transnational and global in nature, establishing "joint talent schemes within the North Atlantic region" and "inviting film professionals to share with us their experiences of setting up film industries in small communities," while also promoting films made in the Greenlandic language (FILM.GL n.d.). FILM.GL, then, is as committed to building networks of production and circulation as it is to addressing questions of language conservation and representation. FILM.GL has lobbied actively for increased funding for film production from the government of Greenland (Naalakkersuisut), efforts that reflect both the global use of cinema and media as tools for postcolonial nation building and branding and the fact that Greenlanders cannot directly access Denmark's generous public support mechanisms for film production and therefore need their own infrastructural support. It is also concerned with marketing Greenland as a production site, to the extent of listing *Dr. Strangelove or: How I Learned to Stop Worrying and Love the Bomb* (Stanley Kubrick, USA, 1964) as a film shot on location at the Thule Air Base, tying Greenland to an international production site through film history.

Circulation and Publics

The growth and development of circulation practices in the global circumpolar North are not inseparable from public and private funding schemes that foster film production in the service of language preservation, cultural revitalization, and economic development. The numerous circulation practices constitutive of global circumpolar cinephilia respond to both a specific set of local needs and a transnational media context. The local needs are defined by few conventional exhibition facilities, small audiences scattered across geographically distanced communities, low bandwidth for cable or digital streaming, and limited public funding to support screenings. Attempts to reconcile these conundrums turn to contemporary digital global media distribution, such as Isuma Igloolik TV's digital distribution project Isuma.tv, launched in January 2008. As of 2022, the site states that it has more than seventy-eight hundred Indigenous videos in more than seventy languages (Isuma TV n.d.). Established in 2018, the ISFI's digital streaming repository Sápmifilm features more than eighty films. Both sites promote worldwide circulation of Indigenous film.

The growth of Arctic film festivals, with production networking, industry days, and development workshops attached to them, is especially significant, and several of these have promoted cross-Arctic Indigenous production. The

FIGURE 1.2. Ice projection at Skábmagovat, the Indigenous Peoples' Film Festival, Aanaar, Finland. Photo by Terhi Tuovinen.

Alaska International Film Festival was established in 2001 and has evolved to include support mechanisms for independent local film and media development. Tromsø International Film Festival, focusing on films made across the circumpolar North, was launched in 1991. Skábmagovat, the Indigenous Peoples' Film Festival in Aanaar, Finland, has been screening films outdoors in a snow theater on the darkest weekend of the year since 1998 (figure 1.2).

Skábmagovat is especially noteworthy for its decades-long commitment to sharing and supporting the growth of Arctic Indigenous film cultures. In Skábmagovat's 2014 festival program, artistic director Jorma Lehtola makes this priority explicit: "For the indigenous peoples of the Arctic, film has become an important tool both in strengthening the identity and communicating with other peoples. The circumstances of production and the resources available vary, but the field keeps expanding" (J. Lehtola 2014). Founded in 1999, Toronto's imagineNATIVE is the world's largest Indigenous film festival. It has played a central role over the past decade to advance Arctic Indigenous cinemas, with many of Arnait's and Isuma's works premiering there, and with focused programs on Sámi film in 2015 and Greenlandic film in 2016. As a programmer of global circumpolar Arctic moving images, imagineNATIVE has been the key organization in the development of global

Indigenous film and media cultures. This has not gone unnoticed, and Netflix is one of the main sponsors of imagineNATIVE since 2020 (Lead Industry Partner), as is Crave (Presenting Partner), a Canadian streaming platform—which demonstrates the increasing global reach and recognition of Indigenous filmmaking.

New festivals continue to arise, including Nuuk International Film Festival and the Golden Raven International Arctic Film Festival in Anadyr, Chukotka Autonomous Okrug, Russia, both of which started in 2017. Launched in 2016, the Arctic Open Festival, in Arkhangelsk, Russian Federation, whose motto is "To Make the Arctic Closer," brings together films from Arctic countries, with a focus on the Nordic countries, Indigeneity, and the Arctic ecosystem. International festivals such as Berlin, Cannes, CPH:DOX, Gothenburg, and Sundance have also foregrounded Arctic Indigenous film production over the past decade, just as major art museums, such as Louisiana in Denmark, the Royal Ontario Museum in Toronto, and the Venice Biennale, have curated Arctic media production, especially by Indigenous filmmakers. Often, individual works shot in the Arctic have been read uncritically as synecdoches, where one work, like *Nanook of the North,* stands in for the whole circumpolar Arctic. The increase in production and the global circulation of global Arctic moving images has brought to the forefront the heterogeneity of both the global region and the diversity of works made by Arctic Indigenous artists and their allies.

ARCTIC TWENTY-FIRST-CENTURY CINEMA AND THE GLOBAL CIRCUMPOLAR NORTH

This book offers an intervention into the old organizational paradigm of Arctic studies built on a North-South vector and determined by the ideologies and power structures of explorers, colonizers, and settlers. Instead, we propose the value of a comparative approach that builds on the foundations of global media studies, emphasizing interconnectivity and complexity, to argue for a heterogeneous circumpolar North that emerges through cinema and media. Yet, at first glance, the chapters may seem like they follow a nation-state or area studies model, because of their geographical orientation. One of the reasons for this structure is that cultural, linguistic, and political differences across the Arctic do have a foundation in the nation-state and colonial paradigms that shaped them. These continue to strongly influence production and circulation, which the principles of media sovereignty and interventionist historiography notably challenge. At the same time, it is our desire to not reinstantiate the nation-state homogeneity that the film and media works challenge. Therefore, the chapters are structured to speak to and through each other, foregrounding the hybrid and porous nature of twenty-first-century Arctic media production and representation.

Chapter 2 examines what we call "new Arctic explorers" and examines twenty-first-century ice imaginaries, beginning with the origins of IMAX arising out of multiscreen technologies to document the Arctic in a spectacularized fashion. It also considers *Chasing Ice* (Jeff Orlowski, USA, 2012), as a means through which the new explorer engages in metrical and masculinist practices to make visible ice melt. Chapter 3 begins in Nunavut, home to the Isuma and Arnait film collectives. We examine Isuma's recent films about the climate crisis and the impact of resource extraction, and works that respond to Canada's Truth and Reconciliation Commission (TRC) and the Qikiqtani Truth Commission (QTC), such as *One Day in the Life of Noah Piugattuk* (Zacharias Kunuk, Canada, 2019). The Cold War and its legacies also play a role as the impetus behind the relocation of Inuit to the very high polar North during the 1950s, and the traumatic effects of this governmental decision, as documented in *Martha of the North* (*Martha qui vient du froid*, Marquise Lepage, Canada, 2009). In chapter 4, we consider feature-length works by Arnait directors Madeline Ivalu and Marie-Hélène Cousineau that reflect the TRC, the QTC, and the National Inquiry into Missing and Murdered Indigenous Women and Girls (NIMMIWG), focusing on their fiction features *Uvanga* and *Restless River* (Canada, 2019) and the documentary *Sol* (codirected by Cousineau and Susan Avingaq, Canada, 2014). We also address the global significance of works such as Arnaquq-Baril's *Angry Inuk* (Canada, 2016) and the political and economic case it makes against the international ban on commercial seal hunting. In both Nunavut chapters, we delineate how principles and practices of media sovereignty and interventionist historiography are mobilized in the works of Isuma, Arnait, and Arnaquq-Baril, and their political significance in Nunavut and Nunavik.

Next, two chapters address recent filmmaking from Sápmi. First, chapter 5 provides the historical context for the rise of Sámi filmmaking since the Álttá (Alta) dam protests in the late 1970s, which laid the foundation for the environmental justice and self-determination tradition in Sápmi filmmaking. We examine the environmental justice and self-determination discourses in films by Nils Gaup and Paul-Anders Simma, as well as the coalitional and ally-based filmmaking by Swedish documentarian Stefan Jarl in his investigation of the Chernobyl nuclear accident and its effects on Sámi reindeer herding. We then examine media sovereignty, interventionist historiography, and visual anthropology in the documentary *Dreamland* (Rachel Gomez Andersen and Britt Kramvig, Norway, 2016) and in *Sami Blood,* the critically acclaimed art cinema work by Amanda Kernell. *Sami Blood* investigates how colonial and ethnographic practices coalign to an extent previously unexamined in Swedish cinema and culture. In chapter 6, we look at recent experimental documentary films by Sámi women, focusing on a range of short and feature-length works that use animation, remediation, and first-person narration to offer alternate perspectives on Sámi history, especially from women's perspectives. Works by Elle-Máijá Tailfeathers, Katja Gauriloff, and Liselotte Wajstedt are addressed in relation to transnational Indigenous politics in Canada and Sápmi, Sámi sovereignty debates and Indigenous feminism, the resettlement of the iron-ore mining town Giron, the impact of resource extractivism in Sápmi, and the history of forced relocation of the Skolt Sámi.

Chapters 7 and 8 examine the rise of filmmaking in Greenland since the Self-Government Act of 2009 and the work of the Greenlandic Reconciliation Commission (GRC, 2013–17). We trace the growth of a Greenland-based film industry, the significance of Greenlandic language politics for film production, and how Greenlandic films circulate domestically and internationally. Chapter 7 provides a history of Greenland films over the past decade, foregrounding their cinematic richness, from horror and bromance comedies to art cinema and experimental documentary. This diverse body of works includes those of experimental media artist Jessie Kleemann, who addresses the impact of the climate crisis of the melting Greenlandic ice sheet; and the marital melodrama *Anori* (Pipaluk Kreutzmann Jørgensen, Greenland, 2018), the first fiction feature film by a Greenlandic woman. In chapter 8, we turn our attention to two significant documentaries addressing the Danish colonial impact on Greenland and the work of the GRC. The first is the rockumentary *Sumé: The Sound of a Revolution* (*Sumé: Mumisitsinerup nipaa,* Inuk Silis Høegh, Greenland/Denmark/Norway, 2014), which combines 1970s found footage with interviews of Greenlanders about the role the rock group Sumé had in forming self-determination movements that led up to Home Rule in 1979. A different perspective is offered in *The Raven and the Seagull* (*Lykkelænder,* Lasse Lau, Denmark/Greenland, 2018), a Danish-Greenlandic allyship collaboration that satirizes Denmark's colonial treatment of Greenland. Both these works are clear representations of interventionist historiography.

In chapters 9 and 10, we turn to twenty-first-century Russia for an examination of the Arctic as a privileged national imaginary that builds on Cold War geopolitical and environmental paradigms. Chapter 9 examines the 2007 global media event in which Russian explorer Artur Chilingarov, via a submersible, planted a flag on the North Pole seabed. We discuss both state-sponsored media events (such as the Arctic Days festival) that advocate increased resource extraction, including on Indigenous lands, and documentary and art films that revisit Cold War and Stalinist legacies of a militarized, colonized, and strategic Russian north, such as Oleg Soldatenkov's *The Polar Prize* (*Poliarnyi priz,* Russia, 2014), Vladimir Tumaev's *White Moss* (*Belyy yagel,* Russia, 2014), and Aleksey Fedorchenko's *Angels of Revolution* (*Angely revolyutsii,* Russia, 2014). In chapter 10, we consider the Cold War legacy of security and environmental issues in the Russian Arctic, examining recent art cinema such as Alexei Popogrebsky's *How I Ended This Summer* (*Kak ya provyol etim letom,* Russia, 2010) and Zvyagintsev's *Leviathan.* We further address the film and media imaginaries of Svalbard, and especially the Russian coal-mining settlements depicted in a number of recent documentaries. We address how the archipelago is mobilized in internationally circulating film and media—as a paragon of global scientific and artistic collaboration for communicating the impact of the climate crisis—and the ongoing (and never truly buried) Cold War imaginaries that continue to shape Arctic cinema in the twenty-first century. In chapter 11, the book's conclusion, we discuss the state of Arctic filmmaking post-pandemic, identifying recurring trends and new directions with regard to Indigenous and interventionist historiography, popular cultures, global media cultures, and environmental and climate change discourses.

_____

TWO

New Arctic Explorers and Twenty-First-Century Ice Imaginaries

FROM METRICAL DOCUMENTARY TO IMAX SPECTACLE

DECREASES IN THE EXTENT, THICKNESS, and coverage of sea ice, and in the size, regrowth, coverage, and density of glaciers, have increasingly been at the forefront of global media representations. How this attention to melting ice shapes discourses of climate change in twenty-first-century Arctic documentary cinema is the focus of this chapter. In the context of Western modernity, there has always been an "anxiety of ice." If today there is too little, in the past there was too much. Polar explorer narratives and Arctic imagery from the late 1700s onward have consistently established ice as dangerous, treacherous, forbidding, and a problematic impediment to the realization of the Western world's ulterior interests (e.g., Craciun 2016). At the same time, discourse in glaciology has neglected epistemological questions about the discipline and how glaciology has ideologically constituted the study of ice. Specifically, as Mark Carey and colleagues note: "Given the prominent place of glaciers both within the social imaginary of climate change and in global environmental change research, a feminist approach has important present-day relevance for understanding the dynamic relationship between people and ice—what Nüsser and Baghel ... refer to as the cryoscape" (Carey, Jackson, and Rushing 2016, 771; see also Sörlin 2015). In rethinking the image of the glaciers and Arctic sea ice in climate change works, the films addressed in this chapter, which we label "new explorer films," attempt to counteract conventional climate science's reliance on quantification, instrumentalization, and abstractification in order to explore new epistemologies of ice. They stand in contrast to the activist Indigenous works we address throughout much of the rest of the book.

The explorer context is a dominant historical explanatory model in Arctic studies. A critical reframing of that model is long overdue. Yet it continues to motivate resource allocation for ongoing colonization of the Arctic. This historical model is reflected in the recurring tropes of Western visual history of Arctic representation. Icescapes are a privileged motif. Paintings, often of a massive scale, capture icebergs, wide glacial landscapes, or a white expansiveness of snow. This tradition is evident in seminal works such as Caspar David Friedrich's classic painting of an expedition ship crushed by ice and rocks, *Das Eismeer* (1823–24); a ship frozen in ice in American landscape painter Frederic Edwin Church's *Aurora Borealis* (1865; see Belanger and Stenport 2017); or photographic images from northern expeditions that position ships dwarfed by ice, as in the still photography by William Bradford in *The Arctic Regions: Illustrated with Photographs Taken on an Art Expedition to Greenland* (1873). These canonical images helped constitute an Arctic seemingly devoid of inhabitants and foregrounding adverse climates, difficult terrain, remoteness, and danger. In addition, some polar explorers who were themselves artists saw the explorer mythology as reductive and problematic. In 1895, Hungarian artist-explorer Julius von Payer implored the British Royal Geographic Society to launch a different kind of artistic expedition to relay the astonishing variety of landscapes, cultures, and light patterns of the Far North in sophisticated formal terms, to counteract the already established aesthetic register that portrayed the Arctic as an inhospitable icescape devoid of life and color (Payer 1895). These aesthetic traditions continue to influence Arctic representational practices as one of the most dominant Western approaches (Hill 2008; MacKenzie and Stenport 2013; Potter 2007).

Perhaps one of the earliest images to come from the Arctic through cinema is that of the rugged male Arctic explorer. Moving images—at first, slide shows or revolving images through panoramas, and by the early twentieth-century *actualités,* newsreels, and documentaries—have played a privileged role in bringing the results of polar and Arctic scientific expeditions to the South. Lecture tours, such as those by Fridtjof Nansen, Robert Peary, Roald Amundsen, Knud Rasmussen, and others, were generally accompanied by richly illustrated photographic slide shows and later, in the case of Rasmussen, films, in which photographic evidence, mounted for immersive purposes, was mobilized to gather support for new scientific expeditions to the North (Diesen 2015). Thus, the connection between explorers, science expeditions, and moving-image representation and exhibition were quickly established, as the emerging film medium spread around the world to bring back information about little-visited places. These images not only brought the Arctic to the world, but also functioned as a means of funding expeditions, including in films ranging from *The Voyage to the North Pole* (Robert W. Paul, UK, 1903), *From the North Cape to the North Pole* (Nordisk, Denmark, 1909), and *A Dash for the North Pole* (Charles Urban Co., UK, 1909; consisting of 1903–05 stock by Anthony Fiala from the Ziegler Polar Expedition), to later North Pole aviation films, such as *Roald Amundsen: Lincoln Ellsworth's Polar Flight 1925* (*Roald Amundsen: Lincoln Ellsworths Flyveekspedisjon 1925*, Paul Berge and Oskar Omdal, Norway, 1925). This kind of filmmaking quickly became well known to cinema audiences, as evidenced by how fast the Arctic exploration films were mocked, self-reflexively analyzed, and undercut. One sees this in fantastical renditions, such as Georges Méliès's *The Conquest of the North Pole* (*À la conquête du pôle,* France, 1912); *avant la lettre* mockumentaries produced to influence public opinion on who reached the actual North Pole first, such as *The Truth about the Pole* (Frederick A. Cook, USA, 1912); and comedic shorts built on the myths of exploration, such as *Lieutenant Pimple's Dash for the Pole* (Fred and Joe Evans, UK, 1914).

During the 1920s, Danish-Greenlandic ethnographer Knud Rasmussen, part Indigenous and an explorer, made multiple films about Greenland, Nunavut, and Alaska that serve as significant counterpoints to Flaherty's approach and the imaginaries those spawned. An embedded ethnographer born in Greenland and part-Inuit, educated in Denmark and supported by the country's National Museum, Rasmussen bridged Indigenous and Western cultures like few others, especially at that moment in time (see Thisted 2011; for an analysis of his cinematic practice, see MacKenzie and Stenport 2020, 2021). Writing a century ago, as the disciplines of anthropology and ethnography were being established, Rasmussen did not pursue a critical reflection on his own practice; indeed, his work has been considered biased and as lacking rigor by many later anthropologists (see Krupnik 2016, 1–30). Yet his interest in using film as a documentary tool continued to shape Danish depictions of Inuit cultures, which later Danish-Greenlandic artists such as Pia Arke forcefully mobilized to critique centuries of gendered and patriarchal representations of "scientific" and "anthropological" examinations of the Arctic, especially evident in video works such as *Arctic Hysteria* (Greenland, 1996) and Arke's 1995 treatise "Ethno-Aesthetics" (Arke 2012; see also MacKenzie and Stenport 2019b; and see chapter 7).

As André Bazin argues, writing a few years before McLuhan was formulating his theory of the global village, the documentary grows out of polar exploration films: "It was after World War I, that is to say in 1920, some ten years after it was filmed by Ponting during the heroic expedition of Scott to the South Pole, that *With Scott to the South Pole* revealed to the filmgoing public those polar landscapes which were to constitute the major success of a series of films of which Flaherty's *Nanook* is still the outstanding example" (Bazin 1967). Bazin goes on to note that "because of the success of the Arctic films," the explorer films exploded in popularity, though these often took on "exotic" qualities (1967, 154–55). Bazin identifies a shift after World War II, a move away from sensationalism to authenticity, which can also be understood as the positioning of the scientist-as-explorer at the heart of the documentary expedition film (see also Cahill and Caminati 2020).

Arctic documentary footage from the late 1950s reflects the fact that the vast northern circumpolar region was central in Cold War geopolitics, which includes science diplomacy (see chapter 10; see also Stenport 2015; Kinney 2013). This development also relates to the Great Acceleration and the growth of global change science, based on science diplomacy that bridged ideologies of both East and West. The Arctic Ocean, as a seemingly stable sheet of impermeable ice, was a key region of analysis and study. Heavily militarized, the Arctic Ocean and the land surrounding it furthermore became dotted with military installations (sometimes covert) that also served as scientific data collection points for meteorology, atmospheric science, astronomy, geology, and glaciology. These practices and technologies relied heavily on the emergent global media network, including the growth of satellite technology and fiber cable networks under the ocean.

As an emergent technological media practice, documentary film reflected these changes, as lightweight cameras with adequate sound uptake could more easily be brought on polar exploration journeys and in-home television sets provided expanded outlets for documentary footage, with the launch of the TV series *The Wonderful World of Disney* (then called *Disneyland*) in 1954. This alignment of documentary, ethnography, science, and geopolitical mobilization for ideological purposes is evident in several of Disney's Oscar-winning "Real-Life Adventures" from the late 1950s, such as *The Alaskan Eskimo* (James Algar, USA, 1953) and *Men against the Arctic* (Winston Hibler, USA, 1955), and in works such as *White Wilderness* (James Algar, USA/Canada, 1958), all of which relied on husband-and-wife explorer team

Lois and Herb Crisler to shoot the material, and all of which were shown on Disney's TV show. In these films, scientists continued to play a key part, though their roles transitioned, in Cold War cinema, from the figure of the intrepid polar explorer to being stationed in the Arctic North for empirical data collection as well as military purposes.

The environmental, science, and geopolitical context of the Arctic exploration documentary has a profound connection to the rise of a global media culture in the early 2000s. New explorer films can be metrical, self-reflexive, or spectacular in their documentary register—and, as one can see by the use of these terms, all are highly mediated. Explorers of yore brought cameras with them to document and empirically prove their explorations (and therefore follow realist documentary conventions about how to represent the world through documentary as transparent, empirical, and "real"). "New" explorers, by contrast, explore with a camera—using an on-screen presence, antirealist techniques (e.g., time-lapse photography), and a self-reflexive approach to documentary. Framed in the history of documentary, the "old" explorers are Griersonian in their approach; the "new" explorers are Vertovian. The new explorer film thereby often combines the figure of the environmentally oriented scientist and artist into a new expeditioner, who undertakes a journey to launch inquiries into climate change, bringing viewers to key sites where anthropogenic environmental effects can be visualized to their greatest impact through self-reflexive or spectacular means.

In addition, many of these new explorer films self-reflexively call attention to both the challenges of using cinematic technology and the lingering aesthetic registers of past cinematographic practices, though this self-reflexivity does not extend to the impact on the climate of travel to make these works. The Arctic is a key setting for new explorer documentaries, as the space remains highly mediated and accessible through moving images. With the rise of mass global travel since the 1980s, many distant and "exotic" places became easier to access. This does not generally pertain to the Arctic, as costs and infrastructure for traveling north remain prohibitive. Scientific expeditions continue to be one of the few ways to finance travel (though in recent years "cruise expeditions" have grown with the melting of ice in the Arctic, especially in the Northwest Passage and around Greenland). In addition, new explorer films set in the Arctic have maintained a privileged status, as film cameras venture where tourists cannot so easily traipse, leaving documentary as one of the central ways to "explore" the region in spectacular, virtual, and metrical fashions.

Virtual tourism of the Arctic through moving images has increased as ice has melted during the twenty-first century. The Arctic has lent itself to this use because of its historical representation as an otherworldly spectacle. One of the central aspects of this virtual spectacularization is IMAX—it can be seen as the art and technology of virtual new exploration. And the Arctic seems to have been made for this technology, especially for locations that few will experience in person. "One of the things that IMAX turned out to be rather good at," notes Graeme Ferguson, one of the technology's inventors, "is to make you feel you are in the picture. When the audience comes out of an IMAX theatre, a large number of people say, 'I felt as if I was there'. . . . That sense of verisimilitude comes from two things: one is the large screen, which engages your peripheral vision, and the other is the very high resolution" (Ferguson 2014, 145; for related arguments about IMAX and Bazin's "myth of total cinema," see Acland 1998, 431; Arthur 1996; Griffiths 2006).

IMAX promotes immersion through the assumption of a near 180-degree view and, very often, 3D, and lessens the possibility for a viewer to maintain an imagined external and impartial point of view. At a moment of proliferating "visual evidence" on multiple screens and multiple digital channels (on YouTube, other social media, streaming services, and beyond), the story IMAX tells about climate change retains an aura of authenticity—an IMAX film is a special occasion to immerse oneself in environments seemingly about to disappear. Our examination of IMAX therefore addresses several notions of the cinematic "new Arctic explorer" in an age of climate change. These include the scientist on screen as an avatar for the audience, the experience of audience members in the theater, and the cultural connotations of climate change and melting ice as "hyperobjects" (Morton 2013), whereby IMAX, because of its scale and spatial compression, becomes a privileged medium.

We concentrate on the traditional IMAX documentary travelogue, circulated mostly in museums, science parks, and other destination venues purpose-built for IMAX (we leave aside the increasing numbers, over recent decades, of suburban IMAX cineplexes and Hollywood IMAX feature films such as *Polar Express* [Robert Zemeckis, USA, 2004]). Venues such as these have a long tradition of exhibiting works relating to scientific inquiry and exploration, by emphasizing, among other things, environmental preservation and sustainability. These venues give the films authority and epistemological weight that they would not have were they exhibited in suburban cineplexes or, for that matter, on television or through social media and digital streaming services.

IMAX is an early example of another documentary and global media phenomenon specifically related to the Arctic. It harkens back to Graeme Ferguson's global, circumpolar multiscreen film *Polar Life/La vie polaire* (Canada, 1967)—produced for Montréal's 1967 world's fair, known as Expo 67, and included in the long-running exhibition that succeeded it, *Man and His World/Terre des hommes*—on which McLuhan worked as a consultant in the development of the technology (Marchessault 2005, 82). The film screened in the "Man the Explorer" Pavilion, which had four subsections: Man, His Planet and Space; Man and Life; Man and the Oceans; and Man and the Polar Regions, indicating the centrality of the Arctic to the Canadian Cold War imaginary, and the utopian dreams of bridging the divides between East and West that underpinned a key aspect of the ethos of Expo 67. Indeed, in *Polar Life/La vie polaire,* the binary between East and West in the popular imaginary was turned on its head through cinematic dialectical juxtaposition of the lives, wildlife, and climates of the global polar North and South. In addition, *Polar Life/La vie polaire* also implicitly addressed the emergent centrality of the Arctic in Canada's Cold War and national imaginaries, as a location to be conveyed as inhabited, protected, and "Canadian"—which dates back to forced Inuit relocation in the 1950s (see A. Marcus 1995)—in the case of any challenge to its autonomy from the USSR or the USA (see also chapter 3).

Polar Life/La vie polaire (figure 2.1) documents the Arctic and Antarctica, offering glimpses of the lives of the Inuit, the Sámi, and the northern inhabitants of Finland, Sweden, Greenland, Norway, Alaska, Canada, and Siberia. It was projected on eleven screens, with the audience seated on a rotating "turntable" that revealed three screens at a time (Ferguson 2014; MacKenzie 2014; Gagnon 2019). "Sublime" images of the Arctic as a space of ice and desolation are contrasted with images of the diverse array of flora in northern Sweden and Finland, challenging the notion that all the Arctic is a frozen, endless wasteland. In its use of multiple screens and atypical images, *Polar Life/La vie polaire* challenges the homogeneity of Arctic representation.

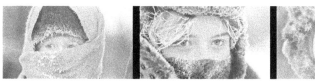

FIGURE 2.1. *Polar Life/La vie polaire* (Graeme Ferguson, Canada, 1967), tryptic, Expo 67, Montréal, April 27–October 29, 1967.

While relying on some of the dominant themes so often found in Arctic and Antarctic films—archival images of Scott, Rasmussen, and Amundsen appear early in the film as a triptych, for instance, along with found footage from Herbert Ponting's *The Great White Silence* (UK, 1924)—the film also conveys a diversity of images that portray the Arctic as a hybrid space with traditional practices alongside those of modernity, depicting both Arctic rock bands ("The Polar Bears") and go-go dancers, juxtaposed with Sámi and Inuit traditional dress. Using its three-screen structure (the number of screens visible to the audience at once), *Polar Life/La vie polaire* presented a diverse array of images that challenged traditional representations of the Arctic.

Moreover, *Polar Life/La vie polaire* foreshadows many tropes found in IMAX films over the past fifty years: aerial shots across vast landscapes; the audience as a new explorer of terrain as yet unseen by most of the world; and the depictions of both poles as a means to unite remote locations, accentuating a universal humanism central to emergent globalization theory and the philosophy behind world expositions. Such transnational, comparative, non-linear, multi-narrative approaches later became central to IMAX.

The Arctic at the World's Fair: A Global Space

Expo 67 was as interested in projecting scenarios of the future as it was in documenting world cultures of the present. Moreover, more so than any previous world's fair, it was fully enmeshed in and mediated by moving images, including the emergent "expanded cinema" (Youngblood 1970). *Polar Life/ La vie polaire* was a central component of this impetus, drawing heavily, as so many of the exhibitions at Expo 67 did, on McLuhan's concepts of the global village and the media being "an extension of ourselves" (McLuhan 1962, 1964). Expo 67, with its slogan "Man and His World," was, like many world's fairs, meant to signify the entire world in one location, including île

Notre-Dame, a human-made island in the St. Lawrence River. Where Expo 67 differed was in its use of moving images to "reach out" globally, with expanded cinema building connections to the wide array of locations from which this world came.

The Arctic, notably, was featured extensively, in disproportionate contrast to its total world population. Similarly, McLuhan's theory of the global village was a catalyst for the emergent global media studies paradigm; his concepts were greatly influenced by his colleague Edmund Carpenter's works on communication and Arctic anthropology (Carpenter and Flaherty 1959; Carpenter and McLuhan 1960). McLuhan offered a diagnostic about the ways in which communication technologies were transforming "remote" locations in the world. Importantly, McLuhan's concept did not argue that the global village led to cultural homogeneity; instead, he saw the global village that emerged from communication technology as one that juxtaposed cultural and sociological practices, bringing cultures together through a recognition of their profound, and at times incommensurable, differences. This ethos was at the heart of Expo 67, and, in terms of the film's radical juxtaposition of diverse and discordant Arctic images, the aesthetics and politics of *Polar Life/La vie polaire*.

Expo 67 put *Polar Life/La vie polaire*'s depiction of the Arctic and Antarctic, with its globalized new explorer ethos, at the center of what could best be called a postmodernist "non-place" (Augé 1995) on the periphery of Montréal, while at the center of the world's attention. The official souvenir guide to Expo 67 was uncannily prescient about the polar regions' centrality for twenty-first-century environmental media imagery: "Another display shows that if the ice of the Antarctic were to melt, seas would rise so much that the shape and size of other continents would change" (Milne 1967, 51). Concern about melting ice in the Arctic began to spread by the 1960s, as submarine measurements of sea-ice thickness were being gathered from the late 1950s onward (e.g., Rothrock et al. 1999). This is also a direct outcome of the increased military surveying of Arctic Ocean ice sheets, and of Cold War geopolitics, which emphasized the polar ice cap as a buffer between East and West. At the same time, science diplomacy among glaciologists meant that information about ice melting was shared between the superpowers, which is the context that underlies the concerns found in a world's fair guide about global interconnectedness.

In this context, it appears that Expo 67's designers understood moving images, in innovative and expanded cinematic forms, to be an appropriate way to convey concerns over environmental impact and societal changes in the global circumpolar regions. Using the world's fair as a site for conveying formal experimentation, *Polar Life/La vie polaire* both foregrounds and questions the status and authority of Arctic and polar explorers. Like many Expo 67 and IMAX films, *Polar Life/La vie polaire* is about different ways of seeing, and the microscopic images of ice projected on a large screen foreshadow much of the "world made visible" aesthetic of IMAX. Indeed, nature in *Polar Life/La vie polaire* is not only stereotypically "polar"; while there are images of ice and snow, there are also aesthetically rich and abstract images of what four-thousand-year-old ice looks like under a microscope.

Polar Life/La vie polaire was an early form of immersive and expanded cinema. Invented by Ferguson, NFB/ONF director Roman Kroitor, and Galt-Cambridge, Ontario, mayor Robert Kerr, IMAX arose out of discussions initiated by Ferguson after Expo 67. The subsequent development of IMAX in 1970 for the Osaka World's Fair in Japan was a direct result of *Polar Life/La vie polaire* and a desire to launch a commercialized version of expanded and immersive cinema (see Feldman 2014, 159–83). This move toward turning IMAX technology into a commercialized form also led to a shift in the hierarchy of the film crew. As Ferguson argues, the difference between multiscreen and IMAX is that with multiscreen projection, the large, single-channel image becomes predominant (MacKenzie 2014). By contrast, in *Polar Life/La vie polaire*'s eleven-projector and three-screen cinematic realm of montage and fragmentation, meaning is largely constructed collectively at the point of projection and with the editor serving a key role (on this transformation, see Youngblood 1970, 352–54). This aligns with Sergei Eisenstein's "montage of film attractions," a method he defines as "the comparison of subjects for thematic effect" (Eisenstein [1924] 1988, 48), though in the case of multiscreen, the images are juxtaposed concurrently and not sequentially.

In IMAX's myth of total immersion, the meaning is constructed by the cinematographer through special cameras, wherein footage is projected seamlessly onto a spherical surface. IMAX thus combines several key components of the new explorer in documentary cinema: bringing a film camera to remote locations to record and capture evidence, and to present the evidence to audiences where few can experience it firsthand. Notably, IMAX technology was originally developed for projection on a screen replicating the form of a globe, thereby incorporating key tenets of the globalization ethos promulgated by Expo 67's interest in "Man and His World" and "Man the Explorer." Given the expense, size, and difficulty of operating IMAX cameras, it is, as Charles

Acland notes, "not surprising to see the IMAX cameras so frequently in the films themselves; technological self-reflexivity relates to the very heart of its business" (1998, 432). The 70mm cameras are large, weigh 249 pounds, and are cumbersome to operate, as they need to be reloaded frequently—holding only four minutes of film—and are prone to malfunction. In that sense, IMAX cinematographers face many of the same challenges as did early polar film photographers, while their efforts are inseparable from and contribute to global media production and circulation.

THE EXPLORER AND THE CLIMATE CRISIS MELODRAMA: *TO THE ARCTIC* AND *WONDERS OF THE ARCTIC*

Two of the best-known twenty-first-century IMAX films mobilizing a new explorer figure while directly addressing anthropogenic climate and environmental change are *To the Arctic 3D* (Greg MacGillivray, USA, 2012) and *Wonders of the Arctic* (David Lickley, Canada, 2014). Made by IMAX veterans, *Wonders of the Arctic* and *To the Arctic 3D* mobilize a range of established IMAX narrative strategies and cinematographic techniques that rely heavily on a new explorer trope. Both films address melting ice and foreground scientists—Western and white—who travel north to be immersed in Arctic environments for research purposes. The scientists are portrayed through their interactions with large photogenic mammals who live on or near ice (notably polar bears and walruses). *To the Arctic* includes scant reference to Indigenous populations, whereas *Wonders of the Arctic* features several Inuit informants who reflect with authority on the implications of climate change and melting ice on environments, societies, and cultures of the North.

Reinforcing their status as "scientific," these Arctic IMAX films screen in museums, science centers, and event venues around the world, selling millions of tickets in the process. When screened in these contexts, viewers are invited to occupy the position of the observing and recording scientists; indeed, the viewer's subject position is given extra authority because of the context of the screening. The films invite the audience to occupy the role of a new explorer, traveling to the Arctic to observe the effects of climate change seemingly firsthand. Because these films play to multiple audiences (both children and adults) for educational and entertainment purposes, their episodic and nonlinear strategies rely on both scientific discourse and emotionally cathartic calls to ethnographic melodrama. This tension in modes of address is part of a long-standing tradition of Arctic cinematic depiction, reconfigured in the new explorer trope to relay anxieties of ice melt.

Scientists are generally configured in both *To the Arctic* and *Wonders of the Arctic* as new explorers, well equipped with top-notch photographic and camera technology to visually capture the effects of climate change. The films' representation of these figures is far from what Paul N. Edwards diagnoses as the vilification of climatologists by global warming deniers during the first decade of the twenty-first century—imagining these scientists as sequestered in labs, crunching numbers, building abstract computer models to be rendered on screens, supposedly far away from empirically observable evidence. Edwards investigates the complex epistemology of climate change, arguing that a simplistic opposition between empiricism and conceptual framework is moot "because *without models, there are no data,*" just as "the models we use to project the future of climate are *not* pure theories, ungrounded in observation. Instead, they are filled with data—data that bind the models to measurable realities" (Edwards 2010, xiii).

In recent IMAX films about climate change set in the Arctic, scientists are active, garner empirical evidence, are hands-on, and are fully immersed in the study of climate change and its effect on the polar ice cap or Greenland's ice sheet. Scientists in *To the Arctic* and *Wonders of the Arctic,* moreover, are portrayed uncritically as positive new explorer role models, in that they are conversant in contemporary theories of climate change, yet are present to observe and garner empirical evidence and convey their findings to the public at large, and especially to school-age children. In that sense, these IMAX films are configured to speak to both camps in the climate change debate by using the figure of the scientist on the ice as a noncontroversial collector of evidence as well as a storyteller.

The Pop Culture of IMAX Arctic Exploration

Recourse is often made to stardom, anthropocentrism, and melodramatic narratives of various kinds in IMAX films. *To the Arctic,* for instance, is narrated by Meryl Streep and features an acoustic soundtrack specially performed by Paul McCartney alongside an alternative take of the Beatles' "Because" (1969) from their archival CD *Anthology 3* (1996). These strategies of mobilizing media stars serve to convey effects of a warming Arctic in a spectacularized manner through an anthropomorphizing cloak. The lives of polar bears play out as a maternal melodrama, framed by Streep's voiceover, as the mother polar bear seeks to feed her cubs and defend them against attacks by hungry, full-grown, male polar bears. This implicitly ties their story of survival to the quintessential Arctic explorer myth—Sir John Franklin's lost expedition in search of the Northwest Passage, during which the crew eventually succumbed to cannibalism.

While many IMAX documentaries employ a Griersonian voice-of-God commentary, mobilized as an objective and impartial authority, *To the Arctic*'s new explorer pathos is one negotiated by Streep as an emotional avatar for the audience. The maternal polar bear melodrama is offset against the rationality and clarity of scientists, pitching effect against empirical evidence and quantifiable statistics about the retreat and thinning of polar sea ice, the melting of millennia-old glaciers, and changing atmospheric conditions (see Stenport and Vachula 2017). There is furthermore a connection between melodrama, science, and ethnographic evidence in the documentary film tradition, which stretches back to the genre's beginnings. This history includes canonical ethnographic films such as *Nanook of the North* (Robert Flaherty, USA, 1922) and *The Wedding of Palo* (*Palos brudefærd*, Friedrich Dalsheim and Knud Rasmussen, Denmark, 1934), which were marketed both as entertainment and as realistic evidence of Indigenous life in the Arctic in the first part of the twentieth century. Both films, however, included depictions of practices and traditions that were no longer current at the time of filming, contributing to a film history of salvage ethnography, which has been significant in Arctic film narration since then. This continues in *To the Arctic,* which contains footage from *Nanook of the North,* with Nanook popping out in 3D toward the audience while devouring raw seal meat. Here, Nanook stands in for both past and present-day Indigenous experiences of climate change. Dislocated from time and space, this sequence in *To the Arctic* positions Inuit culture as never-changing until the effects of climate change destabilize eons-old practices, while disregarding the impact of environmental transformations on the cultural and social conditions of present-day Nunavut.

IMAX Environmental Aesthetics, Technology, and the Scale of Polar Exploration

IMAX is a medium often dedicated to environmental examination. It is also a mode of exploration, as Alison Griffiths notes: "Ascending and conquering are thus tropes . . . most certainly co-opted by IMAX in its iconography and camerawork" (2006, 252). This ethos is furthermore reflected in many film titles by way of their "active prepositions" (2006, 242), as in *To the Arctic*. Moreover, that film's director of photography, Brad Ohlund, affirms the significance of an explorer's perspective for a successful IMAX film: "We're out here in the middle of no place, surrounded by ice. I've always said our job is to put the camera where it doesn't want to be. Now we have the ability to put it in many places it just couldn't be" (*To the Arctic Featurette no. 4*).

The ethos of IMAX films such as *To the Arctic* and *Wonders of the Arctic* indeed conflates scientists with new explorers. Speaking about her voiceover collaborations with *To the Arctic* director Greg MacGillivray, Streep affirms her belief in a strong correlation between science and climate change communication, education, expedition, immersion, and global interconnectedness: *To the Arctic* "is such an intimate look at a vast place and the changes that are being made there have such implications for all of us living in the temperate world" (*To the Arctic Featurette no. 1*). The ideological and historical rhetoric of IMAX is reinforced by Streep: "The great thing about films like this is that they take us places where you can't get into your car and drive [and IMAX has] captured impossible subjects and made them available to us in the most vivid ways" (*To the Arctic Featurette no. 1*).

As part of this visualization process, aerial photography continues to be a standard feature of IMAX, and it is not surprising that *To the Arctic* opens with a "phantom ride" (drawing on Tom Gunning's influential term) in an airplane, looking across the steep edge of a shimmering white Greenlandic glacier that abuts a turquoise ocean (see Acland 1998, 436). Traditionally, footage of Arctic and polar landscapes of ice have lacked depth of field or deep focus: the big open landscape, often with reflections off ice and snow against a large sky, seemed to convey two-dimensional monotony rather than three-dimensional contrast.

As a technology, IMAX (further augmented when projected in 3D) notably redresses this flatness by providing an opportunity for a three-dimensional cinematic form on a scale concomitant with quintessential Arctic icescape motifs. As Griffiths notes, "most IMAX films eschew the pan's horizontality in favor of a perpendicular movement into the frame that evokes the sensation of penetrating space through depth cues. Panoramic vision no longer refers to a lateral sweep associated with the cinematic pan but functions more as a synonym for an overall view" (Griffiths 2006, 240; see also Ross 2012, 212). This cinematographic approach is consistently employed in *To the Arctic,* echoing strategies of aerial polar exploration films such as *Roald Amundsen: Lincoln Ellsworth's Polar Flight 1925.* In the 1920s, the connections between the airplane, as a new mode of transportation, and movies, as a new form of mass media, converged in the aerial cinematography aesthetic. That aesthetic continues to be used by IMAX today, as in the opening shot in *To the Arctic,* which interpellates the spectator, through the IMAX camera and projection technology, as a new explorer flying over the edge of glaciers. Not all recent IMAX films set in the Arctic are explicitly about the figure of the new explorer as scientist, but they still mobilize related tropes for representing the Arctic.

Several Arctic IMAX films function as travelogue—a centuries-old and established genre familiar to most Western audiences—and include it as part of IMAX's "extraordinary form," which is "associated with a special trip" and theaters that are often located in "tourist sites" (Acland 1998, 437). These IMAX films include *Great North* (Martin J. Dignard, Canada/Sweden, 2000) and *The Emerging North* (Alan Booth, Canada, 1986), a ten-minute film made for Expo 86 in Vancouver for the Northwest Territories Pavilion. Both of these films include captivating landscape photography, close-ups of Arctic animals, renditions of Indigenous life in the North, and resource extraction as intrinsic to Arctic visual representation. *Great North* is sponsored by Hydro Québec (the provincially owned hydropower utility), and *The Emerging North* by PetroCanada (at the time, the federally owned petroleum company). These two Arctic IMAX films reflect Acland's contention that "nature tourism [and, by extension, IMAX] is always about being where human traces are not visible, even though this is an ontological impossibility. Access is permitted to an otherwise prohibited environment. Contaminating human forces are held at an ideological distance in order to construct intimate communion with the natural world. And, paradoxically, this often propels more elaborate technological initiatives, of which IMAX is a superior example" (Acland 1997, 295–96).

The Emerging North is a typical tourist/travel IMAX work along the lines of what Acland outlines, as it uses no voiceover, which works well for an international audience visiting a world exposition. It combines traditional views of the Arctic, especially landscape shots and fauna (polar bears, rabbits, muskox, etc.), with tourism (kayaks running rapids) and "new technologies" such as icebreakers, oil rigs, hovercrafts, computers, and radio stations—all in service of resource extraction. At the same time, it intercuts these developments with accounts of cultural practices in the Yukon Arctic. More recent developments seek to take immersion further. For instance, Oculus Rift technology launched the Arctic-set *Polar Sea 360°* (Kevin McMahon, Canada, 2014) as "the world's first 360° documentary." Through this VR platform, a viewer virtually explores the Arctic through a helicopter flight, looking around, 360 degrees, while doing so. Like the immersive works of Expo 67, from which *Polar Life* and later Arctic IMAX films emerge, *Polar Sea 360°* is a vertiginous experience, giving the viewer an immersive exploration. This new technology is thus one of many examples of the kind of virtual tourism that is part of the new Arctic climate change explorer ethos. The Oculus Rift VR technology is especially well suited to display environments, and these aspects have become critical to the platform (see Heuer and Rupert-Kruse 2015). And, like IMAX, Oculus is the ultimate extension of the spectacularization of the Arctic.

ARCTIC ENVIRONMENTS AND THE DIFFICULT ART OF COMMUNICATING CLIMATE CHANGE: *CHASING ICE*

If one of the goals of IMAX, a visually spectacular moving-image technology, is about spatial compression to generate immersive effects, then one goal of Jeff Orlowski's *Chasing Ice* (USA, 2012) is temporal compression to make climate change visible—achieved by a massive sequencing of still images. *Chasing Ice* is a key example of the new explorer film, combining the figure of the scientist-artist-expeditioner with inquiries into climate change in the Arctic. The film, which won awards at Sundance and an Emmy, takes melting Arctic ice as its impetus and its subject, offering a self-reflexive perspective on the visualization of retreating glaciers in remote locations in Greenland, Iceland, and the USA. Through still photographer James Balog's construction of a massive digital archive of time-lapse photography, the film generates compelling and media-friendly representations of retreating and thinning glaciers, seeking to make visible some of the effects of global warming. Balog inverts the figure of the scientist explorer of yore, not documenting an undiscovered, unchanging Arctic landscape, but instead recording its disappearance. *Chasing Ice*'s international popularity stems largely from its revealing, disturbing, and aesthetically captivating photographic documentation of melting ice. The film has also been screened for politicians, including as part of a 2013 Earth Day special screening at President Obama's White House, attended by policymakers and staff. *Chasing Ice* is thus both a historically connected and contemporary representation of ice and an example of the epistemological value of those representations.

Balog began documenting melting ice when he was assigned to photograph Icelandic glaciers for *The New Yorker* in 2005. This led to a commission from *National Geographic* to document glaciers globally, which resulted in the cover story "The Big Thaw" in June 2007 (he published a second series of photos under the title "The Melt Zone" in *National Geographic* in June 2010; see Balog 2007, 2010). In 2007, the year that the Arctic Ocean's ice cover was at its lowest since the beginning of comprehensive measurements, Balog also founded the Extreme Ice Survey (EIS), a long-term program that maintains over forty time-lapse photo sites across the Arctic. Global recognition of the diminished ice coverage also coincided with attention to the International Polar Year of 2007–08. This globally mediated international awareness of melting ice in the Arctic—as seen through NASA and European Space Agency satellite images, as well as Balog's *New Yorker* and *National Geographic* photos—brought visibility to climate change. Indeed, the Arctic became the popular image of anthropogenic climate change (see Christensen 2013).

What gives *Chasing Ice* its power is the transformation of pictorial images into metrical ones, where images function as abstracted scientific data and not as transparent windows onto reality. Built in part around the long-standing tradition of the pictorial documentary, *Chasing Ice* equates sight with knowledge (see, e.g., Burnett 1977). As a reviewer in *Film Comment* notes, "if it weren't for the time-lapse imagery, it would add nothing really new to the debate. But on the basis of that footage alone—astonishing sequences that condense years into seconds—it's a must-see" (Chang 2012). Orlowski himself was driven to document the EIS when he joined as a volunteer. But in much the same way that the line between explorer and scientist had been blurred in the past, the ethos of the film was not necessarily one of science education: "'So much of climate change is described in charts, and graphs, and numbers, but that's not how we wanted to approach it,' says Orlowski. 'We didn't want the film to be science-y. I wanted just enough science to understand the context . . . and no more'" (Thomson 2012, 24). Balog confirms this priority in the film, explaining that the story of climate change "is in the ice." The film bridges the pictorial and metrical traditions of documentary by containing images that resonate with a long-standing set of Arctic image conventions, yet counteracts these as a means of interventionist historiography, transposing them into a metrical form, revealing the new forms of knowledge that lie embedded within them. The film's remit of bridging art and science takes on additional valence. By "making ice the center of the story," climate change becomes visible; ice, the film affirms, is a photogenic subject. Balog's images are composed as fine art photography, conveying ice as luminescent, reflecting and refracting color, gradation, and hue. Similarly, the compelling formal qualities of icebergs and glaciers are sometimes portrayed by Balog as evocative of expressionist art, at other times as conventional landscape pictorialism.

This formalist approach provides a counterpoint to conventional science communication, as a commentary in *Nature Climate Change* foregrounds: "For Balog, the point of the project is both to collect good scientific data and to communicate the issues of glacial retreat to the public using stunning artwork. 'Most of the time, art and science stare at each other across a gulf of mutual incomprehension'. . . . These high-definition time-lapsed photographs of remote glaciers also have research value, helping plug holes in the data" (Jones 2012, 142). Balog thus uses pictorial art to reimagine the dominant metrical aesthetics deployed in other contemporaneous visual media, especially composite satellite data imagery. Satellite imagery emphasizes data quantification and abstraction, while leaving out reference to the aesthetic, and therefore affective, nature of the images, under the guise of objectivity. For example, the records of the lowest Arctic Ocean ice cover to date in 2007 were conveyed through extensive composite satellite data imaging, distributed globally through news media. Though measurement of the extent, age, and thickness of ice in the Arctic Ocean has been going on for centuries, the amount of data generated by scientists and through satellite imagery has, in the early twenty-first century, increased dramatically.

One component that makes *Chasing Ice* epistemologically different from the satellite images of melting ice is scale. "The key idea," writes cognitive psychologist Markus Knauff on the function of the metrical in relation to scale, "is that individuals use visual images to reason." Images "are not epiphenomenal in human inference but have a causal power in our thoughts." Accepting that human reasoning relies on visual images, the "main assumption . . . is that these images represent metrical information about distances between objects within the image in an analogical way, and reasoners scan these metrical images to find new relations that are not explicitly given in the premises" (Knauff 2013, 141). *Chasing Ice* introduces a human scale in the representation of climate change, in contrast to the planetary renditions that dominate satellite composite imagery (see Wormbs 2013). Throughout the film, the white, male explorers are asked to stand in frame, in various photos, to demonstrate a human sense of scale. As an aesthetic and rhetorical effort, these calls to the recognition of human scale also help support the notion that climate change is *anthro*pogenic.

Because of its retrospectively deterministic nature, the nihilism of the Anthropocene—and its Western, imperialist, and patriarchal origins—complicates engagement and action, which ostensibly contradicts the goals of works like *Chasing Ice*. This dilemma or dialectic between determinism and action—including in the role that art plays as a mode of representation and call to action—pervades many contemporary climate change documentaries, and especially those we discuss as "new explorer" works. *Chasing Ice* therefore situates the white figure of the human explorer as central to examining climate change, because a human presence makes the effects and impacts of warming visible. Seemingly small and insignificant in the vastness of the landscape, the figure helps convey to an audience the potentially devastating effects of ice melt.

Yet this use of "scale," and the whiteness of the new explorers, elides the fact that for centuries, Inuit and other northern Indigenous peoples have inhabited glacial environments; implicitly, it is the interjection of the white, male avatar that functions as a means by which to make ice a subject so that an audience can understand the scope and scale of the change at hand. Whiteness—in this case, the white explorer's body—becomes the new metric for the image of climate change, reinforcing the notion of the white explorer as the bearer of empirical knowledge. This trope, especially through the various activities of the National Geographic Society—a sponsor of many polar expeditions—has repeatedly been called into question (Hill 2008; Lúthersdóttir 2015). *National Geographic*'s "Race" issue (Goldberg 2018)—which seeks to redress the NGS's colonialist ethos, and its main publication's role in the whitewashing of "explorer" and "expedition" legacies—is a sign that this history has begun to be openly questioned by the NGS itself.

The Scientist as a New Explorer in the Arctic

Chasing Ice contains many sequences that echo past Arctic explorer films. Balog, trained as a geomorphologist, is both an expeditioner and an artist/photographer. In accordance with such tropes, *Chasing Ice* documents the difficulties that Balog, Orlowski, and their associates experience with the basics of image capturing: mounting cameras in remote locations, shielding the equipment from snow, ice, wind, and rocks, and overcoming technological sensitivity to the climate. These problems are not new. Indeed, Anthony Fiala, one of the first to film in the Arctic during the Wegener polar expedition in 1903–05, complained about the difficulty of operating a camera in the polar region: "Of all my photographic apparatus, the bioscope gave me the most trouble, particularly in the low temperature of spring and early summer. . . . [A]s soon as the instrument became cold, the film broke like fragile glass," the instrument coated "with a film of ice" (Fiala 1907, 141).

Balog, through self-narration, reinscribes many of the old tropes of masculine exploration: the environmental dangers, battling with his health and failing body, the supportive family that awaits his return, his team of relatively anonymous acolytes, and the support of *National Geographic,* which, for Arctic exploration, extends back to the days of Peary and Cook (see Poole 2004, 89–103). New explorer tropes pervade the film. In contrast to many twenty-first-century climate change films discussed throughout the rest of this book, *Chasing Ice* does not engage with the Indigenous peoples directly affected by melting ice in the Arctic or with the effects on anyone around the world. The only participants are North American or European: white men, female family members, and a few scientists.

Centuries-old tropes of Arctic exploration are fully mobilized in *Chasing Ice,* perpetuating exactly what Mark Carey and colleagues claim needs to be refuted: "Most existing glaciological research—and hence discourse and discussions about cryospheric change—stems from information produced by men, about men, with manly characteristics, and within masculinist discourses" (Carey, Jackson, and Rushing 2016, 772). Yet, in contrast to many historic Arctic exploration journeys, the quests in *Chasing Ice* are not about claiming new land, planting a flag, capturing specimens, or pursuing salvage ethnography of Indigenous peoples' cultures. In *Chasing Ice,* Balog is present to engage in a salvage ethnography of landscape and of the disappearance of one of its salient features: Arctic glaciers. This metaphor of glaciers as subject to salvage ethnography points to the manner in which Balog's dialogue refers to them: they are understood to be primordial, primitive, powerful, and quickly fading away into water, not through cultural imperialism, but because of modernity's impact on environments and climate. The glaciers are construed as passive victims of the relentless force mobilized in the name of progress. This echoes Carey and colleagues' argument about the relationship between glaciology, exploration, and patriarchy: "Many natural science fields have historically been defined by, and their credibility built upon, manly attributes such as heroic (often nationalistic) exploration and triumphs over hostile, wild, and remote landscapes" (2016, 777). In regard to *Chasing Ice,* they state that

> Balog may not have chosen this [masculinist] approach, but the filmmakers and media adhere to tropes of masculine vigor, risk, adventurous exploration, and heroic science to attract audiences and validate research, thereby sustaining these masculinist glacier narratives into the twenty-first century. Thompson and Balog's work is impressive to be sure because collecting the data they gathered was no easy feat and they are yielding insights for science and climate change impacts. But read alongside older heroic scientific narratives, the masculinist attributions ascribed to this type of field science remain prevalent over three centuries. (2016, 779–80)

Chasing Ice challenges many aspects of the traditional representation of ice and its melting in climate change media, but it nevertheless mobilizes other gendered, masculinist narratives of exploration and endurance to anchor the images with a patriarchal voice of authority.

The "salvage ethnography" ethos of scientific exploration raises concurrent issues. For example, rhetoric seeking to "save the Arctic" has a long history, having been promulgated by anthropologists and ethnographic filmmakers—Flaherty among them—in the early twentieth century, part of efforts to document and record purportedly vanishing Indigenous cultures (see MacKenzie 2015; Stuhl 2016). In *Chasing Ice,* the object of exploration is to capture vanishing ice. The filmmakers seek to retain a record of the ice's primordial qualities—not cultural artifacts that can be brought south for display in ethnographic museums, but images that can tell stories about a vanishing physical history of earth, its atmosphere, and its geological characteristics. Indeed, *Chasing Ice* contains a compelling, though short, sequence about the function of ice cores drilled from Greenland's ice sheet by scientists who use these cores to garner and analyze data from thousands of years past (see also Martin-Nielsen 2013).

These Greenland ice cores provide an alternate epistemology of earth, containing clues to climatological, meteorological, and glaciological information that may explain the future impact of present-day climate change. Similarly, Balog's captivatingly beautiful still photography in *Chasing Ice* is used to capture small bubbles of ancient air—atmospheric specimens of times past—that erupt through the many billions of water-filled cryoconite holes that dot Greenland's ice sheet. These bubbles release records of the earth's past into the atmosphere, while speeding up the melting process. This confluence of specimens of millennia-old records of the earth (ice cores drilled and air bubbles released) is hauntingly juxtaposed with the process of establishing a massive database of digital images of the very same glaciers that are disappearing. An additional perspective combines the digital records with social memory and the trauma of lost environments: "Implicitly, a strong narrative running through the film reveals that the footage ... is understood as an artifact, as evidence of what once was, as social memory for glaciers that will not exist much longer. Orlowski and Balog believe the glaciers documented are ruins. They are what remains" (Jackson 2015, 486). *Chasing Ice* thus depicts these glaciers as primitive epistemes—as harbingers of knowledge about to be lost. This rhetorical stance is not new for the Arctic, though it is mobilized to new effects in *Chasing Ice* to convey twenty-first-century attention to the impact of melting ice in the Arctic.

The Myth of Transparency and Climate Change Aesthetics

The importance of *Chasing Ice* in the context of science communication about climate change impact is compelling, precisely because of its synthesis of pictorial and metrical representations. The same year Balog began conceptualizing large-scale time-lapse photography as a documentary mode, debates emerged as to whether documentary photography would even be a valuable tool "in the communication of climate change messages," writes Kate Manzo, as scholars "and photographers have noted that the complexity of climate change and its potential and often invisible risks make the issue difficult 'to illustrate photographically'" (2010, 197). These debates foreground the notion that climate change appears too large and too diffuse for accurate and compelling visualization as "hyperobjects" (Morton 2013). They also coincide with a decrease in mainstream media attention to climate change as a result of the 2008 global financial crisis (Pearce et al. 2015). *Chasing Ice* arguably seeks to counteract both of these social imaginaries—the first that climate change cannot be compellingly visualized, and the second that economics trumps the environment. Melting glaciers are consistently mobilized in *Chasing Ice* as the most effective way to bridge these social and epistemological realms.

Chasing Ice, in essence, responds to and seeks to overcome deficiencies in two dominant models of science communication about climate change and its global implications that were in broad circulation during the first decade of the twenty-first century. The deficit model of science communication postulates that publics are uninformed, and that conveyance of more scientific information will predispose these publics to learning, knowledge, and, eventually action. This model, however, leaves out the nonscientific knowledge that publics rely on to make decisions (history, culture, personal experience, etc.), which influences and sways scientific beliefs (see Pearce et al. 2015).

Chasing Ice mobilizes aesthetics to overcome some of these deficiencies; similarly, it critiques the transparency fallacy or "myth of transparency," a default assumption that science communication visuals are unmediated and not symbolic constructs. Indeed, "people—experts and non-experts alike—frequently treat scientific graphics as transparent windows onto reality. Obviously, they are not: anyone familiar with scientific graphics knows how many hours, instruments, and software programs go into creating a good graphic, and how different the finished product looks from the raw data" (L. Walsh 2015, 365). The transparency fallacy similarly obscures nonobjective factors such as fear, emotion, and ideology, and promulgates views of science as elitist, divorced from everyday or embodied experiences, to "work at odds with goals of galvanizing personal and political action to mitigate climate change" (2015, 366).

What is innovative about *Chasing Ice,* as an early twenty-first-century example of science communication, is that it offers self-reflexive insights, not only into the process of visualizing climate change, but also into documentary filmmaking. *Chasing Ice* returns to the scientific root of making still images move: to make visible the invisible through metrical representation. Both Eadweard Muybridge's photographic studies—of horses trotting in 1887 and of humans walking and jumping in 1883, leading to his publication of twenty thousand photographs of human and animal movement in book form in 1887—and the development of Étienne-Jules Marey's chronophotographic device in 1882 were based on the premise of making visible to the human eye the breakdown of movement that could otherwise not be seen (see, e.g., Williams 1981; Braun 1992). *Chasing Ice* inverts this process; if Muybridge and Marey were slowing down time, Balog is speeding it up to the same end: to make visible what is not visible in human experiences of temporality. And like Marey and Muybridge, Balog, through trial and error, must invent new and modified technologies in order for his project to succeed. Balog, then, is part of a long tradition, one extending back to the first attempts to capture movement through the use of still images in order to make visible the world at a different temporality than the one that can be captured by the human eye unaided. As such, *Chasing Ice* also ties into the long history of the animation of still images for the purpose of scientific inquiry. As for both Muybridge and Marey, aesthetics are an important part of that process.

Given Balog's status as a celebrated nature photographer, it comes as no surprise that his photography reproduced in *Chasing Ice* is captivating; quite a few of Orlowski's shots of Balog and crew in landscapes of ice, rocks, and gravel are as well. On the one hand, these images are conventionally beautiful in composition, perspective, shape, gradation, hue, and shade, as they reflect a long tradition of Western landscape portrayal, whether pastoral, spectacular, or sublime. On the other hand, *Chasing Ice* mobilizes a complementary approach that works in tandem with an allegiance to conventional aesthetic registers. In the context of climate change, this is important, because a clear political interest is expressed in these images as well. Partly by refusing to uphold the images' status as autonomous aesthetic objects, the process of massive time-lapse photography is one of breaking the very landscape down to metrical components, visualizing glacial retreat as one discrete bit of information separated from another and emphasizing process rather than a complete visual rendition of supreme formal characteristics. In this way, time-lapse photography becomes an aesthetics of description of the very images that one is enraptured by in terms of their aesthetics.

Chasing Ice is also a film about memory and epistemology—about visually documenting what is about to disappear, securing the significance of ice in history. Having retrieved the computer memory card from one of the remote cameras, Balog is shot against a vast landscape, holding the small object in his hand and addressing the camera, saying that "this is the memory of the landscape." The once ice-covered location is now only present as absent, caught like a fly in amber inside the digital memory of the chip itself—freeze dried, as it were (add electricity and technology, and the landscape will reconstitute and come back to life from the past). One of the main contributions of *Chasing Ice* is that it foregrounds an aesthetics of mass-scale digitization, and thereby makes visible the quantification process involved in digital renditions, but refuses to create a seamless, photorealistic virtual reality. It generates its overwhelming truth claims by visualizing the laborious process of generating massive amounts of data, and making that data accessible to viewers in a way that will have affective impact on audiences.

Chasing Ice does not offer an account of the climate impact of the production of these images (see Maxwell and Miller 2012), including the expeditioners' air travel. Digital camera technology may be thought of as carbon neutral, but it is made with rare-earth metals with limited life spans that create e-waste; cameras are typically shipped abroad or turned into landfill when the initial life span is over; storing digital images on servers consumes energy; and the servers themselves need frequent replacement. Perhaps there is a feeling—as among the leaders from 150 countries who fly to climate change conferences in Copenhagen or Paris—that the good offsets (but does not carbon offset) the bad. But the lack of self-reflexive examination of the impact of producing these images, or at the very least acknowledgment of the impact or the steps taken to offset it, raises questions about the efficacy of making digital films about climate change. Both climate change and digital images are made by people and the technologies they deploy. Indeed, the technologies deployed to fight climate change in *Chasing Ice* are essentially the same technologies that are causing the problem; they parallel and intersect with each other, but without the film's protagonists acknowledging the fact.

THE NEW EXPLORER SPECTACLE AND
THE ANTHROPOCENE PROJECT

Chasing Ice and *To the Arctic* seem to confirm the proposition made in 2000 by Nobel Prize–winning chemist Paul Crutzen that "'we are not in the Holocene anymore' . . . we are in the *Anthropocene*" (quoted in Warde, Robin, and Sörlin 2018, 165). These considerations are central to the art installation, book, exhibition of still photography, and documentary film that constitute *The Anthropocene Project: The Human Epoch* (Canada, 2018) (for an overview of the whole project, see Burtynsky, Baichwal, and de Pencier, 2018). Canadian photographer Edward Burtynsky and filmmakers Jennifer Baichwal and Nicholas de Pencier seek, throughout their extraordinarily ambitious project, to exemplify and critique the concept of the Anthropocene. *The Anthropocene Project* builds on the premise that documentary images have a distinct power, with the artists proposing that "[lens-based] media represent the most technological, the most anthropogenic, of art forms," which is what makes them especially important (Hackett 2018, 14).

The Anthropocene emerges simultaneously with the global growth in visual and mediated cultures: the spread of photography in the late nineteenth century is a product of and represents industrialization, the rise of

Western capitalism, and the movement west and north of settler cultures in North America. Lens-based media is an aspect of polar exploration, certainly. Along those lines, Jennifer Fay's *Inhospitable World: Cinema in the Time of the Anthropocene* (2018) emphasizes that Western cinema is a cinema of fabrication, spectacle, and scale. She argues that "the dominant mode of film production is governed by a simulationist aesthetic, a preference for the artificial, the mechanical, the temporary, and the made over any location or object that is 'naturally' or historically given" (2018, 6), by which "filmmaking occasions the creation of artificial worlds, unnatural and inclement weather, and deadly environments produced as much for the sake of entertainment as for scientific study and military strategy [so that] human worldmaking is undone by the force of human activities" (2018, 4) and where there is "a movement from the artificial studio, in which simulated nature may be totally controlled before and on film, to the anthropogenic planet that is predicted to *defy all prediction*" (2018, 9). While the Anthropocene is not synonymous with the climate crisis, early symptoms of global warming became the epitome of the Anthropocene (see Robin, Sörlin, and Warde 2013, 491–501). In the twenty-first century, the visualization of melting ice, disappearing as if into thin air, became a concrete exemplifier of the impending crisis, often pictured in popular media as an emaciated polar bear drifting alone on the Arctic ocean on a small ice floe.

The Anthropocene Project, global in its undertaking, is a part of the spectacularization of climate change, foregrounding massive terra-transforming projects of extraction, infrastructure, pollution, and human-caused environmental change. *The Anthropocene Project,* moreover, places Burtynsky as a photographic new explorer, akin to Balog, without him appearing as an on-screen avatar. More recently, *I Am Greta* (*Greta,* Nathan Grossman, Sweden/UK/USA/Germany, 2020), the documentary about climate activist Greta Thunberg, her school strike (Skolstrejk för klimatet), and her journey across the Atlantic Ocean in a racing yacht (to eliminate carbon emissions) in order to speak to the US Congress and the UN (see Thunberg 2019), can be seen as a new, next-generation, nonpatriarchal representation of the new explorer.

Chasing Ice, Arctic IMAX cinema, and *The Anthropocene Project* are examples of documentary forms deeply intertwined with the history of polar exploration. At the same time, these films exemplify how documentary representations of the Arctic across metrical and spectacular registers have become popular for engagement with and knowledge of climate change. The documentaries also exemplify the significance of specific locations to represent large-scale environmental change, a notion that resonates with McLuhan's concept of the global village. As we have demonstrated in this chapter, the rise of global media is also intertwined with the Arctic, as global change science, Cold War science diplomacy, and the Great Acceleration all have salient points of connection with the Arctic. These aspects are significant also to twenty-first-century Arctic Indigenous cinema, which circulates globally in ways that further political activism focused on self-determination, the environmental impact of resource extraction and climate change, and connecting local knowledge sites in the Arctic with the rest of the world. These are some of the items we will explore further in the chapters to come.

THREE

Isuma and Indigenous
Media Sovereignty

IGLULIK (IGLOOLIK), NUNAVUT, HAS A POPULATION of about seventeen hundred people. Yet this small hamlet is home to two of the global circumpolar Arctic's most important film and video art collectives: Isuma and Arnait. The emergence of the Isuma and Arnait collectives and the many directors they advanced parallels Nunavut becoming the third territory in Canada, which split from the Northwest Territories after a 1992 referendum. The push toward self-governance began in the 1970s, reflecting worldwide pan-Indigenous political mobilization efforts that also spanned Greenland and Sápmi. The long delay between a first proposal in 1976 and Nunavut's founding in 1999 was not only a result of the Canadian government dragging out the process because of an indifference to Inuit and Indigenous goals, but also reflected that the federal government, dealing with the rise of Québec separation, was uninterested in creating a new semiautonomous part of Canada (Légaré 1998, 274). Canada's various constitutional "crises" abated by the late 1990s, and Nunavut came into being on April 1, 1999, as an outcome of the Nunavut Land Claims Agreement, which was signed on May 25, 1993.

The constitution of the Government of Nunavut (GN) "was as close to fashioning a government on a blank piece of paper as anyone is likely to see" (Hicks and White 2015, 4). While in some ways Nunavut's form of government aligns with others in the Canadian nation-state, there are important distinctions that have directly and indirectly contributed to the rich and unique twenty-first-century media environment of Nunavut. These encompass "its departmental structure, which included such distinctive departments as sustainable development and culture, language, elders, and youth; a commitment to Inuktitut as a working language of government; and an attempt to imbue both public policy and government operations with traditional Inuit values (*Inuit Qaujimajatuqangit*—IQ)" (Hicks and White 2015, 5). Nunavut thus strove to incorporate IQ principles into the structures of government, including consensus building and decentralization, while also recognizing cultural production and Indigenous language as central to governance. For example, administrative bodies typically based in the capital are currently dispersed across Nunavut. Developing such a dispersed, decentralized government capable of upholding Indigenous values and Indigenous Knowledge (IQ principles) hinges on effective communication technology: "The entire Nunavut project, and especially decentralization, was premised on the availability of high-speed electronic communications to connect far-flung offices" (Hicks and White 2015, 261).

However, what was not foreseen was the lack of investment in IT and infrastructure—an ongoing concern—to make internet governmentality feasible: the role of communication technology is nevertheless central to the building of what can be properly understood as a Nunavut identity tied to both IQ and Western models of national and subnational identity (see also Yunes 2016, 98–103). These principles also reflect Benedict Anderson's notion of nationhood as a process emerging through a shared linguistic imagining of one's collective past, mobilized through a medium specific to the moment at hand. This allows a popular narrative of a shared past to emerge, one that is mobilized for a shared understanding of the present: the imagined community (Anderson 1991). In the nineteenth century, the process happened, according to Anderson, through newsprint and the novel. In twenty-first-century Nunavut, it happens through audiovisual media production and circulation. The use of oral traditions through audiovisual media is significant because it strengthens the Inuktitut language, reaffirms oral storytelling practices central to the community for millennia, and revivifies history through the act of sharing these stories over a vast territorial expanse. All of this leads to the further strengthening of an Indigenous imagined community. This process is reflected in the fact that the use of technology is embedded within the Nunavut Constitution Act as a way of building community and the centrality of media in that process.

To put the development of the astonishing range, quantity, and quality of film and video production in Nunavut since its founding into perspective, we return to the GN's emphasis in its charter on language, culture, sustainability, and support of elders and youth. These facets of Nunavut life have been supported by decentralized communication technology and are reflected in the content of many films from Nunavut that often thematically address issues of language, culture, environmental and social sustainability, and the relationship between youth and elders. Through her extensive research on and work with Indigenous media-makers—including Nunavut directors such as Zacharias Kunuk—visual anthropologist and media historian Faye Ginsburg has coined the exceptionally useful concept (already discussed in chapter 1, and explored throughout the book) of Indigenous "media sovereignty," which Ginsburg describes as follows:

> I think of this as media sovereignty, a term I introduce to describe practices through which people exercise the right and develop the capacity to control their own images and words, including how these circulate. Here, I draw on a classic legal definition of sovereignty as the possession of authority over an area, extending this more typical idea of political authority over a land and populace to the possession of technical, cultural, political and creative control over media produced by indigenous peoples and about their lives. This approach dialogues with the discourses of native North American intellectuals emergent since the mid-20th century. (2016, 583)

The connections between means and modes of production are central to media sovereignty. The concept also resonates strongly with the priorities of GN and with the principles of media collectives such as Isuma and Arnait, both of which we discuss in detail in this and the following chapter.

Media sovereignty goes beyond the aesthetic, thematic, and cinematic concept of "visual sovereignty," as it was first developed by Tuscarora artist and scholar Jolene Rickard in 1995 (Rickard et al. 1995) and subsequently reformulated by a number of practitioners and Indigenous scholars (see Robinson and Martin 2016). For Kristin L. Dowell, visual sovereignty is about the act of production: "the articulation of Aboriginal peoples' distinctive cultural traditions, political status, and collective identities through aesthetic and cinematic means" (2013, 1). To address histories of both popular documentary and "scientific" representation of Indigenous peoples by anthropologists and ethnographers, Michelle Rajeha describes visual sovereignty as "what it means for indigenous people 'to laugh at the camera'" and "to confront the spectator with the often-absurd assumptions that circulate around visual representations of Native Americans, while also flagging their involvement and, to some degree, complicity in these often-disempowering structures of cinematic dominance and stereotype" (Raheja 2007, 1160).

Raheja goes on to note that visual sovereignty is a productive "space between resistance and compliance wherein indigenous filmmakers and actors revisit, contribute to, borrow from, critique, and reconfigure ethnographic film conventions, at the same time operating within and stretching the boundaries created by these conventions" (2007, 1161). It is important to note that Raheja's influential concept engages directly with an *ur*-text of and for Inuit visual representation: Flaherty's *Nanook of the North* (USA, 1922), which has informed nearly a century of filmmaking in, about, and for the Arctic and against which many of the films of Isuma and Arnait must be read (for the history of the remediation of Flaherty's film, see MacKenzie 2015). At the same time, media sovereignty should be understood not exclusively as a reaction against settler colonialism, but also as a way to demonstrate how non-Indigenous and non-Nunavut-based funding sources and circulation mechanisms have been a strong component of Nunavut's film culture, as these have supported one another.

From the late 1990s onward, concerns surrounding media sovereignty in Nunavut have often centered on alternative modes of production (access to inexpensive gear) and alternative modes of distribution (videotape, DVD, streaming) in order to circulate works beyond the necessity of a single channel or a public screening room. This not only allows greater access, but also foreshadows the movement of media works into the home, along with other kinds of new/alternative venues, leading to a surge in screening and community discussion. These priorities resonate with a long history of the National Film Board of Canada/Office national du film (NFB/ONF) and its support for filmmaking in Arctic Canada and beyond by "putting cameras into the hands of people." These practices emerge from the "Challenge for Change/Société nouvelle" ethos of the late 1960s and early 1970s—that people can document their own everyday experiences (with directors functioning, in the first instance, as "facilitators") as well as their own understanding of the histories that have shaped their communities (see Waugh et al. 2010). This richness of storytelling connects Nunavut to the long history of oral communication as a community-building practice and a means to document history. It also reinforces GN's emphasis on connectivity across the dispersed territory and the inclusion of every remote village and hamlet in its community. Cameras were and continue to be a means of interventionist historiography across Nunavut, serving as a conduit, often an inexpensive one, through which Inuit reclaim their own communal and personal histories. In this way, responding to *Nanook of the North* through media sovereignty and employing Inuit actors and production crews (especially in the case of Kunuk's reworking of the ethnographic writing of Knud Rasmussen) actively remediates history in the Inuktitut language and places Inuk identities and embodied experiences at the forefront of the narrative.

Distinctive of contemporary Nunavut media sovereignty is that the filmmakers use many different forms and genres of production—documentary, experimental media, television, fiction, digital imaging, ethnography, and animation—in order to reach a variety of intersecting local and global audiences. For both economic and aesthetic reasons, Super 8 film, home video technologies, and consumer digital technologies are all incorporated into the aesthetic choices made by practitioners. Arctic Indigenous media, then, works outside of dominant modes of image-making to engage in a form of cultural counterprogramming that embodies Nunavut's media sovereignty. Media sovereignty thereby moves beyond production itself to include the circulation, distribution, and reception of works. Media sovereignty in the Arctic is often trans- and inter-Indigenous in the context of a global media circulation, but it is still mobilized at the local level. In this and the next chapter, we examine a range of works from across the Canadian circumpolar North as political documents with impact and resonance across local and global spheres of influence.

THE HISTORY OF MEDIA SOVEREIGNTY AND NUNAVUT, THE INUIT BROADCASTING CORPORATION, AND THE ABORIGINAL PEOPLES TELEVISION NETWORK

The advent of satellite technology, celebrated in Canada as a means by which to unite a geographically large and dispersed country, was greatly debated in Inuit culture (see Roth 2005). These concerns are implicit in the GN charter through its emphasis on language, sustainability of communities, and connectivity. Indeed, despite the connectivity that satellites allowed, including in the field of medicine in the Far North, many communities resisted having images from southern Canada and the United States broadcast into their communities. This rejection was not an act of cultural isolation, but an attempt to avoid colonization. The problem, therefore, was not with modern technology but with its content. This led to the rise of IBC (Inuit Broadcasting Corporation) in 1982. It was North America's first Indigenous-language network and the world's first Indigenous media project broadcast by satellite. The principles of IBC subsequently informed filmmaking collectives such as

Isuma and Arnait and arguably influenced how the imagined community of Nunavut was eventually constituted. By the 1990s, the IBC positioned television and satellite technology as one of the many technological advancements that were fully incorporated into Inuit culture. In contrast to satellite broadcasts developed by CNN, MTV, and Sky in the United States and Europe during the 1980s, which included sleek new production standards in the interest of fostering global for-profit connectivity, IBC broadcasts were narrowcast and low-tech. The program content was also consistent with the larger political and cultural aims of Inuit rejection of outsider satellite programming, in that it sought to enhance cultural and historical recognition of local and Indigenous practices and engage in media sovereignty to support the principles of a Nunavut community imagined through media production (see MacKenzie and Stenport 2019a).

IBC's trajectory is indicative of larger Canadian Radio and Television Commission (CRTC) policy issues surrounding diversity of representation. The history of the cable channel APTN (Aboriginal Peoples Television Network) is instructive. APTN grew out of the federal government's interest, in the late 1980s, in establishing a designated northern channel, TVNC (Television of Northern Canada), which launched in 1992 (see Alia 1999; Hafsteinsson and Bredin 2010). As Lorna Roth argues, TVNC became a vehicle "to figure out ways to encourage its audiences to think along pan-Northern lines—beyond the local and regional to a more global perspective" (Roth 2005, 191). IBC programming screened on TVNC, and this channel straddled competing demands: catering to a northern audience and becoming an intercultural mediator and educator, engaging a North-South audience (2005, 198). TVNC was successful in reaching southern audiences through satellite signals. It showed that balancing "linguistic *narrow*casting and cultural *broad*casting" was possible (2005, 198). In 1998, the transition from TVNC to APTN was put in motion with the understanding that APTN—launched in 1999—would become a national TV channel and be included in all basic cable packages. The overall aim was establishing adequate representation of Indigenous populations within the Canadian public sphere, as "one of the three Founding Nations of Canada" alongside English and French Canadian settlers (2005, 204).

Incredibly, APTN, one of the world's most successful publicly funded Indigenous channels—with national and international reach—was based on television programming, production support, and distribution methods originating in the Arctic. The sparse population in a vast territory supported the development of various communication channels, by which IBC, TVNC, and APTN have served significant functions in fostering and encouraging local Indigenous moving-image production. To understand Canadian Inuit moving images, one must see them as distinct from Indigenous North American filmmaking writ large. While First Nations Canadians are already surrounded by white, Western, settler culture, the geographic distance between the South and the North means that concerns about assimilation in the Arctic have traditionally not been as strong. The draw of some of these technologies for Indigenous production in the North is precisely to bridge the expansive geographic spaces of the Arctic.

REMEDIATION, COLLABORATION, AND THE NFB/ONF

The extraordinary efforts of the NFB/ONF, over almost eight decades, to document Arctic and Indigenous populations cannot be overestimated when it comes to the conceptualization of Inuit media sovereignty. Funded by the Canadian government, the arm's-length agency promoted a range of documentary, fiction, and animation production in the North, and it brought the Arctic South to the rest of Canada and to the world. This legacy of representation and extraordinary body of work has been instrumental in supporting a robust practice of Indigenous filmmaking in Nunavut since the 1990s. For example, after Telefilm Canada offered only $100,000 for the making of Kunuk's *Atanarjuat: The Fast Runner* (Canada, 2001), funding was provided by the NFB/ONF. Although the NFB/ONF had ceased making feature fiction films by 2001, Kunuk successfully argued that documenting Inuit mythology was a form of documentary filmmaking (see Alioff 2001). The NFB/ONF has nevertheless been criticized for producing colonial images of the Arctic and working in extractive ways to make documentaries—up until the late 1960s—without Inuit agency. *Unikkausivut: Sharing Our Stories,* a 2017 NFB/ONF initiative, addresses both the agency's legacy in the Arctic and how that legacy and history is being continuously reformulated, remediated, and reimagined as part of an ongoing process of media sovereignty. The project is a comprehensive DVD and online streaming collection of NFB/ONF Canadian Arctic and Inuit films and includes over sixty NFB/ONF films that are considered "the most important worldwide, that represent all four Canadian Inuit regions (Nunatsiavut, Nunavik, Nunavut and Inuvialuit)" (National Film Board of Canada 2017b). *Unikkausivut: Sharing*

Our Stories contains representations far beyond the territorial boundaries of Nunavut, including Nunatsiavut, which crosses the boundaries of Labrador and Northern Québec; Nunavik in Northern Québec; and Inuvialuit in the Yukon and Northwest Territories. The NFB/ONF further notes that its more than one hundred animated films and documentaries about Inuit constitute "a unique audiovisual account of the life of the Inuit—an account that should be shared with, and celebrated by, all Canadians" (National Film Board of Canada 2017b). With this body of work, the NFB/ONF has produced more Arctic documentaries than any other organization worldwide.

Unikkausivut: Sharing Our Stories is an example of a collaborative, coalitional act of media sovereignty. Developed in tandem with the Inuit Relations Secretariat of Aboriginal Affairs, Northern Development Canada, and the Government of Nunavut's Department of Education, and supported by various Inuit organizations, the effort made most films available in Inuktitut, which speaks to the importance of repatriation in acts of media sovereignty. The films in the collection are divided into four categories, reflecting the dominant paradigms of Arctic and Inuit films made by the NFB/ONF since its earliest years: works produced from (1) 1942 to 1970, labeled ethnographic documentary films; (2) 1971 to 1977, described as early collaborative works and the initial stages of Inuit filmmaking; (3) late 1970s to mid-1990s, defined as films by non-Inuit filmmakers defending the rights and values of Inuit peoples and advocating an affirmation of Inuit culture; and (4) 1999 to 2011, films reflecting the emergence of "a true Inuit cinema"—films made by Inuit for Inuit. This trajectory delineates the history of documentary representation and activism.

The circulation of this massive set of documentaries and Inuit imagery proved challenging; though the GN has maintained its priorities in digital connectivity across Nunavut, the reality is that remote communities have less dependable internet connections. In response, a boxed set of DVDs was provided to remote Inuit communities; the GN distributed the first set of twenty-four films to every school in Nunavut. The circulation of these images, including in educational contexts, speaks both to a long history of colonial images returning to the Arctic as documentary evidence and to the remediation and repatriation questions that this process engenders. NFB/ONF film commissioner Claude Joli-Coeur outlines the rationale for including early ethnographic productions in *Unikkausivut: Sharing Our Stories*: "That is part of the richness of our collection, to see the evolution of how Inuit culture was represented. We have films that were originally made by white people, with a European approach. And later, we see white people who have transformed themselves through the cause of the Inuit, becoming almost Inuit themselves. We are now making films with Inuit creators" (Hamilton 2011).

Unikkausivut: Sharing Our Stories has been marketed by the NFB/ONF as an educational tool for outsiders, in keeping with the NFB/ONF's mandate to "interpret Canada to Canadians and to other nations" (National Film Board of Canada 2014) and as an archive for the Inuit to learn about their history and cherish linguistic and cultural preservation. This aspect is especially important for Inuk filmmaker Alethea Arnaquq-Baril, who affirms that the early documentaries included in the collection serve as a valuable way to acknowledge the richness of Inuit history and culture. These older films, she notes, "are copied and passed around and people catch the occasional screening over the years—they're definitely cherished up here. A lot of people who watch them, they have family members in them. . . . They're watched with as much excitement as anything else on TV" (CBC News 2011). This repurposing and remediation of historical material thus has the potential of becoming an act of media sovereignty on the part of Inuit viewers.

NFB/ONF's ongoing production support for locally produced works resonates with the agency's priority to get media into the hands of the people in order for them to tell their stories as they see fit. The collectively made, short-format digital films in *Stories from Our Land 1.5* and *2.0* (Canada, 2010–14), made in collaboration with the Nunavut Film Development Corporation (NFDC), exemplify additional aspects of media sovereignty practices, especially by foregrounding community building through the collective aspect of filmmaking, through workshops and production support for emerging Inuit filmmakers. This collaboration between the NFB/ONF and NFDC further developed into a short film program for emerging Nunavut filmmakers, *Stories from Our Land 2.0*—which allowed participants to create a documentary story up to ten minutes in length (see MacKenzie and Stenport 2019a). There is no standard theme, genre, or aesthetic for these films. Instead, the project celebrates contemporary Inuit filmmakers and their diverse approaches to representing and self-documenting life in the Canadian Far North.

Isuma and Media Sovereignty

The developments of Isuma and Arnait (the latter to be discussed further in chapter 4) are closely tied to the previous production and distribution practices of the NFB/ONF, IBC, TVNC, APTN, and NFDC, as these publicly funded organizations have helped foster a vibrant moving-image storytelling tradition in the Canadian Arctic. Both collectives build on the significance of oral history for Inuit cultures and emphasize the communal, process-oriented, and cultural aspects of filmmaking. Isuma is the world's first majority Inuit-owned independent film and video production company, and its overarching concern is to produce counter-ethnographic accounts of Inuit cultures and to promote Inuit language and cultural knowledge production within the Arctic. As Asinnajaq argues, their process foregrounds how Isuma members "have empowered and fought to work in a paradigm that matches the values of the collective." Asinnajaq notes "the way those values are present, both in the films onscreen and in the many ways of working and living off-screen; and the way they maintain the social support structures that have kept Inuit strong over thousands of years" (Asinnajaq 2019; see also Evans 2010, 122–39). Combined with its emphasis on collaborative processes and explicit rejection of mainstream hierarchical filmmaking and production structures, Isuma functions as a self-governing and sovereignist film and media production company.

Isuma's processes, practices, and priorities have effectively called into being Inuit media sovereignty through its extensive output, numbering over forty major works and countless shorter productions: shorts, features, fiction, documentary, multimedia, video art, installations, television, animation, and more. These priorities include ongoing emphasis on and innovation in digital sharing, circulation, and distribution to Inuit and global Indigenous communities, including through their own extensive website (isuma.tv), the Digital Indigenous Democracy (DID) project, and a globally accessible dedicated iTunes platform. Maintaining funding for Isuma's vast array of projects over thirty years has required massive efforts, including its reemergence from bankruptcy in 2011. Securing stable funding has been especially challenging because of Canada's complex public/private partnerships for film production, especially for feature films. Given all this, the political commitment to Inuit processes and perspectives and the deliberate attention to the goal of media sovereignty, which is integral to all these efforts, is even more astounding.

In a 2010 essay, Kunuk raises a series of questions about what kind of role media making can play for the Inuit: "Can Inuit bring storytelling into the new millennium? Can we listen to our elders before they all pass away? Can we save our youth from killing themselves at ten times the national rate? Can producing community TV in Iglulik make our community, region and country stronger? Is there room in Canadian filmmaking for our way of seeing ourselves?" (Ertel 2010). Today—more than three decades after Isuma was founded by Kunuk, Paul Apak Angilirq, Pauloosie Qulitalik, and Norman Cohn in 1990—the impact of Isuma for, within, and beyond Nunavut is undeniable, and the questions posed by Kunuk continue to bear relevance and guide the collective's work and current discussions of media sovereignty.

Isuma's work can be organized into three significant strands. The first strand is the collective's use of historical reenactment to Indigenous political ends, bringing Inuit life and stories to the screen. This strand includes works such as *Angakusaujauq: The Shaman's Apprentice* (Zacharias Kunuk, Canada, 2021), winner of the award for best Canadian short at the Toronto International Film Festival, which functions as an animated reenactment of Inuit mythology. The second strand consists of Isuma's more directly political works that speak to the contemporary moment. These documentary works include *Exile* (Kunuk, Canada, 2009), *Qapirangajuq: Inuit Knowledge and Climate Change* (Kunuk and Ian Mauro, Canada, 2010), *Our Baffinland* (2013), and *Silakut: Live from the Floe Edge* (Isuma, Canada, 2019). The third strand runs through all of Isuma's work and is centered on the politics of digital exhibition and circulation, which includes process-oriented practices about access. Most considerations of Isuma address their best-known histor-ical-reenactment fiction feature, *Atanarjuat: The Fast Runner,* which won the Caméra d'Or at Cannes in 2001 (see Huhndorf 2008; Evans 2010; Bohr 2015; Chartier 2018).

Historical reenactment through docudrama provides a throughline for the entire body of Isuma work, wherein the diegeses offer historical accounts framed by contemporary relevance. Key examples include Isuma's thirteen-part narrowcast television series *Nunavut* (*Our Land,* Canada, 1994–95), a significant achievement that documents local traditions before "moderniza-tion." It also serves as a precursor to practices seen in *Atanarjuat* (which can be read as a documentary of Inuit mythology, in the manner Kunuk pitched it to the NFB/ONF), *The Journals of Knud Rasmussen* (Kunuk and Cohn, Canada/Denmark, 2006), *Searchers* (*Maliglutit,* Kunuk and Natar Ungalaaq, Canada, 2016), and *One Day in the Life of Noah Piugattuk* (Kunuk, Canada, 2019).

As a critical component of Isuma's foundational work in media sover-eignty, this first strand of works also provides an interventionist historiogra-phy because it is non-isolationist, connecting the Iglulik group with other Inuit and Indigenous communities. It also supports language preservation in Inuktitut and maintains key historical practices, such as handicraft, hunting, and dog-sled transportation, keeping these forms of knowledge alive as part of prop making and setting and providing opportunities for the community to maintain expertise in these practices. An important component of the trajectory of the historical-reenactment docudrama films has been an increasingly more explicit focus on political content and activism, including the impacts of settler colonialism, Christianity, forced relocation, and sexual abuse within Inuit communities. Later films, such as *Searchers* and *One Day in the Life of Noah Piugattuk,* indicate that these issues can be addressed more openly now than was the case during Isuma's first years as a collective, perhaps because of the public knowledge brought about by the Truth and Reconciliation Commission of Canada (TRC). These docudrama practices provide connections to past colonial filmmakers while also creating new modes of documentary-based media sovereignty.

THE POLITICS OF EXTRACTION AND THE CLIMATE CRISIS

Inuit Knowledge and Climate Change is a significant documentary in a number of ways. First, it is one of the few Nunavut-originated and Indigenous works that directly address the impact of the climate crisis in the Arctic. Although one might assume that climate change activism through documentary storytelling is a staple genre of Arctic Indigenous documentary, it is not. As Candis Callison argues, among Indigenous coastal communities in the North American Arctic, there is a resistance to climate change discourse. One local politician of Kotzebue, Alaska, for instance, remarked to Callison that "climate change . . . we don't really talk much about that. It's more something they talk about on CNN. It's out there. It's not what *we* talk about" (Callison 2014, 44; see also Wright 2014). In this context, *Inuit Knowledge and Climate Change* stands out as a deliberate political intervention. It directly addresses the impact of the climate crisis in Iglulik and nearby regions and does so in ways that fully resonate with the established practices of the Isuma collective. These include emphasis on insider perspectives, Inuktitut language use, community engagement, collaboration, and process-based practices, alongside multiple modes of circulation. The work exists in different running lengths, has been screened in a variety of forms—including at the United Nations Conference of the Parties to the Convention on Biological Diversity, known as "COP 15," in Copenhagen in 2009—and has moved from single-channel screenings to satellite interactive broadcasts to streaming through platforms live on the Isuma website and iTunes (on Isuma and COP 15, see MacKenzie and Stenport 2019c).

In *Inuit Knowledge and Climate Change,* Kunuk and Ian Mauro, a geographer, interview Inuit elders and Arctic inhabitants about the changes they see taking place on a daily basis to address a salient set of issues. The sparsely populated Arctic environments are among the most polluted on the globe, despite being the most remote regions from sources of industrial and agricultural pollution. Burning of fossil fuels that release greenhouse gases into the earth's atmosphere happens disproportionately in the South, yet the Arctic region is warming faster than anywhere else on the globe. As toxins travel through the air and via ocean currents to become lodged in the basin of the Arctic Ocean, ice and low temperatures further bind these slow-to-degrade pollutants, which eventually make their way into the wildlife, fish, and potable water that sustain local populations. *Inuit Knowledge and Climate Change* addresses these issues through its first-person stories of the effects with great poignancy, offering counterclaims to some Western scientific accounts of climate change. According to *Inuit Knowledge and Climate Change,* Indigenous knowledge, experiences, and perspectives are routinely ignored in attempts to understand climate change.

Through the use of what seems, in the first instance, to be a fairly traditional documentary form, *Inuit Knowledge and Climate Change* reverses the roles and assumptions with respect to Indigenous and Western perspectives on the Arctic, on science, and on climate change as an act of media sovereignty. This reversal allows for Inuit to take on the voices of authority. The form through which this authority is conveyed is important. First, the Arctic is immediately presented as populated; there is no pretense of the region as an empty space with few inhabitants, in contrast to many other Arctic climate change and explorer films, such as *To the Arctic 3D* (Greg MacGillivray, USA, 2012) and *Chasing Ice* (Jeff Orlowski, USA, 2012) (see chapter 2). Interviewees are local hunters, community activists, artists, and inhabitants of Nunavut communities, from Iglulik to Frobisher Bay. Stories and information are predominantly conveyed in Inuktitut (with English subtitles) and medium-length headshots. Subjects are positioned in local environments; interview questions and "voice-of-God" commentary are absent. The film's message is meant to be both political and urgent, as Kunuk notes: "Over the years, nobody has ever listened to these people. Every time [the discussion is] about global warming, about the Arctic warming, it's scientists that go up there and do their work. And policy makers depend on these findings. Nobody ever really understands the people up there" (Dixon 2010). The perceived lack of engagement with, or interest from, Western scientists is a central issue for the interviewees in *Inuit Knowledge and Climate Change*.

The film had a significant impact in Canadian news media, which speaks to Kunuk's intent in making a political intervention and ensuring that Inuit voices were heard. The Canadian Broadcasting Corporation (CBC) gave it a great deal of coverage in 2010, with particular focus on one of the most contentious claims made by elders in the film. They stated that the earth seemed to be shifting on its axis, an analysis that was based on the placement of the sun and the stars in the sky. Remarkably, the CBC's coverage was not dismissive, and so a pan-Canadian audience was able to hear these voices, framed through the voice-of-authority of the CBC, even if they never saw the film. The elders' accounts were initially summarily dismissed (Schuppli 2020, 281–82), with scientists claiming that the Inuit most likely perceived atmospheric change and "interpreted" the visual data poorly. Yet, in 2016, scientists created data visualization illustrating the ways in which global warming and climate change have slightly shifted the axis of the earth through atmospheric heating, thereby making a "discovery" that had already been observed and reported by Inuit elders, and dismissed by the media, six years earlier (Hall 2016).

Inuit Knowledge and Climate Change builds on climate crisis activism throughout Nunavut, undertaken during the first decade of the twenty-first century. A key component of this work was the petition in 2005 to the Inter-American Commission on Human Rights by Sheila Watt-Cloutier, international chair of the Inuit Circumpolar Council from 2002 to 2006. The petition sought "relief from human rights violations resulting from the impacts of global warming and climate change caused by acts and omissions of the United States" (Watt-Cloutier 2005). Watt-Cloutier's petition was one of the first political actions to posit a clear link between the climate crisis and negative impacts on Inuit lives, livelihoods, and customs, and to situate these negative implications in the context of climate justice. In her memoir *The Right to Be Cold* (2015), Watt-Cloutier takes a comprehensive stance on the connections between cultures, communities, climate, and globalization in the Far North. Naomi Klein, in her review of the book, puts it this way: "If the ice disappears, or if it behaves radically differently, then cultural knowledge that has been passed on from one generation to the next loses its meaning. Young people are deprived of the lived experience on the ice that they need to become knowledge carriers, while the animals around which so many cultural practices revolve disappear" (Klein 2015). Watt-Cloutier addresses how interconnected these issues are and that they are inherently about sovereignty, justice, and human rights, especially when linked to extractive and polluting industries, such as mining and shipping, which are impending across Nunavut: "How is it that the extraction industry is going to work better for us? This is one heck of a risky business we're getting into as a means of pulling ourselves out of poverty" (Watt-Cloutier 2015, xx).

The political recognition of Watt-Cloutier builds on several media works that address the climate crisis in the Arctic. As such, her leadership in producing the documentary *Sila Alangotok: Inuit Observations on Climate Change* (IISD, Canada, 2000) is especially significant as an example of Inuit media sovereignty. The film garnered attention during the first Climate Justice Summit conducted in parallel with the United Nations Climate Conference (COP 6) in The Hague in 2000 (Watt-Cloutier 2015, 197–98, 319–20). *Sila Alangotok* is likely the first Inuit work of climate justice media sovereignty ever made (it was cofunded by the Northwest Territories, a number of Canadian agencies, and the Walter and Duncan Gordon Foundation). Shot in Sachs Harbour, Northwest Territories, the documentary features talking-head participation by a number of Inuit elders, as well as interviews with Western scientists conducting a research project that explicitly integrates Indigenous knowledge and local observations on the climate with the impact these environmental changes are having on fishing, hunting (subsistence and livelihood), community and infrastructure (permafrost), and health (sunshine and melanoma risk). The film offers an especially striking commentary on the climate crisis, suggesting it is the unpredictability of conditions that is most concerning.

While *Sila Alangotok* does portray the scientists as collaborators, there is also a Eurocentrism at play, with scientists making statements such as "It is just so pleasurable to listen to people who have this knowledge," as if it were surprising that the Inuit know something about their own environment. The instability of the changing nature is one of the key points that the film returns to again and again: travel on ice is perilous, and knowledge of when animals will be migrating based on the state of the weather has also become much harder to discern. The film combines first-person interviews with sessions run by scientists to discuss Indigenous knowledge—not just about how the climate has changed, but also about what the future-oriented effects of these changes mean. While *Sila Alangotok* incorporates Indigenous knowledge, *Inuit Knowledge and Climate Change* reverses the lens, making Inuit knowledge central to understanding the changing climate.

It is therefore important to note that Kunuk's work was not produced and released in a vacuum, but paralleled rising concerns of climate change in the Arctic and the growing awareness of the concept of the Anthropocene. *Inuit Knowledge and Climate Change, Sila Alangotok,* and Watt-Cloutier's human rights petition emerged in riposte to Western "climate determinism" (see Hulme 2009) and offer counter-imaginaries that address the cultural and scientific debates surrounding the Anthropocene. Kunuk's and Watt-Cloutier's priorities emphasize that stories of climate justice should travel, including as part of COP 6 and COP 15, and should be digitally available anywhere in the world, while staying true to the community collaboration that undergirded the documentary ethos of these productions. Furthermore, these films forcefully demonstrate what we know is a misnomer: the Anthropocene is not about universal human causality; Inuit, like many Indigenous groups, have not brought about the climate crisis through the burning of greenhouse gases. Instead, the climate crisis in the Far North can be understood as an example of continuous settler colonialism.

TRC CINEMA, ORAL TESTIMONY, AND THE TRAUMA OF FORCED RELOCATION: *EXILE, MARTHA OF THE NORTH,* AND *ONE DAY IN THE LIFE OF NOAH PIUGATTUK*

The history of the effects of settler colonialism takes center stage in what we call Truth and Reconciliation cinema. Settler colonialism in the mid-twentieth century was perhaps most strongly experienced by Inuit through the forced-relocation orders mandated by the Canadian government. Several recent works by Isuma and other filmmakers have addressed the social and cultural impact of the relocations during the 1950s and 1960s. Kunuk's films about relocation in his home region of eastern Nunavut, Qikiqtani, echoes testimonials presented in "Nuutauniq: Moves in Inuit Life," a report in *Qikiqtani Truth Commission: Thematic Reports and Special Studies 1950–1975* (Qikiqtani Inuit Association 2013). The work of the Qikiqtani Truth Commission (QTC) addresses the forced relocation that took place in communities such as Grise Fiord, Resolute Bay, Iglulik, and Kivitoo, and it includes multiple testimonials similar to those presented in *Exile, Martha of the North* (*Martha qui vient du froid,* Marquise Lepage, Canada, 2008), *Kivitoo: What They Thought of Us* (Zacharias Kunuk, Canada, 2018), and

One Day in the Life of Noah Piugattuk. The report confirms that the state erroneously perceived Inuit consent to the relocation plans, as the Inuit didn't verbally protest, but instead often met the implementation of these measures with withdrawal and silence.

The QTC concludes that "Qikiqtaalummiut [Inuit in *Qikiqtani*] experienced a mix of voluntary, pressured and forced moves between 1950 and 1975, usually in response to government priorities. The federal government's primary goals were often contradictory. On the one hand, it wanted to keep Inuit self-sufficient through hunting or wages. On the other hand, it wanted to ensure that they lived in government-created permanent settlements where it would be cheaper to provide education, health and other government services" (Qikiqtani Inuit Association 2013, 45). The report goes on to note that "Qikiqtaalummiut suffered what scholars have called 'domicide' (the killing of one's home) when they left the land. . . . It represents the loss of independence and replacement of a way of life" (2013, 45). This "domicide" echoes through the works of both Kunuk and Lepage.

Made the year before *Inuit Knowledge and Climate Change,* Kunuk's *Exile* features testimonials by Canadian Inuit who were moved from Northern Québec to the High Arctic in 1953–55 to create the image of Canadian sovereignty, though the Inuit had in fact left these lands hundreds of years earlier (for an in-depth discussion, see A. Marcus 1995). This early-1950s relocation to the High Canadian Arctic became a focal point of the Cold War, with symbolic and material initiatives undertaken to strengthen ideological ownership of the Arctic. In Kunuk's forty-eight-minute documentary, Inuit presently living in Resolute Bay, Grise Fiord, Inukjuak, Iqaluit, and Ottawa tell their stories of relocation, including Martha and Renée Flaherty (who are descendants of Robert Flaherty, director of *Nanook of the North*). In total, members of thirteen different families are interviewed, offering complementary yet distinct views on the relocation and its profound and ongoing effects.

In terms of Western documentary practice, *Exile* can best be understood as an observational documentary, with the caveat that unlike in most works in the observational documentary mode, Kunuk is not an outsider gazing in. As oral testimony, the work functions within an Indigenous idiom as an act of media sovereignty by providing participants agency over their own accounts of history and the way these stories ought to be told. Meanwhile, its aesthetic can be "translated" into proximate forms of Western documentary for cross-cultural understanding. Its main mode of circulation has been through Isuma TV, garnering over forty-six thousand views by 2022, and Kunuk's status as a "major" filmmaker garnered press coverage and cultural capital for *Exile.* In the Canadian context, because of his international recognition and awards for *Atanarjuat: The Fast Runner,* Kunuk's voice is one that is granted attention by the dominant Canadian public sphere to which many other Indigenous artists have far less access.

Martha Flaherty's story of relocation (she is also interviewed in *Exile*) is told in the NFB/ONF film *Martha of the North.* Made a year before *Exile, Martha of the North* includes interviews with many of the same participants featured in *Exile,* but it differs markedly from that film's focus on verbal testimony and talking heads. Made by a Québécoise director through a multi-year script and story collaboration with Martha Flaherty, the film is included in the NFB/ONF's DVD box set *Unikkausivut: Sharing Our Stories* and is "classified as an Inuit production" (Bertrand 2019, 296). Distinguishing features include numerous historical-reenactment scenes as well as found footage, including photography and moving images from and related to *Nanook of the North.* The Flaherty family speak in harrowing detail about the trauma inflicted by the relocation, including loss of community, family, and language. The found footage included in *Martha of the North* also signals how stereotypical visual representation of northern Indigenous populations might have made it easier and more palatable for the Canadian government to mislead the Inukjuak community about the relocation north. To outsiders, the dramatic differences in climate, winter light, and ecosystems for hunting and gathering would have been offset by the pervasive visual imagery that had constructed a monolithic image of the Arctic and Inuit life by that time.

Martha of the North also differs from Kunuk's productions, not necessarily because it was funded by the NFB/ONF (many Isuma works, including *Atanarjuat: The Fast Runner,* have had Canadian government funding) and made in French and English rather than in Inuktitut, but because of its extended focus on women's lived realities. Thematically, the film is based on the journey motif to retell the experiences of its protagonist as in some ways exceptional, which includes the fact that Martha is Flaherty's granddaughter and went on to become a well-known interpreter and translator, as well as an important spokesperson for Inuit rights through the Nunavut Land Claims process and the women's rights organization Pauktuutit. Other aspects of Martha's story are foregrounded as communal, including her traumatic experiences of residential boarding school. The aesthetics of *Martha of the North* signal a higher production value than *Exile;* it is lit and shot in ways that emulate European art and heritage cinema, in historical reenactment scenes and through the adherence to historical and cultural accuracy in props and costumes.

Made in four versions—feature-length versions in English and French and one shorter version in each language for instruction ("campus" versions)—*Martha of the North*'s intended audiences include Indigenous and non-Indigenous viewers, whereas *Exile* is intended to be viewed by Inuit communities. The film circulated at festivals and was shown on CBC. As such, *Martha of the North* is a production entirely in line with the NFB/ONF mandate "to interpret Canada to Canadians and to other nations." As Karine Bertrand argues, community screenings "generated overwhelming reactions from the people who had lived through the relocations," and the documentary was a "prime reason why the government finally offered public apologies and recognized the damage done to the families that were separated from their land and relatives" (Bertrand 2019, 296). Soon after both films were released, and after years of resistance, Canada's minister of Indian and Northern affairs apologized for the government's actions, while claiming he didn't know what the government's motivation was for moving the families (Campion-Smith 2010). Like *Exile, Martha of the North* is political and engaged in interventionist historiography that calls to light the systematic violence enacted upon Indigenous communities by settler colonialism.

Kunuk's fiction feature *One Day in the Life of Noah Piugattuk* presents a different but related examination of Inuit relocation. Instead of reflecting historically on the "High Arctic" relocation—a high-profile geopolitical move by the Canadian government—*Noah Piugattuk* thematizes a standard mid-twentieth-century governmental operating procedure: the coerced displacement of Inuit from their land to settlements. Kunuk's film offers a metonymic snapshot of one contact between an Inuk hunter and a government representative in charge of relocation in the area, with the experiences of one individual standing in for the experiences of the community as a whole. The film is set during one day in 1961, in Kapuivik, on North Baffin Island, where after breakfast with his wife, Noah Piugattuk (Apayata Kotierk) sets off on the ice to hunt. The work builds on almost thirty years of Isuma practices in significant ways. As a fiction feature, it engages community-based filmmaking, working with a local Inuit crew and making a contemporary film about history in order to keep these past events alive in the present.

Isuma has a long history of documenting Inuit communities in a digital archive of stories told by elders. Some of these recordings are looked upon not as "films," but as a way of maintaining oral testimony. Some become complete "works," as is the case with *Exile* or *Kivitoo,* the fourth episode of their "Hunting with My Ancestors" series. *Kivitoo* combines found footage and oral testimony that document elders reflecting on their forced displacement in 1963 to Qikiqtarjuaq, fifty kilometers to the south.

In *Noah Piugattuk,* Isuma's oral archive of interviews with elders is drawn upon as the basis of the film's narrative. Aiming, in part, for circulation within the global media market, Kunuk transforms "oral testimony" from the Isuma archive into what the end credits stipulate is a script "based on a true story." Turning back to its oral testimony traditions, the film ends with the actual "oral testimony" of Noah Piugattuk, recorded a decade earlier, revealing the Indigenous origin of a work that can be construed by international audiences as an exemplar of Indigenous art cinema.

One Day in the Life of Noah Piugattuk foregrounds a fact known to Inuit across the Arctic: traveling by dogsled over frozen ice- or snow-covered ground is efficient and effective across vast lands. Meeting other travelers while hunting is a critical way of bartering and sharing news across the scarcely populated territory. The open expanses are contrasted with the opening and closing shots of the Piugattuk domestic space. Filmed mostly with still camera and long shots, formal aspects of the interior scenes replicate the stillness of the external landscapes, though one is cramped and cluttered and the other open and expansive. The interior scenes feature Noah and his wife, both of whom are mostly silent, while the wife's repeated coughing foreshadows, one can assume, a tuberculosis diagnosis. The dominant sound in these scenes comes from the wall clock, tick-tocking into the future. Once relocated, the closing sequence suggests, the family will have access to medicine as well as scarce staples such as sugar and coffee. But the move off the land is not construed as positive.

The figure personifying settler colonialism, patriarchy, capitalism, and the government is "Boss," whose English is translated by Evaluarjuk (Benjamin Kunuk), who serves the role of interpreter and cross-cultural diplomat seeking to bridge two worlds that are seen as diametrically opposed (figure 3.1). Noah's seemingly subdued reactions and his silence in this moment are significant: he does not explicitly challenge the demands made but does not consent, either. His unspoken protest is important—Kunuk's interest in authentic language use resonates also when words in Inuktitut are not uttered. The lack of real consultation and the misconstrual of Inuit reactions to white or governmental authority was synthesized by Rosemarie Kuptana

FIGURE 3.1. Noah Piugattuk (Apayata Kotierk) and "Boss" (Kim Bodnia) in *One Day in the Life of Noah Piugattuk* (Zacharias Kunuk, Canada, 2019).

in 1993 when she testified to the Canadian Royal Commission that for Inuit to present "a challenge to the authority of the Qallunaat [white people] or defiance of their requests was almost unthinkable" (Kuptàna 1993). For Kunuk, arguably, this interchange demonstrates Indigenous agency and perspective just as much as a spoken protest in Inuktitut, translated into English, would have done.

Danish actor Kim Bodnia plays "Boss," which provides an important connection to Kunuk's earlier fiction film about settler-Inuit contact, *The Journals of Knud Rasmussen*. In that film, Bodnia played the Danish-Greenlandic ethnographer and explorer Rasmussen. Kunuk reinterpreted a critical sequence in Rasmussen's treatise on pan-Inuit cultures, languages, and practices synthesized from the Fifth Thule Expedition (1921–24) and told to audiences worldwide in *Across Arctic America* (1927). In *The Journals of Knud Rasmussen,* Kunuk retells the 1922 meeting between Rasmussen and the shaman (*angakkuq*) Aua, presenting the encounter as a forced choice between adopting Christianity or remaining true to Indigenous spirituality. The meeting of traveling westerners and Inuit on the ice is also featured in the films made about the Fifth Thule Expedition, including *With Dogsled through Alaska* (*Med Hundeslæde gennem Alaska,* Leo Hansen and Rasmussen, Denmark, 1927; see MacKenzie and Stenport 2020, 2021). This film contains numerous sequences shot in present-day Nunavut where

Rasmussen and cameraman Leo Hansen are pictured in scenes not dissimilar from the meeting in *One Day in the Life of Noah Piugattuk*.

The connections between Isuma and Rasmussen are subtle, but they signal through Bodnia a cinematic continuity across time periods. The strategy also connects to the legacy of Rasmussen in Kunuk's oeuvre, including the TV series *Our Land* (Canada, 1993–94), which, like the earlier NFB/ONF *Netsilik* series (1965–69), was largely based on the ethnographic descriptions by Rasmussen and others participating in the Fifth Thule Expedition. The scene of Noah and "Boss" meeting in *Noah Piugattuk* should be understood in the context that no more than four decades have passed since Rasmussen traveled across Inuit lands from Greenland to Alaska on dogsled. At the same time, Bodnia's presence confirms Isuma's iterative process and the maintaining of connections between films, just as the contested legacy of Knud Rasmussen is reaffirmed.

While these works directly address the issues raised in the QTC report, more generally, in terms of the Canadian imaginary, they can be understood as part of a post-TRC Canadian cinema (including not just Inuit filmmakers, but Indigenous filmmakers throughout Canada). The TRC was overseen by the parties of the Indian Residential Schools Settlement Agreement, which was launched June 2, 2008, and concluded on December 16, 2015. The forced relocations were somewhat better known in the Canadian consciousness than the abuse of Inuit in residential schools, though there were Indigenous protests against the schools as early as the 1960s, and some residential schools remained open in Canada as late as 1998. The trauma of residential schooling is a key part of the TRC's findings. Forced relocation paralleled the dislocation of the residential school system, both actions functioning as forms of domicide, which was central to the Inuit experience of residential schools, where they faced certain forms of abuse distinct from those suffered by the First Nations to the south. The conclusion of volume 2 of the TRC's final report ("The Inuit and Northern Experience: The Final Report of the Truth and Reconciliation Commission of Canada") states:

> The harm done by residential schooling in the North remains. . . . When they returned to their communities, they were estranged from their parents, their language, and their culture. Many of their parents, the generation still in a state of shock from the upheavals of the 1940s through the 1960s, could not knit their communities back together again. The removal of children added to the damage already done by other economic and demographic changes. (Truth and Reconciliation Commission of Canada 2015, 186)

In post-TRC cinema such as *One Day in the Life of Noah Piugattuk,* the suspense is non-diegetic; viewers can infer things will end up badly for Noah and his family. We know this because of documentaries such as *Exile, Kivitoo,* and *Martha of the North,* but also because of the deep investigative and community-engaged TRC process and ensuing reports. Kunuk's and Lepage's works can be understood as catalysts for both the QTC and TRC processes, signaling the political significance of Inuit media sovereignty and the ways in which media works function as interventionist historiography that shapes public and political life.

The TRC is, as the term *reconciliation* implies, a process. Settlers in Canada must reimagine and come to terms with what they have done and, in many cases, continue to do. The works of Isuma, in particular, make this process possible through their use of multiple modes and forms to reach a variety of audiences. Some works, such as *Exile* and *Kivitoo,* speak first and foremost to Inuit communities, though they are also mobilized through digital platforms. Others, such as *Noah Piugattuk,* address—alongside Inuit viewers—Canadian and global audiences through a form of art cinema. Here, Isuma takes the mandate of the NFB/ONF "to interpret Canada to Canadians and to other nations," using it to its own ends in order to place settler Canadians in the position of reconciling the State's own past actions against the Inuit and Indigenous people more broadly. *Martha of the North,* produced at arm's length by the state through the NFB/ONF, circulates bilingually, both on public television and in schools, as an educational tool. To begin the process of reconciliation, the stories and histories of the acts of the Canadian nation-state need to be revisited through multiple modes and from multiple perspectives, allowing Indigenous people to have ownership over how their past experiences are told, and enabling settler Canadians to recognize themselves in the actions of the settler nation-state.

Local and Global Circulation: Isuma's Community Building across Multiple Networks

Circulation is a key component of Isuma's community-building priorities and its politics, and an integral component of Indigenous media sovereignty. Indeed, questions of circulation are as important as questions of production and representation. These priorities have led the collective to build, on the website isuma.tv, one of the most robust digital platforms and global networks of any Indigenous film collective. This is no small feat, considering the unreliable or partially nonexistent internet in the Far North. In the works we have discussed in this chapter, the politics of circulation reflect and intersect with many of the topics—climate change, resource extraction, relocation—along with the production circumstances of the works themselves. A key example of this constellation of representation, production, and circulation can be found in *Inuit Knowledge and Climate Change,* which circulated in a number of ways beyond single-channel exhibition. The completed version of *Inuit Knowledge and Climate Change* has had a significant film-festival and internet audience, viewed nearly 120,000 times on Isuma's website, while DVD distribution has primarily been isolated to library holdings.

Isuma pursued a politically activist line in response to an opportunity to attend COP 15 in Copenhagen in 2009, when the film was invited to screen as part of the United Nations University's *Indigenous Voices on Climate Change Film Festival* at Denmark's National Museum. Kunuk and codirector Mauro declined to attend, noting with irony that the hundreds of COP 15 delegates who had flown to Copenhagen would increase net planetary carbon emissions: "'We don't need to travel there and burn more fossil fuels,' Kunuk said. 'We can use digital technology to share what we have found with the world' through 'a live internet broadcast'" (CBC Arts 2009). Instead, they stayed in Nunavut, used the Isuma website to stream the twenty-three festival films, and moderated a global discussion about climate change impact on Indigenous communities from every continent.

This initiative was part of Kunuk's ten-day media event, the internet-streamed *Live from the Floe Edge* in December 2009, running concurrently with COP 15. Prior to *Live from the Floe Edge,* Kunuk and Mauro also held live editing sessions for the fifteen-minute version of *Inuit Knowledge and Climate Change* as a "call-in internet show," where anyone could comment and provide feedback on the evolving film, foregrounding the participatory nature of the project. These efforts were part of a six-month "internet campaign" called *Countdown to Copenhagen,* which also included talks by Sheila Watt-Cloutier, such as her Baldwin-Cartier Lecture on climate justice. Isuma encouraged alternative screening venues, from student centers (Hart House in Toronto) to art galleries (Urban Shaman in Winnipeg) to living rooms, in a call to action for "a creative happening/salon," declaring that "we need some grassroots (cross-continental) energy to start going viral!"—with the goal of eventually achieving "a network of 1000 local-to-global screening groups around the world watching *Live from the Floe Edge*" by December 2009 (PlanetFriendly 2009).

Inuit Knowledge and Climate Change also screened as part of the *Carbon 14: Climate Is Culture* exhibition at the Royal Ontario Museum in Toronto in 2013–14. By placing Indigenous voices alongside those of scientists in a museum setting, the exhibition provided the imprimatur of cultural capital to Isuma's work. The catalog for *Carbon 14* describes the exhibition as exploring "the growing global issue of climate change through the eyes of scientists, artists and cultural informers. Art and science come together in this engaging and provocative exhibition" (*Carbon 14* 2014, 18). In this exhibition, Kunuk and his collaborators exhibited three works, all of which mobilize different visual platforms. Along with *Inuit Knowledge and Climate Change,* which screened as a single-channel work, the exhibit also included *Our Baffinland,* which used GPS-enabled cameras to relay effects of increased resource extraction and toxin release on wildlife as experienced by hunters on Baffin Island near the Mary River Mine, run by the Baffinland Iron Mines Corporation. The work's title refers to the Baffinland mine, not to Baffin Island itself. By using "Baffinland" as the title and modifying it with "Our," the work reclaims the land and the mine as the Inuit's own.

The larger project, *Our Baffinland Atlas,* is an open-access virtual map with links to over a dozen site-specific videos, photos, and interviews about the effects of pollution and resource extraction. Kunuk and Mauro give voice to those in the Baffin Island community and bring to light public concern and dissent around mining and its expansion. *Our Baffinland* makes explicit use of early twenty-first-century everyday technology: the iPad. This device, while prioritizing personal screening, also allows for more interactive engagement than a large-screen video projection. The catalog describes the project as follows: "Using a new generation of GPS-enabled cameras, this spatial media project explores the place-based knowledge of Inuit elders and hunters, and their experiences in a changing and increasingly industrialized Arctic" (*Carbon 14* 2014, 42). The use of the handheld device allowed attendees to navigate the space in their own way and individually decide how many links to click and in which order, moving from video to photos to interviews. This interactive process is distinct from the monological mode of address found in single-channel works, and it allows viewers to construct their own experience. This also reflects media sovereignty principles as it accentuates the need for multiple voices to understand and participate in the debates around climate and resource extraction on Baffin Island.

If *Our Baffinland* can be seen in the ways described above as interactive, *Kobe* (Canada, 2012), Kunuk's 3D collaboration with Félix Lajeunesse and

FIGURE 3.2. *One Day in the Life of Noah Piugattuk* at the 58th Venice Biennale, May 11–November 24, 2019.

Paul Raphaël, which was also screened at the Royal Ontario Museum, can be better understood as immersive. Shot in stereoscopic DCP 3D and screened with glasses, the work explores the Nunavut landscape and Inuit cultural practices through the avatar/protagonist Kobe, a ten-year-old Inuk. Edited without commentary in the style of an observational documentary, the work complements *Inuit Knowledge and Climate Change* and *Our Baffinland* by offering viewers the feeling of being "in" Nunavut, allowing for a visual-sensory feeling of connection. The use of a young boy also provides a contrast to the voices of the elders in the other works, pointing to the future.

The practices of circulation developed over two decades were fully deployed in Isuma's exhibition in the Canadian pavilion at the 58th Venice Biennale in 2019 (figure 3.2). The six-month exhibition consisted of three major projects, based on illuminating "Canada's forced relocation of the Inuit people in the 1950s and the power of the media today to reclaim history and see the best in us" (La Biennale di Venezia 2019). Along with screening the premiere of *One Day in the Life of Noah Piugattuk* (which played in the pavilion throughout the Biennale), the exhibition also included *Silakut,* which was a program of webcasts "from Nunavut to the world" and *Isuma Online,* an online curation of Isuma and Indigenous works from thirty countries and multiple languages.

Connecting back to Isuma's media intervention during COP 15 when Kunuk ran the identically titled live internet event *Live from the Floe Edge,* the Biennale efforts highlight a decade of work and emphasize the relationship between climate change and resource extraction in Nunavut. *Silakut* consisted of ten episodes streamed live during the summer of 2019, which were archived on the Isuma website, then edited and released as a single-channel documentary under the same title. The single-channel version of *Silakut* compiles aspects of the various episodes, delineating not only the community debates, but also the process through which they take place. Each of the streamed episodes of *Silakut* covered community discussions about relocation, mining, Indigenous history, and music—making public, on a global scale, conversations about the politics and culture of Inuit life that are rarely given a voice outside of local communities. In the ten-episode series, mining and resource extraction were of particular importance, especially the Mary River Mine, which was first discovered in 1962 and was purchased in 1986 by Baffinland.

Treaties have existed since the 1990s in relation to socioeconomic Indigenous rights on Baffin Island and the resource extraction at the mine: "Signed in 1993, the NLCA provides that the Nunavut Impact Review Board (NIRB) act in the environmental, cultural, and socioeconomic interests of the Nunavummiut. Aside from the James Bay Agreement, this is one of the most elaborate regulatory regimes of resource development on Indigenous territory in the world, and the NIRB is perhaps the best model of direct engagement that exists" (Scobie and Rodgers 2019, 233–34). With the increasing visibility and awareness of the climate crisis in Nunavut, views on resource extraction have shifted along with the changing climate. *Silakut* gives voice to this growing dissent by Inuit inhabitants of Baffin Island, many of whom do not feel they have a voice in the decisions being made in the community. By streaming these images at the Canadian pavilion at the Venice Biennale, *Silakut* becomes sanctioned both as Canadian nation-state discourse and as art. *Silakut: Live from the Floe Edge* can also be understood as an important example of post-TRC cinema through its media sovereignty and reimagining of modes of inclusion in the Canadian nation-state.

MEDIA CONVERGENCE AND DIGITAL ARCHIVES

One of the ways in which Isuma's diverse practices of circulation form a new kind of media sovereignty is through, as Faye Ginsburg notes, media convergence: "As Indigenous media has grown more robust over the last two decades—in part through the increasing convergence of media forms—this synthesis of media technology with new forms of collective self-production has much to offer Indigenous communities as they redefine themselves to their future generations, as well as the broader world, in the twenty-first century" (2019, 270). Indeed, for Ginsburg, one of the key aspects of media sovereignty is the fact that it takes place in the "digital age" (2016, 581), which allows for various forms of convergence and migration between platforms. Isuma's interventions in the global mediascape through *Silakut* at COP 15 in 2009 and the Venice Biennale in 2019 build on their deliberate interest in promoting a global circumpolar and Indigenous media coalitional solidarity. A key initiative in this vein is the "Digital Indigenous Democracy" (DID) project, which was started in 2013 by Norman Cohn and Zacharias Kunuk. DID links radio, television, digital, and social media materials from Inuit, Sámi, and other Indigenous cultural producers.

As an act of media sovereignty, Isuma emphatically does not want their work to live in a vacuum or in isolation from others. This impetus translates into the creation of a living DID archive, where documents from the past are not simply collected, but where new documents, as they are produced, engage with the past, present, and future. These kinds of digitally networked and media convergence efforts have not, to date, been greatly associated with Indigenous filmmaking in other parts of the world. Isuma's interest in maintaining a DID archive in a rapidly changing media environment, and the effort and labor it takes to maintain such an archive, constitutes an activist and interventionist approach that challenges a capitalist system of planned obsolescence and its market-driven obsession with the new. Yet Isuma has not been immune to capitalist pressures, having reconstituted itself multiple times over the decades, including declaring bankruptcy and defaulting on a loan with a Montréal-based Nunavut investment firm in 2009. The firm did not have a cultural investment in Nunavut and, as such, had a low stake in the cultural value of the company. Despite the significance and success of Inuit Indigenous filmmaking and Isuma, economic and infrastructural limitations continue to persist in the Arctic.

Given these financial pressures, Isuma's prioritizing of multimodal and media-convergent circulation is integral to media sovereignty and emerges as deliberately political in nature. As a whole, their work functions as locally situated texts positioned for a global, transnational, and diverse set of audiences, including other Indigenous communities, Western scientists, policymakers, and the general public, which reframes the work through its various modes of circulation, screening, streaming, and distribution. In these contexts, Isuma's multichannel works and living digital archives intersect with a wide range of dominant and alternative discourses on the Arctic, including climate change, resource extraction, sovereignty, and truth and reconciliation. These practices are acts of media sovereignty in that they not only give voice to Inuit people, but also place them at the center of various cultural and political platforms, such as publicly funded news, international conferences, galleries, and international exhibitions. If dominant forms of media writ large—such as the internet, broadcast television, film festivals, cinema distribution, streaming services, and museums—can be seen as both monological and hierarchical, the media sovereignty practices of Isuma—appropriating and retooling these forms to their own political and cultural ends—redress these inequities, while throwing into question the supposedly normative nature of dominant political and cultural modes of communication and representation.

————

FOUR

The Arnait Collective, Feminist Practice, and Inuit Self-Determination

LOCATED IN IGLULIK (IGLOOLIK), NUNAVUT, the Arnait Video Production Collective is the first women's Inuit filmmaking collective. Founded in 1991, Arnait builds upon and connects to the production, distribution, and exhibition strategies deployed by Isuma while providing an important countermovement to that collective's more pronounced focus on men's lives. Arnait engages in the collective production of films and videos from Inuit women's perspectives, where politics of recognition and representation often foreground gendered inequities of settler colonialism, seen also in the lives of children and youth (for a history of the Arnait collective, see Cache Collective 2008; Evans 2008, 172–90; MacKenzie and Stenport 2016; McGough 2019). Like that of Isuma, Arnait's early body of work parallels the rise of Inuit political mobilization and the founding of the Government of Nunavut (GN) in 1999. Similarly, like those of Isuma, Arnait's more recent works have addressed many of the issues explored in the report "Nuutauniq: Moves in Inuit Life," part of *Qikiqtani Truth Commission: Thematic Reports and Special Studies 1950–1975* (Qikiqtani Inuit Association 2013), and in the residential schools report issued by the Truth and Reconciliation Commission of Canada (2015). These issues include forced relocation, mental health, and suicide. Isuma's and Arnait's bodies of work both illustrate and helped shape the QTC and TRC conversations during the first decades of the twenty-first century. What sets Arnait's work apart, however, is the focus on storytelling through the underrepresented voices of women.

Through political and governmental processes, gender roles are both constructed and reaffirmed, such that Inuit and "Indigenous discourses and colonizing policies include boundaries of exclusion and silence that entrap men and women differently" (Altamirano-Jiménez 2008, 128; see also

Kuokkanen 2019; NIMMIWG 2019). These processes were also part of the founding of the GN, where a gender parity clause for parliamentary representation was eventually voted down. Isabel Altamirano-Jiménez argues that gender inequities in the Inuit self-determination process are evident across several different social and cultural layers. These include a denigration by male leadership of colonialism's unequal effects as less important "women's issues" and the foregrounding of male-centered identities, practices, and economic priorities, such as hunting and resource ownership, as central to the nation-building or self-determination process; to this end, "land claims processes are embedded in the colonial representations of women as landless and domestically placed" (Altamirano-Jiménez 2008, 131).

These assumptions also carry through in the QTC's reports, which do not address gendered experience of the relocation process during the period 1950–75, except for one section about women's domestic labor. Arnait's works should be read as a reaction to these inequities and can be put into context with Altamirano-Jiménez's assertion that "the contestable issue of gender remains submerged in political struggles emphasizing self-determination, cultural difference, and experiences of material and social inequalities" by which "Indigenous women's voices remain 'muted'" (2008, 129). Arnait's working principles and examined topics resonate with a different modus operandi, namely, that "Indigenous traditions portrayed by women generally place women at the centre of communities, families, and political and cultural practices including the participation of the collective in achieving balance and consensus" (2008, 130). Arnait's emphasis on women's community-oriented work thus functions as a political strategy in addition to one that is important in cultural and aesthetic terms.

Arnait's productions, Dianne Chisholm argues, can be read as resistance and resilience activism through the lens of Gerald Vizenor's (2008) concept of "survivance," in that it comes "to flourish in contemporary media despite and in response to colonialism's systematic suppression of oral traditions" (Chisholm 2016, 211; see also Bertrand 2017, 2019). There is an additional critical component implicit in survivance, though not explicitly emphasized by Vizenor: the oral and cultural traditions of women have always been further silenced by colonialism, imperialism, patriarchy, and capitalism compared to their male counterparts. Addressing this double silencing in multiple ways—through content, practice, circulation, and international feminist solidarity work—is part of Arnait's extraordinary legacy and contribution, not only to Indigenous cinema and media, but also to world cinema, as well as to Nunavut and Canadian culture, politics, and history. At the same time, like that of many Indigenous women activists and artists, Arnait's work does not follow the road map of Western feminism. Indeed, "Indigenous women are complex figures to feminists, not only because of these women's double racial and gender identities, but also because [their] actions and political positions seem to point in contradictory directions" (Altamirano-Jiménez 2008, 129; see also Green 2007).

True to their emphasis on thematic complexity, community involvement in production and circulation, cross-cultural negotiation, multimodal delivery, artistic experimentation, and the use of Inuktitut (alongside English and French) and Inuit oral storytelling practices, Arnait's body of work is not uniform, though all works prioritize the points of view of Inuit women and employ women-only or women-dominated production crews. Arnait has directed, produced, and coproduced over twenty works since its founding, including four fiction features, one documentary feature, two television series, twelve short and mid-length documentaries, short experimental media works, one short and one mid-length fiction film, web-based digital media, museum and gallery exhibitions, and works of animation.

Arnait's cofounders, Marie-Hélène Cousineau, a Franco-Québécoise from Montréal, and Madeline Piujuq Ivalu, an Inuk from Iglulik, collaborated on the majority of these productions over the thirty years of Arnait's existence between its launch in 1991 and its closing in 2021, also including a number of other Iglulik-based women film practitioners such as Susan Avingaq, Mary Kunuk, Martha Maktar, and Lucy Tulugarjuk. The collective has been extensively recognized in Canada, with Cousineau and Ivalu honored as 2014 Women in Film by Telefilm Canada at the Toronto International Film Festival (TIFF).

In addition to their centrality to Inuit and Canadian cinema and public culture, Arnait operates in a global context. Their productions have garnered extensive international recognition. For instance, *Before Tomorrow* (*Le jour avant le lendemain,* Marie-Hélène Cousineau and Madeline Piujuq Ivalu, Canada, 2008) had a two-week run at Film Forum in New York, competed at Sundance, and won the award for best first Canadian feature at TIFF. Other feature-length films, such as *Uvanga* (Cousineau and Ivalu, Canada, 2013), *Sol* (Cousineau and Susan Avingaq, Canada, 2014), *Tia and Piujuq* (Lucy Tulugarjuk, Canada, 2018), and *Restless River* (Cousineau and Ivalu, Canada, 2019), have circulated widely through film festivals and on Canadian streaming platforms such as Crave. Arnait's work has a global reach through its digital archive and repository, with many works freely available on its website, www.arnaitvideo.ca, at isuma.tv, and through Isuma's dedicated iTunes channel. The collective operated internationally for most of its thirty-year history as part of its commitment to feminist media collective collaboration. This includes travel and international exchange between Inuit and Greenlandic elders in Nuuk in 2013, which focused on sharing Indigenous knowledge by reviving premodern communication routes of goods and cultural transfer (Arnait Video Productions n.d.[b]). Arnait also partnered with Ojo de Agua, in Oaxaca, Mexico, a grassroots collective that promotes Indigenous communication through local media development. Arnait toured with them in February and March 2009: "In each place, a combination of workshops, seminars, cultural performances and screenings were held so that people could meet and get to know each other in different types of spaces. Our hopes were that people could compare cultural experiences, find out who is working on similar things." Arnait asked the question: "What will come from extending these networks between north and south? We don't really know. But this has been an effort to set up a network, a conversation that spans periods of time and that bypasses national boundaries" (Cousineau 2010).

Arnait's three-decade production history spans a range of concerns pertaining to feminist filmmaking practice that extend far beyond the preconceived assumptions of national or Indigenous film practice; to this end, their works constitute a critical but heretofore often overlooked contribution to Indigenous and world cinema and feminist politics.

ARNAIT'S ORIGIN: COLLECTIVE PRACTICE AND WOMEN'S LIVES

A collective process is key to Arnait's origins and agenda, as Cousineau, Ivalu, Avingaq, and other members have continually affirmed (see McGough 2019; *Ikuma: Carnet de Tournage* 2008). The "originality of Arnait Video Productions' works," Cousineau explains, stems from "a production process that is in harmony with the lives of the women involved in each project. Our production values reflect the cultural values of the participants: respect for community events, for Elders, for hunting and fishing seasons, for certain traditions belonging to particular families, among others" (Arnait Video Productions. n.d.[a]). This emphasis was part of the foundation of the collective and reflects the gendered, everyday experiences and historical practices of Inuit life in Iglulik, as Cousineau reports in a 1996 interview: "I find that activities are separated by gender in this community. . . . I realized that the gender thing didn't work like it worked in my culture. That was obvious. Men were doing things together and women were doing things together and they were both working for the same purpose. . . . I was interested in women's problems and women's culture" (Fleming 1996, 13–14). Susan Avingaq reflects on her decision to join Arnait after hearing a call for women's participation on local radio: "Also I could participate, even if my husband did not want me to. What struck me [in that announcement] was that women and men could do the same things, that we are equals" ("Conversation" 2008, 31). Madeline Ivalu affirms that discriminatory attitudes remained in place well after Arnait had established itself; in 2007, "when we began the project *Umiaq,* the men tried to discourage me because I was a woman" ("Conversation" 2008, 32).

Arnait's early films, such as *Attagutaaluk (Starvation,* Arnait Ikkagurtigitt Collective, 1992), focused on oral recording of historical events through women's perspectives. These productions were as much documents as they were documentary videos. The collective's aesthetic strategies expanded as they developed a variety of goals for their video and film production. One of the best-known early video works is *Qulliq (Oil Lamp,* Susan Avingaq, Madeline Ivalu and Marie-Hélène Cousineau, Canada, 1993), which depicts, in real time to show the exact procedures, a historically central women's task: lighting a seal oil lamp in a newly built snow house. The soundtrack consists of the women (Avingaq and Ivalu) singing as they prepare the lamp. Pedagogy, authentic language use, and practical documentation thus coalesce in the video, further strengthening Arnait's explicit goal of simultaneously archiving, supporting, and creating women's activities—from lighting the qulliq to making videos—and imbuing these with artistic intentionality. *Qulliq* also stands as a reminder that Arctic ice cultures were different only thirty years ago. Climate change effects were not yet as noticeable then. Local and global discourses around the crisis of diminishing ice cover and increased temperatures were not front and center in the media and in politics. In hindsight, Arnait's snow- and ice-focused works come to function as confirmation, reminder, and memorialization of the centrality of cryocultures for Inuit, what Sheila Watt-Cloutier calls "the right to be cold" (2015). Add to that the environmental aspects of Indigenous self-determination and the work done by women to that effect, and works such as *Qulliq* assume heightened political significance.

In this way, *Qulliq* and related short Arnait works from the 1990s and early 2000s foreground women's activities, which do not take center stage in the works of the Inuit Broadcasting Corporation (IBC) or Isuma. To this end, Arnait's works reframe and prioritize stories, practices, norms, and values that would otherwise "remain 'muted'" (Altamirano-Jiménez 2008, 129). Using their voices as a political tool was central to Iglulik women wanting to participate in the Arnait project, as it would allow these little-told stories to be more widely known through television broadcast. Arnait's early works provided the film education and production design expertise that laid the ground for later feature-length fiction works. Many of these early works were also experimental, filmed without a script through collaborative processes, including computer and hand-drawn animation, reenactment, and the mixing of multiple technologies. Arnait's *Show Me the Map, Part I: A Changing World* (Marie-Hélène Cousineau and Carol Kunnuk, Canada, 2010) and *Show Me the Map, Part II: People Can Stand Up* (Cousineau and Kunnuk, Canada, 2010) both examine the impact of climate change and iron ore extraction on Baffin Island. Moreover, the ice melt allows for greater acceleration of resource extraction as multinationals prospect for the increasingly accessible oil that lies underneath, thereby portraying Arctic environments in new and changing ways. The two-part documentary also addresses debates in Inuit communities about emergent uranium mining. Some organizations have supported increased resource extraction, while other grassroots groups argue against these developments, demonstrating that the strain between environment and economy is a complex one in Nunavut. These works engage in different forms of media sovereignty through the use of multiple platforms, from online streaming to exhibition in gallery spaces, to engage with a variety of communities.

Arnait's media sovereignty and its emphasis on Inuit women's lives has multiple layers and operates across many political registers. In addition to the ways they connect the global to the local, Arnait filmmakers assert their role in a national political context—their films and videos are repeatedly described as "Canadian." Material on Arnait's website emphasizes the group's mandate to preserve the cultural heritage of Nunavut—and most of their early works are exclusively in Inuktitut—while also educating all Canadians about the North, inviting them "to value the voices of Inuit women in debates of interest to all Canadians" (Isuma TV 2009). This sentiment both firmly positions Arnait within an explicit Canadian tradition and foregrounds the fact that—at least in terms of Indigenous Inuit women—other media organizations have not done so.

The language politics of Canada are important for understanding Arnait's priorities in intervening in the national discourse. For the thirty years they worked together, Cousineau and Ivalu shared fluency in neither Inuktitut (Ivalu's main language) nor French and English (Cousineau's main languages). The Arnait partnership attests to an ability and a desire to collaborate artistically and politically without a shared or dominant language. Translators do work on scripts and mediate public appearances in the South, and all of their films predominantly, if not exclusively, feature dialogue in Inuktitut. Furthermore, the European (French-English) language debates that have dominated Canadian federal and provincial politics for the past sixty years do not apply and have little resonance for Arnait's works. This is a political intervention by Arnait, effectively asking the settler-dominated public to move beyond debates shaped by colonialism and patriarchy, and to embrace complex and multimodal linguistic and cultural expression that might otherwise be silenced. Similarly, the collective's attention to distinctive cryo-environments throughout their oeuvre integrates a different understanding of "the Arctic" into Canadian and international politics.

SETTLER COLONIALISM, ARNAIT, AND THE TRC: *RESTLESS RIVER, BEFORE TOMORROW,* AND *UVANGA*

These priorities of practice and politics inform Arnait's TRC-oriented feature-length films *Before Tomorrow, Uvanga, Sol,* and *Restless River,* each of which addresses the intersections of settler colonialism and the experiences of Inuit women and Inuit youth. Engaging different pivotal moments of settler colonialism through the eyes of women protagonists, *Before Tomorrow* (set in the 1840s) is a cryoculture film—ice, snow, and Arctic winter are central to the plot. *Restless River* (set in the 1940s–60s) is also a historical narrative that thematizes the bonds between women, children, Arctic environments, and lived experiences of the land while involving community reenactment of cultural practices. The other two TRC films are set in contemporary Iglulik and address the suicide pandemic in Nunavut, *Uvanga* foregrounding intergenerational conflict and trauma between a son and his mother while *Sol* examines the legal system, prejudice, and societal constraints related to suicide and mental health. All four films engage, in different ways, the priorities and practices of the QTC, the TRC, and the National

Inquiry into Missing and Murdered Indigenous Women and Girls (NIMMIWG 2019). *Before Tomorrow* was conceptualized and released at the start of the QTC (2007–10) and TRC (2008–15) processes, *Uvanga* and *Sol* were released during the height of the TRC's work, and *Restless River* concluded the trilogy as the NIMMIWG report was released. The three fiction films, *Before Tomorrow, Restless River,* and *Uvanga,* form a trilogy. The films create an interconnected ecosystem of related stories, experiences, and actions. In turn, the form of the trilogy becomes an important interventionist historiography, serving as critical analysis and reflection of how colonialism has shaped the lives of women and youth over time, with the second two films of the trilogy integrating more English into the dialogue.

Like the truth and reconciliation process and the ensuing reports, the Arnait films share ways of addressing not only what has happened in the past, the effects of settler colonialism, the inequitable and misguided governmental regimes, and the cultural mistrust and miscommunication that ensued, but also how these events reverberate and exist in the present, and how to find ways forward. If the TRC and the QTC not only account for the past, but also make recommendations in terms of reconciliation, we can see in Arnait's works a similar investment in making films that can be understood as ways of addressing the past and present of settler colonialism, its effects, and ways forward. These films share with the NIMMIWG an emphasis on gendered inequality, violence against Indigenous women, and women's experiences of colonialism.

Restless River is a film about an Inuit woman's experience of settler colonialism, the militarization of the Arctic, and the Canadian government's resettlement program. Set in Kuujjuaq, the largest city in Nunavik, Northern Québec, the film has an important historical context. During World War II, the American presence in the Canadian Arctic grew across the region, including a US Coast Guard weather station and a US Army airfield on the shore of the Koksoak River in Kuujjuaq (then Fort Chimo) in Northern Québec in 1941–42. *Restless River* begins during this period, with the lingering American cultural presence as a metaphorical backdrop: Elsa (Malaya Qaunirq Chapman) and her young friends are Hollywood movie–obsessed, unlike the older Inuit generation. The rape of Elsa by an unnamed American soldier shortly before his departure from the American-held Fort Chimo provides the narrative impetus, as Elsa's body is invaded by white, settler patriarchy. The outcome is a son named Jimmy, whose light complexion, blue eyes, and blond hair serve as a symbol of North American colonialism and international geopolitics, which have affected numerous Inuit populations, including through the High Arctic relocation program during the Cold War. *Restless River* engages this still little-known part of Canadian-US history through its effects on one Inuk woman.

Codirected and coscripted by Cousineau and Ivalu, the film is based on Franco-Manitoban/Québécoise author Gabrielle Roy's 1970 novel *La rivière sans repos* (translated into English as *Windflower*), also set in Kuujjuaq. *Restless River* can be seen as an instantiation of post-TRC cinema. It subtly changes the trajectory and agency of Elsa, from the sense of tragedy and defeat that, Cousineau states, suffuses the character's arc in the novel (see Montpetit 2019) to one that allows her, in the film, despite the colonialism she experiences, to gain a sense of ownership and agency over her life at the narrative's conclusion. This trajectory is also a key component of another of Cousineau's post-TRC films, *Angelique's Isle* (codirected by Michelle Derosier, Canada, 2018). The film's similarities to *Restless River* include the representation of strong, female, Indigenous agency, embodied in the survival skills developed by Angelique, a young Anishinaabe woman (Julia Jones), trapped on an island on Lake Superior through the winter of 1845, during the copper rush, with her increasingly demented and violent voyageur husband. Both *Angelique's Isle* and *Restless River* use English as the settler's language.

The fact that *Restless River*—shot and set in Northern Québec, with provincial SODEC (Société de développement des entreprises culturelles) funding, and based on a French-language novel—foregrounds English over Inuktitut is important. In Québec at the time, English (in relation to both French and Indigenous speakers) was the language of power, whether used by Hudson Bay Company employees (who were often English), by members of the American military, or by Catholic priests (who were almost all French but often used English as a lingua franca with Indigenous peoples). In the process of adaptation, Cousineau argues that the use of English reflects the settler context of the time, with film offering a verisimilitude not possible in the novel (Montpetit 2019).

In *Restless River,* the trap that Elsa faces is one not of geographic or linguistic isolation, but of encroaching colonialist modernity and the changes that settler presence brings about. This is evident both in how she supports herself through wage labor as a maid to white settlers—who are trying to replicate the trappings of middle-class suburbia in Québec's "wilderness" outpost—and in how she seeks to bring up her child on the land with

FIGURE 4.1. Grandmother (Madeline Ivalu) and Jimmy (Nick Serino) in *Restless River* (*La rivière sans repos,* Marie-Hélène Cousineau and Madeline Ivalu, Canada, 2019).

extended family, even though her mother and grandmother caution against her acts of rebellion (figure 4.1).

Although Elsa's cultural preferences and her body are colonized, and Westernized, Cousineau notes that the character of Elsa is deliberately construed to exemplify these changes: "We can see the changes around her, in her community, and the internal conflicts the changes cause her. She comes from a more traditional life. This is truly a pivotal time for this community in Nunavik. She crosses it [from traditional life by] raising her child" (Berckvens 2019). These changes speak to the changes in the Arctic's geopolitical and environmental landscape wrought by militarization.

Restless River is also part of the process of truth and reconciliation, with the NIMMIWG as frame of reference. Cousineau states: "Canadians in general do not really know the history of the North. It is an opportunity to experience the life of this woman from the inside, to make a connection between cultures, to make the 'other' less 'foreign', to understand another way of doing things, another culture. Right now, with the Commission of Inquiry into Murdered and Missing Indigenous Women, it is more essential than ever to go and see the experiences of these women" (Berckvens 2019). But *Restless River* is not a story of missing and murdered women per se. Rather, it can be called a TRC film for the ways in which it recognizes and celebrates Inuit women's resilience and the significance of matrilineal bonds. The concluding scene shows Elsa out on the land, on her own, explaining to two young girls of the community that she is happy with her life, and that after facing pressure from the parish priest to find a husband and build a heteronormative family, she still wishes to remain alone.

This is not a new theme in Arnait's work. Indeed, there is an obvious and political point of connection between *Before Tomorrow* and *Restless River:* the relationship between settler colonialism and women's bodies, especially forced sex. Adapting the film from a novel set in Greenland called *Før Morgendagen* (1975) by Danish writer Jørn Riel, Cousineau and Ivalu set *Before Tomorrow* in the 1840s, making it a "first contact" film. This despite the absence of settlers on screen—they are nonetheless present from the very beginning of the film's narrative. The film uses Inuktitut exclusively, another example of Arnait's interventionist historiography, whereby historically significant processes should be told in the original language. One of the oral stories told to the community in *Before Tomorrow* by a visitor, Kukik (Tumasie Sivuarapik), is the Inuit account of meeting settlers for the first time. Along with disparaging stories of the European men's drunkenness, and humor at their expense, is the tale of Inuit men—presumably plied with alcohol—acquiring tools, specifically knives and needles, by agreeing to European men's request to have sex with Inuit women. The narrative assumes a patriarchal foundation to aspects of Inuit culture, and that access to women's bodies could be bartered for goods at the time. The invisible settlers haunt the narrative, with their destructive power manifest in the use of women's bodies to acquire needles and knives, which the film implies was the moment of transmission of smallpox, a disease that eventually kills most of their community while the two main characters, grandmother Ninioq (Ivalu) and grandson Maniq (Paul-Dylan Ivalu), are away on a nearby island. The two persist on their own for some time, guided by Ninioq's shamanistic storytelling, living out the end of their lives on their own terms through traditional practices, until Ninioq makes the decision to let them both die.

Before Tomorrow recognizes and memorializes the centrality of cryoculture to Inuit life. Ninioq notably chooses to let the two of them freeze to death, with the final sequence in black and white suggesting a certain kind of transcendence of the material world. As such, we can also read *Before Tomorrow* as a climate change allegory, emphasizing "the right to be cold" as central to Inuit self-determination. The formal composition of the landscape in winter light is stunning in the second half of the film. This environmental context is also what brings forth the myths and stories told by Ninioq to her grandson. Thus, constitutive aspects of Inuit culture operate on multiple levels in the film; in hindsight, the recognition of the fragility of ice becomes an added layer in the work's richness.

Released at the moment of the initiation of the TRC and QTC processes, *Before Tomorrow* seeks to bridge the past, present, and future through women's oral storytelling by referencing colonial trafficking and the abuse of women, along with the suggestion that Inuit men were at times complicit in this process, whether through rape, prostitution, or murder. This history has had long-standing negative effects on Inuit lives (those of both men and women) and should be understood as central to Arnait's mission to bring these gendered aspects of settler contact to light. A review in the *New York Times* remarks that *Before Tomorrow* is the one feature-length drama associated with Isuma that "focus[es] on women's roles as storytellers and repositories of folk wisdom; its perspective might be described as Inuit feminist" (Holden 2009). As Dianne Chisholm argues, the "film mediates the reappearance of the dead before foreign powers lay their ghost to rest" and confronts "contemporary audiences with spectres of Inuit life that resist being ghosted by the colonial archive. At the same time, they [Arnait] *revive* traditional culture so that it endures in collective memory and contemporary culture with visionary prospects of renewal" (Chisholm 2016, 212, 226; see also Bertrand 2015). The continuity of Inuit ways of living connects *Before Tomorrow* with *Restless River,* including the latter's examination of colonial rape, maternal relationships, and the future for youth. In contrast to Maniq in *Before Tomorrow,* whose premature death is directly related to the negative effects of settler contact, Jimmy's trajectory offers potential reconciliation between his Inuit and settler identities.

Like *Before Tomorrow, Restless River* keeps a focus on the next generation. To this end, the character of Jimmy provides a focal point for many of the rapid mid-century changes in Inuit life that the QTC report outlines. As a baby and toddler, he stays with his grandmother while Elsa works as a maid in the settlement, precluded from bringing her son into that family's house. Elsa then moves farther out on the land to join Isaki, an extended-family member (also called "Uncle" and played by Etua Snowball) who becomes her partner. As a young boy, Jimmy brings Elsa and Isaki together through the learning of traditional male Inuit skills like hunting and living off the land—central to Inuit knowledge and Inuit culture's connection to, and interaction with, local environments. When Jimmy contracts pneumonia—and attracts the attention of a government representative, who affirms that he needs modern medical care and must then return to school—he and Elsa move back to the settlement. They live in what one may presume is government housing; Elsa works as a seamstress, never returning to Isaki.

In the final section of the film, Jimmy is at first presented as a talented but demoralized teenager, whose whiteness is juxtaposed with the dark hair and complexion of his Inuit friends. After an argument with Elsa, when he asks her for details about his father that she cannot provide (such as his name), Jimmy runs away. In the end, we learn that he has realized his ambition to become a pilot, as he calls through shortwave radio to his mother, "Uncle" Isaki, and extended family from the skies. The language transitions in these two scenes are significant. In the first scene, Elsa addresses her son in Inuktitut, he responds in English, and she then switches to English, demonstrating that the cultural bond between them has been weakened. In the second, though, Jimmy speaks Inuktitut in his shortwave message, reconciling himself with his heritage and family by not using the settler language. The fact that he is flying is also significant, giving him a perspective on the Arctic environment that is the most familiar one for westerners: the aerial view of the landscape so well known to both the military and filmgoers.

Restless River expands on two recurrent plotlines in Arnait's oeuvre. The first focuses on women's embedded knowledge. The film includes several scenes with Madeline Ivalu, playing Elsa's grandmother, who functions as a voice of matrilineal guidance while providing a strong sense of continuity between Arnait works. While the film's focus is on Elsa, *Restless River* also examines the domestic life of her white settler employer, Ms. Beaulieu (Magalie Lépine Blondeau), whose ennui, sense of alienation, and, at times, shame are left implied but unexplained, signaling the gender inequalities in white settler culture, while simultaneously addressing the oppressive settler power dynamic that Beaulieu holds over Elsa.

The second plotline recurrent in Arnait's work is the experience of Inuit children and youth through the lens of settler colonialism, often guided by strong maternal figures. *Restless River* refuses to stigmatize Jimmy's biracial identity as a child of colonial rape; Elsa instead prioritizes the need for Jimmy to experience living off the land in the Inuit way, thanks to a strong, positive male figure. The film also postulates that moving to the settlement is both a saving grace (penicillin and hospital care) and a loss, as we see teenage Jimmy and his Inuit friends drinking and skipping school instead of learning skills such as hunting. Moreover, while Jimmy, in an act of racial passing, escapes to the South where he can gain access to education—presumably in part because of his white appearance—the film implies that his Inuit male friends will stay, perhaps succumbing to alcoholism. Finally, *Restless River* shows that the matrilineal bond is weakened by colonialism: Jimmy's internalized anger against the colonial rape is directed toward the victim, Elsa. He leaves then to resume communication at a distance and through a technology vital for remote communities (a shortwave radio), while he flies above the land.

Restless River is a story of how relocation and settler colonialism impact women's lives and matrilineal bonds; as such, it serves as an important companion piece to Kunuk's *One Day in the Life of Noah Piugattuk,* presenting a version of what happened to many Inuit communities in the mid-twentieth century and emphasizing how these lived realities may play out differently for women and men. "Uncle" Isaki in *Restless River,* for instance, refuses to relocate; yet this resistance is implied to be short lived, and that option is not even open to Elsa and Jimmy. On the other hand, *Restless River,* as a post-TRC film, signals ways forward as an act of reconciliation. As with most of Arnait's work, reconciliation is conceptualized less as a grand political statement and more as a forward-looking and redemptive perspective on the lives of Indigenous youth and family that is constantly evolving against the imposition of settler colonialism. Jimmy's escape, to return as a disembodied voice, and Elsa maintaining her strength in her independence offer a redemptive narrative that outlines positive ways forward in light of the NIMMIWG findings.

The third film in the trilogy, *Uvanga,* is set in twenty-first-century Iglulik. Many of the themes introduced in *Before Tomorrow* and *Restless River* reoccur in different forms in *Uvanga.* These include the impact on Inuit lives under settler colonialism; matrilineal relationships to boys and young men; and multilayered connections between the past and present, especially the haunting aspect of dead or absent fathers. *Uvanga* also engages in participatory community production, with many of the cast being inhabitants of Iglulik (Madeline Ivalu plays a grandmother, for instance); uses both Inuktitut and English, interweaving oral storytelling and lessons of intergenerational learning; integrates depictions of hunting and living off the land; and is a character-driven piece. Moreover, the film examines global issues that are not exclusive to present-day Nunavut. Cousineau affirms: "The characters . . . were not inspired by anyone in particular, but there are many families like theirs in the world today: separated, mixed-blood children discovering their roots and identities; Grandparents connecting with newly found Grandchildren; and adults trying to mend broken relationships. This story could have taken place anywhere, but the one we are telling takes place in the

North in a remote community on Baffin Island" ("Director's Note" n.d.). Achieving a realistic portrayal of everyday life in Iglulik through a representative group of characters was important to the film's genesis, Cousineau argues in interviews (e.g., Barnard 2014). *Uvanga* should therefore be read as a political act of media sovereignty that mobilizes art-cinema in service to issues addressed in the QTC and the TRC: premature death and suicide, especially among Inuit men; unequal economic opportunity; alcohol and drug abuse; limited mental health services; and multilayered and complex relationships to local environments.

As in several other Arnait productions, aspects of southern and Québécois culture are integrated into the Indigenous production circumstances and themes to establish a cross-cultural, hybrid, and multidimensional cinematic universe, which resonates with and partly reflects the production circumstances, as Cousineau and Ivalu bring complementary perspectives as codirectors. Brenda Longfellow argues that Arnait's works and the reception thereof are characterized by "'multicultural fluency,' as [Carl Bessire] puts it, which may, in some cases, be attributable to the involvement of non-Native collaborators but which, in any case, produces texts that are polyvalent and eminently adaptable to complex and distinct interpretive communities" (Longfellow 2019, 324).

Cross-cultural themes are an integral part of *Uvanga*'s plot. As Montréalaise Anna (Marianne Farley) returns north with teenage son Tomas (Lukasi Forrest) to revisit Tomas's family, the past comes under increasing scrutiny as Tomas, Anna, and the community are forced to revisit the circumstances surrounding the death of Tomas's father, and Anna's then partner, Caleb. The film's opening does not spell this out; like Tomas, the viewer is left on the outside to learn who is who in the community, as well as what the road and object of discovery will be. The plot, narrative strategies, and production circumstances thus support one another to form an integrated whole. *Uvanga,* then, is about the journeys of all the characters in the film to reveal contemporary life in Iglulik and the experiences of cross-cultural learning.

Like *Before Tomorrow* and *Restless River, Uvanga* foregrounds matrilineage and embraces the notion that maternal relationships are complex and multifaceted, yet central to the lives of young men. In *Uvanga,* the main maternal character (Anna) is white; her biracial son has a different relationship to Inuit culture than Maniq in *Before Tomorrow* or Jimmy in *Restless River.* A central aspect of *Uvanga*'s plot is the child learning about Inuit traditions, including becoming a hunter and adept at living off the land. This is

FIGURE 4.2. Sarah (Madeline Ivalu), Caleb's mother and Tomas's grandmother, hugs a family member in *Uvanga* (Marie-Hélène Cousineau and Madeline Ivalu, Canada, 2013).

an important environmental connection. The film reconfigures the paradigm as a gesture to present-day cross-cultural and linguistic complexity and the mobility in and out of Iglulik. Moreover, the status of the absent father is not technically an abandonment, since Anna left the relationship with Caleb early on; yet Caleb "disappeared" into substance abuse and suicide. Instead of Caleb, Tomas's half-brother Travis (Travis Kunnuk) becomes his guide to Inuit culture, with the audience learning about Iglulik life alongside Tomas. As the grandmother of Tomas and Travis, Ivalu becomes the film's emotional anchor, mediating numerous, complex interpersonal and intercultural relationships, while mourning her son's death and the destruction that alcohol and substance abuse has brought on her community (figure 4.2).

Here, Arnait is deliberately expanding its cinematic universe to include stories of contemporary relevance, demonstrating how Inuit stories resonate for a range of local and global audiences. The film intertwines settler and white culture with Inuit and Indigenous culture, serving as a document of the TRC process and emphasizing a composite picture of settler colonialism through the journeys of its many characters toward reconciliation of past tragedies with hopes of equitable cross-cultural relationships in the future.

In contrast to *Before Tomorrow* and *Restless River, Uvanga* further embraces its contemporary setting through its cinematography, production design, and location. While set in a luminescent Arctic summer landscape, the film's visuals do not construct a transcendental aesthetic of "natural beauty" or "pristine landscapes." Viewers are consistently asked to refute such spectacularization. The environment is first presented from a plane's arrival at the airfield, then through the smudged windows of a car, where run-down houses and buildings are foregrounded. Evidence of settler colonialism, including debris, dirt, and ramshackle housing, is integral to the film's presentation. Characters are notably positioned with their backs to the stunning coastal visuals, as though these visuals are there to satisfy the needs of a "tourist" viewer. These cinematographic strategies are indicative of the film's attempt to be a composite picture of present-day Iglulik. *Uvanga* is not a period piece, nor is it a community-based historical reenactment. Instead, the community creates the characters who inhabit Iglulik, and we see *their* town. The film portrays and presents their reality—spatial, emotional, familial, environmental—as an integral part of its diegesis.

Uvanga's realistic portrayal of present-day Iglulik and its complex cross-cultural relationships, a legacy of settler colonialism, is a summative piece of Arnait's fiction feature trilogy. In these works, women are the tellers of stories and the anchors of their families and communities. They are represented as individuals bestowed with agency and power. Their stories are integral to illuminating the gendered politics surrounding the founding of Nunavut and the historical inequities brought to light by the QTC and TRC processes. Arnait's media sovereignty consists not only in their extraordinary two decades of collective and communal film and video making, but even more so in their insistence on the agency of women in shaping how the past is known, the present enacted, and the future imagined, locally and globally. As such, Arnait's significance—and the worlds brought to life through their trilogy of fiction films—should not be underestimated.

ACTIVIST TRC DOCUMENTARY: ARNAIT'S *SOL*

As an inquiry into the murder or suicide of an Inuk man, Arnait's documentary *Sol* can be seen as an undertaking related to the tasks and reports of the TRC, QTC, and NIMMIWG. The film contributes to a post-TRC process in Nunavut, with Cousineau and Avingaq charting a path forward to address the causes of suicide in Nunavut, while also imagining the potential for a different future. The task of *Sol,* then, is not simply to address the vastly disproportionate suicide rate, but to speak of it publicly, to reduce the silence and shame that surrounds the epidemic. *Sol* is centered on the life and death of Solomon Uyurasuk (1986–2012), a young Inuk performer. The film focuses on whether Sol committed suicide or was killed by the Royal Canadian Mounted Police (RCMP) while in custody. In 2013, the suicide rate in Nunavut was 13.5 times higher than the national average for Canada (CBC

News 2014). As Michael Kral writes, the "Inuit youth suicide pandemic began thirty years ago and is now a major public health problem" whose causes cannot be addressed by a traditional, Western psychiatric or sociological approach of identifying "individual risk factors" but must instead be understood from "a community, cultural, and historical perspective" in which suicide is constituted as "a symbol of social suffering" (Kral 2019, 2). Kral's work, which builds on decades of embedded research with youth in Nunavut and Nunavik, also emphasizes that resolution and reconciliation—he uses the term *reclamation*—will be grounded in community agency, local health programs, and knowledge that reflects Indigenous self-determination movements and sovereignty (2019, 104–42). *Sol* speaks to these priorities and addresses a multiplicity of perspectives in the questions it raises about youth suicide, including what Kral describes as destructive "coping mechanisms for lives fractured by colonial machinations" (2019, 86).

Sol premiered to public acclaim at the imagineNATIVE Film + Media Arts Festival in October 2014, and a slightly edited version was released theatrically in December of the same year. Set in Iglulik, *Sol* foregrounds a range of aspects of contemporary life in Nunavut, including changing intergenerational relationships, unemployment, lack of access to educational opportunities, mental health and substance abuse, poverty, violence, and suicide. While in many ways it is a realist documentary, Cousineau and Avingaq also draw on the elliptical narrative structures of Arnait's fiction films, along with the experimental temporal aesthetics of some of their early videos. This thematic and formal complexity is significant and reflects twenty-first-century developments in documentary practice. As Longfellow notes, Arnait's work is "highly responsive to the transnational field of video art and experimental film" (2019, 324). The world portrayed in *Sol* positions Nunavut as globally connected, and as deeply interconnected with, and shaped by, the politics and culture of the South. Against this backdrop, *Sol* is far more concerned with asking questions—difficult questions that are often left unasked—than with providing didactic answers.

Beyond suicide, *Sol* also foregrounds many other aspects of twenty-first-century life in the North. This complexity is conveyed by the integration of numerous cultural and artistic practices that span Inuit and global pop culture, foregrounding especially the use of Inuktitut. In an interview, local residents address a concern that influences from the South do not allow young Inuit to know who they are, perhaps leading to a drastic rise in substance abuse and suicide. But as in Isuma's *Inuit Knowledge and Climate Change*

(Zacharias Kunuk and Ian Mauro, Canada, 2010), multiple Inuit voices offer a variety of points of view on changing cultural influences. *Sol* includes an interview and in-house performance by an Iglulik-based Inuk hip-hop musician who protests against the suicide epidemic, not by invoking traditional cultural practices, but by writing and recording a hip-hop song, appropriating the disenfranchised musical mode of protest of Black North Americans to intervene in the North.

In *Sol*, Nunavut is presented as an interstitial cultural and political space. A key example of this is the film's use of the art and performances of Artcirq. Along with Sol's life and death, this "social circus project," of which Sol was a member, is a key focal point of the documentary (another Artcirq member, Joey Ammaq, also died under suspicious circumstances in 2014). A local Inuit youth circus, cofounded in 1998 by Guillaume Ittukssarjuat Saladin (who grew up in Québec and Iglulik) in response to two youth suicides, Artcirq is partly funded by Isuma. It brings together polyvalent aspects of culture and combines modern circus arts like clowning, acrobatics, and juggling with traditional Inuit cultural practices such as throat singing, traditional games, and drum dancing.

Along with performing live, Artcirq has made a series of videos and documentaries over the past decade and has released three albums. They performed as part of the Isuma exhibition at the 2019 Venice Biennale, and, as documented in *Sol*, they have traveled and performed globally. Their video works cover a range of areas in performance and cross-cultural practice. These include an account of the group's formation, *Artcirq* (Natar Ungalaaq and Guillaume Ittukssarjuat Saladin, Canada, 2001); a documentary about Inuit adapting to climate change, *Pitanqangittuq* (Guillaume Ittuksarjuat Saladin, Nicolas Tardif, and Félix Pharand Deschênes, Canada, 2010); two public service announcements (PSAs) about littering in the Arctic directed by Saladin, *Dump Tent* (Canada, 2011) and *Raven* (Canada, 2011); a fiction film about substance abuse and violence, *Issaittuq* (Bruce Haulli and Kenneth Rasmussen, Canada, 2011); and a rockumentary about the surviving members of the first Inuit-language rock band, *Northern Haze: Living the Dream* (Derek Aqqiaruq, Canada, 2011). All of these works either document aspects of performance or include performances by Artcirq members. The works engage in media sovereignty through mobilizing video to both make works that address current political and cultural issues in Nunavut and facilitate the involvement of youth in addressing these issues. Many of the works, such as *Amazonia* (Saladin, Canada, 2007), address cross-cultural Indigenous initiatives undertaken by Artcirq—in this case between Amazonia in Peru and Nunavut. This travel is part of a larger frame of reference about cross-cultural communication and intercultural exchange, and travel back and forth to Nunavut is a central trope in *Sol,* the works of Artcirq, and the international collaborations of Arnait.

Along with the social and political concern about suicide and the possibilities of intercultural collaboration, *Sol* and Artcirq demonstrate a commitment to media sovereignty, both in production and in the use of multiple media platforms to spread their messages. Both *Sol* and Artcirq can be seen, in part—beyond their artistic production—as producing PSAs that mobilize a set of websites and additional information about suicide and abuse prevention, which, in the case of *Sol,* are designed to accompany the film (*Sol* n.d.). As Cousineau noted in an introduction to the film at the 2015 Visible Evidence documentary conference in Toronto, when *Sol* screened in Iglulik and other communities in Nunavut, it did so to full houses, followed by community meetings seeking to address both causes and symptoms of the high suicide rate. Yet Cousineau also acknowledged that the film's inconclusive stance on the cause of Sol's death was highly intentional on behalf of Arnait; they sought to demonstrate how Sol's story can be seen as an analogy for many of the interrelated, underlying complexities of twenty-first-century life in Nunavut. This approach allows the film to convey a nuanced, and at times contradictory, presentation of life in the North.

As Bill Nichols argues, documentary films can be sorted into different models, though the boundaries between these models are fluid (2017, 105–07). *Sol* follows this pattern of fluidity, blending what Nichols calls the advocacy, history, and testimonial models (2017, 106). For Nichols, models are forms of discourse that preexist and live outside the documentary form. These models influence and intersect with what he outlines as the six modes of documentary, which are media specific yet fluid. *Sol* intersects with three of Nichols's six modes: the participatory, the reflexive, and the performative (2017, 108–09). *Sol* draws on the participatory through the ceding of knowledge and story to the inhabitants of Iglulik. It is reflexive through its use of self-questioning techniques about what happened to Sol in RCMP custody. It is performative through the presence of both filmmakers as on-screen participants raising questions, and through what Nichols calls direct, experimental encounters (2017, 107). As part of the participatory mode, the structure of the film contains many interviews conducted after Sol's death—most often with friends and relatives, but also with public officials—alongside news reports of local and Nunavut politics. It also shows flashbacks to Sol's childhood and youth as captured in home movies, in TV footage—he appeared with his grandmother Rachel Uyarasuk in Isuma's *Nunavut* (*Our Land,* Canada, 1995)—and in videos of Artcirq performances and tours. *Sol's* use of mediated interview techniques is important in terms of the performance mode; codirector Cousineau cannot speak Inuktitut, and so the film leaves her out of the diegetic picture, while sometimes including her in frame through a Skype window or similar media platforms. Cousineau functions as a cultural translator for settler viewers, while simultaneously acknowledging the difficulty and necessity of the actual intercultural work—in language, Indigeneity, culture, status, age, and class—that the film engages.

As part of the reflexive mode, *Sol* can be seen as a self-reflexive media history of the Isuma and Arnait film and video collectives and the video practices of Artcirq, demonstrating that in Iglulik, moving-image storytelling has become central as an act of media sovereignty. *Sol* integrates sequences from well-known Inuit TV and cinema productions into narratives, providing a running commentary on the extraordinary record of visual image production in the community since the start of IBC in the early 1980s and onward, through the establishment of Isuma and Arnait in the early 1990s, with Artcirq's launch in the late 1990s being yet another powerful example. Therefore, one of the guiding themes at the heart of *Sol* is the ever-changing relationship between Inuit culture and settler colonialism—and the repurposing of settler culture and technology to Inuit cultural, communal, and political ends. As codirectors, through the representation of cultures on screen and the use of various forms of digital technologies and modes of production, Avingaq and Cousineau acknowledge the (at times) contradictory values and perils of asymmetrical intercultural communication and exchange.

In the questions it asks about the Nunavut suicide epidemic over the past three decades, *Sol* also contributes to the important reclamation work that Kral (2019) outlines as central to a resolution. In the case of Arnait and the work of Artcirq that it documents, it is clear that reclamation, in Kral's sense, includes both the formulation of and access to Inuit-developed and community-oriented health as a form of contemporary Indigenous knowledge and strengthened self-governance, as Kral argues. *Sol* also functions as a form of community activism and sensemaking through art, creativity, performance, and cross-cultural innovation and the development of multiple and hybrid artistic forms.

Arnait's role in addressing the suicide pandemic of Nunavut also lies in its emphasis on Iglulik's and Nunavut's global interconnectedness, its support of Solomon Uyurasuk, and Artcirq's exploration of these relationships through artistic practice that is, at the same time, an act of Indigenous political and community activism and media sovereignty to combat youth suicide. Arnait's post-TRC films, such as *Restless River, Uvanga,* and *Sol,* foreground the collective's media sovereignty and community practices as part of an ongoing TRC process, wherein art, creativity, and the imagination of new futures are all central to the collective's practice and to the films it creates. The diegetic function of *Restless River, Uvanga,* and *Sol,* as texts, makes them key examples of what we have called post-TRC cinema, challenging audiences to reconcile the past and present effects and actions of settler colonialism.

BEYOND ARNAIT: ALETHEA ARNAQUQ-BARIL'S ENVIRONMENTAL AND GEOPOLITICAL ACTIVISM

One of the best-known current practitioners of Inuit documentary political activism is director and producer Alethea Arnaquq-Baril, owner of Unikkaat Studios in Iqaluit, Nunavut's capital. As the home of the Nunavut Film Development Corporation, Iqaluit is the administrative center of Nunavut film production. IBC has a production office there, and in 2015 the Nunavut Media Arts Centre opened, which houses the Inuit Film and Video Archive. Growing up with Isuma, Arnait, and NFB/ONF films shot in the Arctic, Arnaquq-Baril recognizes Kunuk's significance for the growth of Inuit film and media sovereignty in Nunavut. In addition to maintaining a strong, and often social justice–focused, social media presence on Facebook and Twitter, she has addressed contemporary political issues in Nunavut directly in several of her films. *Tunniit: Retracing the Lines of Inuit Tattoos* (Canada, 2010) presents the director's travels with Inuk lawyer and artist Aaju Peter across Nunavut as they interview Inuk women about the demise of the practice of facial tattoos, a tradition abolished by colonialism. In *Aviliaq: Entwined* (Canada, 2014), made for imagineNATIVE's Embargo Collective II project, Arnaquq-Baril made the first Inuit lesbian-themed film as a piece of fictional reenactment, affirming the director's dedication to challenging cultural and political taboos (for more on Embargo Collective II, see chapter 6). She coproduced the fiction feature *The Grizzlies* (Miranda de Pencier, Canada, 2018), which, alongside *Sol,* addresses youth suicide in Nunavut.

Arnaquq-Baril is best known for her critically acclaimed documentary *Angry Inuk* (Canada, 2016), an NFB/ONF coproduction with Unikkaat Studios (for more on *Angry Inuk,* see Sakakibara 2018; Vanstone and Winston 2019; Burelle 2020; MacKenzie and Stenport 2023). *Angry Inuk* seeks to confront facile animal rights activism as an act of neocolonialism by westerners, foregrounding the economic, cultural, and political stakes of seal hunting for Canadian Inuit. It outlines how the ban on commercial sales of seal products disallows Inuit repurposing of practices, skills, and natural resources that are integral to their communities' survival and adaptation to the global market economy. The documentary takes issue, for reasons both political and cultural, with the animal rights arguments made about seal hunting by NGOs, including Greenpeace; by celebrity activists like Pamela Anderson and Brigitte Bardot; and by supranational and intergovernmental organizations such as the European Union, which banned trade in seal skins first in 1983 and again in 2009—a complete ban was upheld in the European Court in 2015 after it had been challenged by the Canadian and Norwegian governments (see Fakhri 2017; Hennig and Caddell 2017). The film continues Arnaquq-Baril's collaboration with Aaju Peter, and the two function as the film's protagonists (figure 4.3). As Peter notes about the European Union's seal product import ban: "They want us to be like sick little Eskimos who are stuck on the land and go out in our little Eskimo clothes with a harpoon. . . . They will not let us sell commercial products. It's a form of cultural colonialism. A journalist in the Netherlands called it the Bambification of the Inuit, like we're in some Disney movie" (McLaughlin 2010, cited in Wolfe 2019, 284). Peter's prominence as an Inuit activist voice has led to a third documentary, this one about her own life and the unexpected death of her son, the forthcoming Nunavut/Greenland coproduction *Twice Colonized* (Lin Alluna, Canada/Greenland, 2023).

As Arnaquq-Baril notes in voiceover commentary, Nunavut is one of the most food-insecure locations in North America; though subsistence hunting of seal is allowed, such hunting does not generate the kind of income needed to buy groceries flown in from the South to support the growth of a new generation of hunters. It may even make Inuit more supportive of opening up fragile environments for resource extraction, which generates cash-paying jobs.

In *Angry Inuk,* Arnaquq-Baril argues that the seal economy can and should be conceptualized as so much more than "traditional hunting," since garments made from seal skins—like the ones made by Peter and shown in

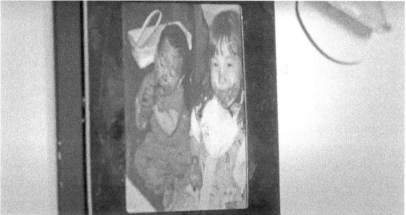

FIGURE 4.3. Aaju Peter and Alethea Arnaquq-Baril discussing sealing, animal rights, and geopolitics in *Angry Inuk* (Alethea Arnaquq-Baril, Canada, 2016).

the film—are a form of locally sourced, ecologically manufactured, and environmentally sustainable clothing that the slow-fashion and animal rights movements *should* be able to stand behind. Seal, moreover, is a more climate-conscious product than synthetic outdoor-wear made of petroleum-derived synthetics. As Louise Crewe (2017) notes, garments—their production, circulation, wearing, and memorialization—exist within a geopolitical (and, in this case, neocolonial) system and are not an innocent or frivolous commodity lacking "real" political and economic implications. Contributing alternative angles to this geopolitical and international relations history is part of what Arnaquq-Baril seeks to do through *Angry Inuk*.

Angry Inuk engages in extensive media activism as well as hybrid documentary practices, which include first-person commentary, animation, and the integration of social media as political and aesthetic tools. The film begins by introducing the director as an activist and on-screen presence, linking her personal family history and community in Iqaluit to the geopolitics of the seal product ban and the role of Western animal rights activists. *Angry Inuk* makes three important claims in its opening sequences. The first is the assertion that animal rights media campaigns by organizations such as Greenpeace have removed Inuit from having a stake in the global sealing industry beyond mere subsistence hunting and local use, even though the export of seal products could be an important global trade commodity for the Inuit. The second is that animal rights media campaigns have focused on protesting the slaughter of photogenic white harp seal pups. This has led to a near-universal anti-sealing sentiment, even though the harp seal is not an endangered species and is hunted only in Labrador and not by Indigenous communities. The third is that the agenda pushed by Bardot and other famous animal rights activists since the 1970s has not been sufficiently challenged by the EU or nation-state governments, which, in the view of Arnaquq-Baril and many others, have mounted little effective opposition to decisions that impact Arctic Indigenous hunters in Canada, Alaska, Greenland, and Russia.

In *Angry Inuk,* Arnaquq-Baril's voice and presence function as a counterpoint to other women in the film, all white and presumably middle class, who work for animal rights organizations. Some of these women self-silence and will not defend their views in the documentary; others hide behind pretaped interviews included in the film, so that their points can be made but go unchallenged by Arnaquq-Baril's questioning. What Arnaquq-Baril's self-narration of all their various withdrawals points to is the fact that women who gain a voice in the public sphere (again, in this case, white women) have a responsibility to defend their points of view and to recognize and acknowledge the fact that their views can be refuted by disenfranchised Indigenous women. These women also need to admit that their own politics are complicit with colonial frameworks that suppress and overrule Inuit political priorities and sovereignty over resources, and recognize that because of their access to donations, campaign funding, and lobbyists through manipulative ad campaigns, they are equipped with overwhelming cultural and economic capital for their activism to set the global agenda.

Angry Inuk locates the seal hunt as part of an inequitable global economy. This purpose also connects to the film's deployment of digital activist politics, demonstrating how, despite all the tropes of Arctic remoteness and the fact that Nunavut is a far-off land in the eyes of most Europeans, EU decisions made in Brussels affect the Inuit politically, culturally, and economically. Arnaquq-Baril deploys a multitude of media platforms—including video, digital media, social media, Skype, and found footage from sources as varied as YouTube and Library and Archives Canada—to demonstrate global connectivity. This archival investigatory process happens within a digitally networked public sphere, and a key aspect of Arnaquq-Baril's activism in this regard is her promotion of the #sealfie media campaign on Twitter, a response to Ellen DeGeneres's 2014 Oscar-ceremony selfie and animal rights tweet against seal hunting. Through media shares, that selfie tweet raised about $1.5 million for the International Humane Society (Athens 2014, 43–46).

The #sealfie countermovement was, in part, a way to demonstrate the importance of sealskin to the Inuit, while promoting it as a fashion statement. As noted in the film, after the throat singer and performer Tanya Tagaq posted a picture of her newborn infant bundled up beside a recently slaughtered seal, there was trollish backlash. *Angry Inuk,* as an act of Indigenous media sovereignty, foregrounds the contested nature of the image of the seal as signifier. As propagated by Greenpeace, the popular image of the white-furred pup clubbed for its pelt by white settlers has only one connotation. Arnaquq-Baril both exposes the ideological and colonial underpinnings implicit in this image and reframes the significance of the image of the seal, on Indigenous terms and within an Indigenous frame of reference, as a form of media sovereignty, including through the use of media activism such as the #sealfie campaign and the integration of social media campaigns into the documentary.

Arnaquq-Baril's film emphasizes that all aspects of the seal hunt are significant in the context of global economics and politics—as well as central to contemporary Inuit culture and society—and that women's political activism is critical for bringing about change that aligns with the priorities of Indigenous sovereignty. These aspects of Arnaquq-Baril's practice demonstrate the significance of interventionist historiography. While the post-TRC works of Isuma and Arnait discussed in this and the previous chapter stem from and address a number of issues examined in the QTC report, as well as in the foundational documents of the GN, Arnaquq-Baril documents political mobilization by and for Inuit in a global context.

These works point to a wide array of formal and political approaches to interventionist historiography and exemplify the Arnait collective's decades-long practice of fostering Indigenous media sovereignty, foregrounding the specificity of Arctic environments, and advancing the principles of what we call TRC cinema. Working through both collectives and arm's-length governmental agencies like the NFB/ONF, the works engage in acts of media sovereignty to advance questions of political self-determination, often closely related to Indigenous language use, environmental priorities, and what Watt-Cloutier deems a fundamental human right: "the right to be cold." The works of these Indigenous women not only foreground a variety of approaches—from narrative cinema to documentary, from experimental work to web-based streaming—but also offer a complementary counternarrative to the works of groups such as Isuma, highlighting the diversity of voices in Nunavut and Nunavik in regard to Indigeneity, the environment, and the role of gender in media sovereignty. These approaches are also significant to other Arctic Indigenous film and media making.

FIVE

Sámi Media Sovereignty and Interventionist Historiography

ENVIRONMENTAL, EXPERIMENTAL, AND ARCHIVAL POLITICS

IN THIS CHAPTER, WE FOCUS on works of interventionist historiography, especially works that serve environmentalist, political, and societal purposes beyond their depiction of historic events or issues of the recent past. As such, these films function as an explicit counterpoint to the silencing of Sámi voices within the Scandinavian historiographical tradition as well as in visual anthropology and ethnographic filmmaking across the global circumpolar North. The use of Sámi languages and the emphasis on oral storytelling—as framing devices, in on-screen text, or in the form of self-reflexive critique—are especially critical. While we point to strands of filmmaking that are different and complementary, we also trace the continuities and contextual points of engagement between them, pertaining to the ways in which these films are deliberately activist examples of interventionist historiography in relation to Indigenous self-representation and self-determination. These films challenge settler colonialism and ideologies of the Nordic welfare states, including environmental sustainability and resource extractivism, while relating these to self-governance debates. Specifically, we address how these recent works provide a counterpoint against a long cinematic history of outsider representations about Sámi that tended to exoticize, infantilize, primitivize, and silence the group into a homogeneous whole (see J. Lehtola 2000; Mecsei 2015; Skarðhammar 2008; Cocq and Dubois 2019; Stenport 2019). While all these strands intersect, one of the major strands we focus on is the growing body of works made by women that address gender politics in the Sámi context, bringing to light narratives, points of view, and perspectives distinct from and complementary to those of the first wave of male Sámi auteurs.

From the mid-1980s onward, key works of interventionist historiography were made by a pioneering generation of male Sámi filmmakers, especially

Nils Gaup and Paul-Anders Simma. Not only are these filmmakers pioneers in the Sámi context, but their works are also foundational to the emergence of transnational and global Indigenous cinemas. Gaup's fiction feature *Pathfinder* (*Ofelaš/Veiviseren*, Norway, 1987), one of the first instances of "Fourth Cinema," reached audiences well outside Sápmi, including through its nomination for an Academy Award for best foreign film—the first Indigenous production to receive that distinction. Māori filmmaker Barry Barclay's 2003 manifesto, "Celebrating Fourth Cinema"—in which he introduced that distinctive term, *Fourth Cinema*—appeared as Indigenous documentary (especially autoethnography) and fiction film (especially art cinema) began to be seen as both part of and in tension with debates around global media and visual anthropology's self-reflexive shift during the 1990s (Barclay 2003; see also Columpar 2010; Ginsburg 1995). The term *Fourth Cinema* does not speak to the history of ethnographic representations of Indigenous peoples; instead, it builds on the decolonization goals inherent in Third Cinema. Moreover, Barclay's concept of Fourth Cinema is articulated a posteriori, championing anticolonial and decolonial strategies as a constituent part of what we define as interventionist historiography. Fourth Cinema, as an emergent global movement, Barclay postulates, will become part of an internationally circulating, critically recognized, financially viable, and distinct cinema. Barclay notes that, at the time of his manifesto, "Indigenous cultures are outside the national orthodoxy. They are outside the national outlook. They are outside spiritually, for sure. And almost everywhere on the planet, Indigenous Peoples, some 300 million of them in total, according to the statisticians—are outside materially also. They are outside the national outlook by definition, for Indigenous cultures are ancient remnant cultures persisting within the modern nation state" (Barclay 2003, 10).

It is worth noting that Barclay includes two Arctic historical narratives that speak specifically to this definition: *Pathfinder* and Kunuk's *Atanarjuat: The Fast Runner* (Canada, 2001). Based on myths retold through oral storytelling practices in the face of settler colonialism, these well-known fiction films address Indigenous relationships to the land and the distinctive Arctic environment of snow and ice, which have shaped these cultures, myths, and spiritual worldviews for millennia. Building on Barclay's call to establish a Fourth Cinema, *Pathfinder* and *Atanarjuat* counteract ethnographic cinema's pervasive and insidious representation that "situates indigenous peoples in a displaced temporal realm" (Rony 1996, 8) and outside of globalization frameworks. Barclay's goal, then, is to assert Indigenous cinema's autonomy with respect to representation of local environments and practices, while also asserting that these representations should be seen as constitutive of globalization movements, and not as separate from or insignificant to them. The importance of this manifesto—and the central role that Arctic Indigenous works played in its formulation—is not simply its diagnostic function in 2003, but the ways in which it charted out the futures of Fourth Cinema, many of which have materialized. In the context of Sápmi, these include the filmmaking trajectory that *Pathfinder* and earlier Sámi documentary works spearheaded, as well as the founding of the International Sámi Film Institute (ISFI) and film festivals such as Skábmagovat and Tromsø. This emergent constellation over a twenty-year period speaks to the processes of funding, production, circulation, and distribution that make Fourth Cinema a form of interventionist historiography, challenging the processes and functions of the First, Second, and Third cinemas that came before.

ENVIRONMENTAL JUSTICE DOCUMENTARY IN SÁPMI

There is nevertheless a substantial history of Sámi film and media predating the release of *Pathfinder*. Sámi-language media production has been important for Sámi self-determination and political agency for over seventy years. Launched in 1946 by NRK (Norwegian Broadcasting Corporation) via radio broadcast, there are now multiple units of public radio and television broadcasters that produce news, documentaries, children's programming, and television series: Finland's YLE Sápmi; Norway's NRK Sápmi; Sweden's SR Sameradion and SVT Sápmi; and Russia's Kola Sámi Radio, Murmansk. The largest of these, with over a hundred employees, is NRK Sápmi. The units also coproduce and collaborate on programming. Sari Pietikäinen notes that the Sámi media are "diverse but uneven" and "function in a complex and paradoxical terrain of language endangerment and revitalization and of political struggle and negotiation of Indigenous Sámi rights" (Pietikäinen 2008, 201). This characterization of Sámi-language broadcasting both builds on the commitment to public service radio and television in the Nordic countries and reflects the significance of film production as an integral part of the Sámi cultural and political revitalization movements of the late 1970s and the 1980s. As Pietikäinen confirms, "the mere existence of Indigenous media signals the vitality and cultural legitimacy of the Sámi community" (2008, 203).

Not only did these productions counteract earlier ethnographic representations to instead foreground Sámi agency in the contemporary moment (in Sámi languages), but they also reflect the rise of international environmental movements during the 1970s, including the UN's World Commission on Environment and Development, led by Norwegian prime minister Gro Harlem Brundtland. While environmental activists in Stockholm and Oslo protested nuclear power plants and nuclear proliferation, thereby endorsing hydropower as an innocuously "clean" and "natural" energy source, these issues took on different and local significance in Sápmi. Hydroelectric dams on Indigenous lands became a galvanizing political flashpoint because of their impact on reindeer herding, subsistence hunting, and fishing. The scars that dams create in the fragile ecosystems of tundra and mountains visualize oppressive colonial relationships that were beginning to be articulated in political discourse at the time.

Several films exemplify constituent components of this movement. First, the transnational coproduction *Ja de bodii dulvi* ("And Then Came the Flood," Finland/Norway/Sweden, 1976), by Sámi documentarians Kalle Mannela (YLE, Finland), Johs. Kalvemo (NRK, Norway), and Per Niia (SVT, Sweden), expressed deep-seated criticism of water resource management in Sápmi. The collaboration among the three Scandinavian broadcasters signals the significance of constituting Sápmi as a transnational region, whose environments and Indigenous practices form a continuum broken by nation-state borders and the colonial histories that imposed Western governance structures onto preexisting Indigenous communities. Indeed, Sámi media producers at the time collaborated to raise awareness of environmental inequities, using the resources of the national governments (YLE, NRK, and SVT) for political and media sovereignty purposes. A similar work of international significance is Norwegian Arvid Skauge's *The Boy from Lapland* (*Ante,* Norway, 1975). First made as a six-part NRK TV series, it was recut to feature length in 1976. Sold to twenty-three countries, the work became one of the first mainstream media critiques of Norway's policies on "Norwegianization" of the Sámi.

Even more importantly, the widespread Sámi opposition to the Norwegian government's building of the massive Álttá (Alta) dam (1979–81) led, as Coppélie Cocq and Thomas Dubois argue, to "a strategy of image making that proved a powerful tool for gaining attention and redirecting public discourse in directions favorable to Sámi" (2019, 27). Contemporary film showed, in unprecedented ways, the Sámi people with agency and interest

not only in shaping representations of themselves as an Indigenous population, but in challenging the actions and policies of the Norwegian nation-
state. The docudrama *Let the River Live!* (*La elva leve!,* Bredo Greve, Norway,
1980) is especially significant in its impact on local and national debates. As
of this writing, Ole Giæver's fiction feature *La elva leve* ("Let the River Live,"
Norway, 2022) was produced with support from the Norwegian Film
Institute and a budget that, in the Scandinavian context, is large: 40 million
kroner. As a historical drama, it features a woman protagonist (played by
Sámi actor Ella Marie Hætta Isaksen) who discovers her Sámi heritage
through involvement in the political protests. The Álttá protests of the late
1970s, and their representations on screen, galvanized community involvement and spurred revival of Sámi Indigenous culture, including the expansion of NRK Sápmi and the establishment of theater groups and cultural
organizations, especially in Guovdageaidnu (Kautokeino). This provided a
training ground for many Sámi actors and directors, especially as part of the
Beaivváš Sámi theater group, which started as an independent ensemble
under the leadership of Nils Gaup in 1981. These community-strengthening
efforts meant expansion of access to the Sámi language through cultural
building efforts in arts, performance, and higher education, including as part
of the Sámi University College, established in Guovdageaidnu in 1989. The
strong growth during the 1980s of Sámi culture initiatives in Sámi languages
must be seen not only as a complement to "traditional" political mobilization, but as leading those efforts and strengthening the cultural foundations
from which to examine, question, and protest against centuries of settler
colonialism.

Notably, the Sámi parliaments in the three Scandinavian countries were
established as an outcome both of environmental justice activism and
of the language and culture revitalization movements. The Norwegian
Sámi Parliament opened in Kárášjohka (Karasjok) in 1989, the Swedish Sámi
Parliament opened in Giron (Kiruna) in 1993, and the Finnish Sámi
Parliament opened in Aanaar (Inari) in 1996. These entities have been criticized as inefficient at advancing self-determination, because they replicate the
settler-colonial and nation-state superimposition of non-Indigenous governance structures onto Indigenous populations, including the marginalization
of non-reindeer-herding Sámi communities in decision-making processes
(Kuokkanen 2019, 100–07). The first wave of Sámi media sovereignty from
the period of the late 1970s through the first decade of the twenty-first century focused on environmental justice, with films made by men and mostly

about activist men. The environmental justice movements and the emergent political structures included a range of women activists, but media sovereignty works of the time largely ignored those contributions, arguably influencing how Sámi Indigenous political movements have since been represented in Sámi activist works.

*Representing the Suppressed Past and Addressing
the Contested Present: Interventionist Historiography
and Gendered Sámi Political Agency*

As political mobilization expanded across Sápmi, several groundbreaking films shaped social, cultural, and political awareness of Sámi experiences, both within and outside of Sápmi. Two of these films were released in 1987 and share the environmental justice impetus that distinguished earlier Sámi documentaries: Gaup's *Pathfinder* and Stefan Jarl's *The Threat (Hotet/ Uhkkádus,* Sweden, 1987). Gaup's ensemble piece features Beaivváš Sámi actors, friends, and family and is filmed entirely in the Guovdageaidnu region. The film emerged through the work of the Sámi ČSV organization, started in the 1970s and inspired by Native American and First Nations activism, with an explicit goal to be both a political and a living cultural movement. Spurred by the first Sámi Rights Commission in 1980, the film foreshadows the significance of media for later TRC films in Nunavut and Greenland (see chapters 3, 4, and 7). As one of the early examples of an Indigenous fiction film entirely in an Indigenous language, emerging at least a full decade before global Indigenous cinema became a recognized concept, *Pathfinder*'s significance in film history and global media cultures cannot be overestimated.

Set in precolonial times in Sápmi in the winter, *Pathfinder* retells a Sámi foundational, centuries-old, myth about a young man outwitting vicious intruders to save his community, though following Global Hollywood action film conventions in the style of the western (see Iversen 2005, 2019). Containing some aspects of autoethnography, especially in the historical reenactment of traditional practices, the film is a clear allegory of Nordic and Russian colonialism in Sápmi. Brutal, inarticulate, and marauding, the invaders, known as Tchudes, are costumed in black and offset against a white-clad pacifist Indigenous group of organized activists who mobilize their embodied knowledge of the environment, climate, animism, and shamanism (see Cocq and Dubois 2019, 88–96). *Pathfinder* emphasizes a reindeer-herding community's agency and their forceful and empowering connections to their environment, and foregrounds the sustainability of Indigenous practices in living off and with the land. At the same time, the film posits a premodern social system outside of what Rauna Kuokkanen describes as "structural inequalities" within the *siida* (the Sámi community). Established "patriarchal and sexist policies," Kuokkanen argues, became formally embedded within the 1978 Reindeer Herding Act, which considers reindeer herding "both inside and outside Sámi society as synonymous with men's activities," though women continue to be deeply involved in the economic, political, and social aspects of herding (Kuokkanen 2009, 501). *Pathfinder*'s success, like that of Kunuk's *Atanarjuat*, stems partially from a utopian portrayal of a precolonial past of stable and separate gender roles apparently aligned with the demands and subsistence opportunities of the Arctic environment. Yet Kuokkanen further argues that "for many Indigenous women, the struggle for the integrity of the land is indistinguishable from the struggle against patriarchy" (Kuokkanen 2019, 44). Drawing on a range of examples from Sápmi, Kuokkanen affirms that, in general, resource extraction tends to be more favored by men (they are more frequently employed in mines, logging, infrastructure, and hydropower projects), while women tend to express concern about the negative outcomes of development, including pollution and environmental degradation, but also the threat to social relations, increases in gendered violence, and disruption of cultural practices, including Sámi language use. Kuokkanen further notes "a problem when [Indigenous] land rights are premised on male activities of hunting, fishing, and reindeer herding" (2019, 169) and women resource development activists "are either excluded from or decline to engage in formal political structures within their communities" (2019, 44). *Pathfinder,* emerging ten years after the Reindeer Herding Act and as a tribute to Sámi political mobilization during and after the Álttá dam protests, eschews any gesture to the complicated internal politics within the Sámi self-determination and self-governance movements, where women politicians and environmental activists have been partially sidelined.

Pathfinder and *Atanarjuat* foreground environmental justice and the rights of Indigenous groups to live on and in self-determination of their ancestral lands. Oral storytelling in Indigenous languages is central to advancing this political claim in both films; similarly, both are part of the kind of emergent Indigenous production and circulation context that Barclay outlines as a potential means to gain international audiences through film festivals and alternative screening venues. Gaup's and Kunuk's momentous works are thus political allegories of Indigenous agency and unity, where, however, acknowledgment or examination of gender politics is absent. Their works mobilize fiction filmmaking as a vehicle of media sovereignty and interventionist historiography, recalling earlier documentary and ethnographic works: for Kunuk, the *Netsilik* series (Asen Balikci and Quentin Brown, Canada, 1967) produced by the NFB/ONF; and for Gaup, the influential Sámi-language environmental justice documentaries of the 1970s. What this demonstrates is that the fiction features that emerge as foundational components of Fourth Cinema are built upon a larger foundation of Indigenous works of media sovereignty and interventionist historiography that have a long history.

EMBEDDED ETHNOGRAPHY AND NUCLEAR
CATASTROPHE: *THE THREAT*

Shot around Giron on 35mm by one of Scandinavia's best-known and internationally recognized documentary filmmakers and released the same year as *Pathfinder,* Stefan Jarl's *The Threat* is the first Sámi-Swedish collaborative ethnographic feature-length documentary. It is also a piece of environmental justice interventionist historiography in solidarity with the work of Gaup. Translated from Swedish into English as *The Threat* and from Sámi into English as *Encroachment,* the film reflects a mid-1980s political interest in Indigenous land rights, cultural revitalization, and increasingly vigorous debates about the destructive impact of logging, mining, and hydroelectricity. *The Threat* must be read in the context of extremely limited Sámi self-representation (or quasi-self-representation) on Swedish film screens in the mid-1980s, especially as voiced by women. Foregrounding statements by reindeer-herding Sámi Lillemor Baer, along with those of her husband Lars Jon Allas, *The Threat* confronts what Kuokkanen calls the myth of the "Strong Sámi Woman" (2019, 172–74). Mobilized during the political protests of the 1970s and '80s to allow for a differentiation between Indigenous social systems and those of the colonizing Nordic settler states, the myth promoted a notion that Sámi culture and societal organization is inherently more egalitarian than that of the Scandinavian mainstream, at least in part because of its environmentalist ethos (we discuss the interconnection between gender politics and environmental justice activism later in this chapter and in the next). While the film's voiceover (read by well-known Swedish performer Torsten "Totta" Näslund) superimposes a conventional (at times) framework of "tradition versus modernity," Lars Jon's and Lillemor's statements counteract this supposedly authoritative narrative, offering a much more nuanced and complex assessment of late twentieth-century Sámi life, challenging dominant ethnographic master narratives.

The Threat is an example of Jarl's commitment to social critique and collaborative ethnography. Since his films of the 1960s and '70s about Swedish youth cultures, Jarl consistently pursued, in the 1980s and '90s, what would become known as embedded ethnography or multi-sited ethnography (for a detailed definition of multi-sited ethnography, see G. Marcus 1995; for a critique and partially redemptive approach, see Lewis and Russell 2011). Embedded ethnography is based on forms of collaboration between ethnographers and their subjects, whereby the ethnographer often becomes a part of the lived experiences of the subjects and the ethnographic tale itself. Similar strategies are used in Jarl's other films about Sápmi, *Jåvna: Reindeer Herdsman in the Year 2000* (*Jåvna—Renskötare år 2000*, Sweden, 1991) and *Land of the Lapps* (*Samernas Land/Sami Atnam,* Sweden, 1994). Critically acclaimed for its visual lyricism and timely topic (though it did not fare well at the Swedish box office), *The Threat* was released internationally and has continued to circulate on the festival circuit.

The Threat examines a traumatic moment in Sápmi history, with roots in a long history of environmental colonialism (figure 5.1). The film was originally intended to be the embedded ethnographer's account of encroachment upon reindeer-herding practices, but the Chornobyl nuclear catastrophe in spring 1986 radically changed the circumstances of filming and the political orientation of the film, as Jarl explains in an essay about it (Nilsson 1991). Partly traditionally realist in its presentation of talking heads juxtaposed with expansive nature shots and up-close shots of reindeer, *The Threat* represents the interiority of characters expressed as on-screen testimony, and mobilizes a poetic aesthetic, especially through the use of ominous recurring shots of reindeer carcasses piled one on top of the other or airlifted by helicopter for safe disposal, with the sound and image of helicopter rotor blades against gray skies, accompanied by experimental synthesizer music.

The Threat is unique in its portrayal—visceral and uncompromising—of the collective trauma of the Chornobyl incident on Sápmi. As the spread of radioactivity reached most of northern Sweden, reindeer, fish, berries, and other sources of Sámi livelihood and sustenance became contaminated. Modifying the focus of the documentary to address the trauma of radioactivity on land, in waters, in the air, in animals, and in humans, Jarl abandons

FIGURE 5.1. Environmental destruction and its aftermath in *The Threat* (*Hotet/Uhkkádus*, Stefan Jarl, Sweden, 1986).

his role as auteur to transition narrative agency to Lillemor and Lars Jon (figure 5.2), while integrating explanations of Indigenous knowledge and Sámi belief systems and politics. In addition, Jarl's film can be read as powerful geopolitical and Cold War commentary; its fast production schedule and release in spring 1987 makes it a counter-document to the USSR's attempts to suppress information about the radioactive material. Moreover, the film makes clear that Arctic Indigenous populations continue to be caught in the middle of the Cold War's oppositional politics—as demonstrated in films such as Marquise Lepage's *Martha of the North* (*Martha qui vient du froid,* Canada, 2009; discussed in chapter 3) and Ivalo Frank's *ECHOES* (Denmark/Greenland, 2010; see chapter 7).

Notably, *The Threat* demonstrates that not all ethnographies made by outsiders about Indigenous Arctic populations represent "bad ethnography" (see Groo 2019). In this film, Lillemor and Lars Jon are protagonists who speak extensively on screen, while the filmmaker functions as an ally and a facilitator, not necessarily controlling the narrative. In this way, *The Threat*'s statements by Lars Jon and Lillemor function as Sámi autoethnography embedded within an outsider film. Lars Jon and Lillemor forcefully indicate throughout that northern Sweden is not "wild"—it has been populated and colonized for centuries. In addition, Lars Jon articulates an autoethnographic

FIGURE 5.2. Lars Jon Allas and Lillemor Baer in *The Threat* (*Hotet/Uhkkádus,* Stefan Jarl, Sweden, 1986).

account that puts Sámi culture front and center, affirming that a Sámi worldview is proudly different, logical, well organized, and highly functional. The impassioned statements by Lars Jon are political and philosophical in ways that extend far beyond the Chornobyl impact. The true significance of Lars Jon's and Lillemor's statements is that through this film, the voices of nonofficial Sámi spokespeople—especially those of women—are brought to the forefront. Their statements provide a forceful illustration of an insidious settler colonization of their lands, with the invisible cesium-137 functioning as an alien invader. Jarl's film is one of the first Swedish films that present Sámi as authorities over their own recent past, challenging the nation-state's implicit adherence to the Great Acceleration theories that have shaped colonialism in Sápmi during the preceding decades.

Indigenous Practice, the Great Acceleration, and the Anthropocene

In addition to its relevance for Sámi environmental justice and self-determination politics with respect to the Swedish nation-state, *The Threat* provides a forceful illustration of what, only a decade after its release, became a hotly contested set of concepts: the Anthropocene, a term coined by Paul Crutzen and Eugene Stoermer in 2000, and the accompanying Great Acceleration paradigm (see Steffen et al. 2015). Controversial among humanities scholars because of its anthropogenic, Western Enlightenment, white, imperialist, and patriarchal roots (see Haraway 2016; Yusoff 2019), the assumptions of the Anthropocene and the Great Acceleration are forcefully questioned by Lars Jon and Lillemor Baer in *The Threat*. Focusing his critique on the presupposed need for eternal growth and industrialization (what the Great Acceleration illustrates), Lars Jon argues that the Sámi wanted neither nuclear energy nor large-scale extractive infrastructure. Sámi practices are not wage labor, nor are they part of capitalism, he affirms, contrasting Sámi and European approaches to the environment and the clash between cultures and belief systems that ensue: "Sámi philosophy is first taking care of animals; never humans first." Examples from reindeer pastoralism guide his assessment: "Right now, the reindeer want to move to the winter pasture in the forest, but we can't have them go there because of the cesium-137. They do not understand that, so two world views are clashing. We have been living well up here, outside of the European system for centuries; now we are confronted with that way of life, against our will. Our work is not work, it is a vocation; it is our way of living." Decentering the assumed priorities of Western modernity and industrialization in the age of the Anthropocene, Lars Jon's statement comes exactly midway through *The Threat*.

Recurring interview sequences with Lillemor foreground a complementary aspect of the film's implicit Anthropocene critique and the Great Acceleration's impact on Indigenous lives. She states that in past years with few reindeer, "we were able to work ourselves out of the situation; by fishing, picking berries, farming, or other means." An unequivocal sense of resilience and resourcefulness characterizes these statements. At the same time, she notes with despair that "everything has been pulled away from under us. There is nothing left." The nuclear contagion exacerbates all other encroachments over the past two centuries by railways, highways, hydroelectric plants, logging, and mining companies, though, she states, the impact of these are more gradual: "But with this event, it is taking out everything. We have no choices anymore." In tandem with the Anthropocene and Great Acceleration critique, Lillemor further reflects on the trauma that the cesium-137 levels are creating. She articulates this as *Trauerarbeit* (German, "work of mourning"). Everyday family practice, connections to land and animals, environmental knowledge, memory, and recollection are disrupted to the breaking point, rupturing bonds that have guided her life and that are core to her identity. "Fishing, it is like a memory now. Is that what the practice will be reduced to? To a memory? What if I had known last year during the fall fishing season that it might have been the last. My sisters and I inherited that duty from our mother, that we would do the fishing for the family. . . . What if I had been thinking last year that this might have been the last time." In an extended sequence, Lillemor works through a traumatic set of memories, documented in the moment. Jarl's long take, with only the bare minimum of questions asked, is strongly reminiscent of other film and media works on topics ranging from the Holocaust to the Anthropocene. Like speakers in those works, Lillemor verbalizes real-life trauma while on camera. As such, Jarl's film is an important piece of collaborative interventionist historiography and an Arctic environmental justice documentary.

TRANSNATIONAL SÁPMI: PAUL-ANDERS SIMMA AND NILS GAUP

Like Nils Gaup, Paul-Anders Simma had a transformative impact on Sámi cinema. Made across Sápmi with actors and crew in multiple languages— North Sámi, Finnish, Norwegian, Russian, and Swedish—Simma's varied oeuvre tells alternative transnational histories of the region. A remarkably dynamic director, Simma has produced a number of realist and political documentaries about different aspects of contemporary life in Sápmi, along with narrative fiction and experimental work. Through his production company Saamifilmi OY (Safi) and in collaboration with other companies and public television in Norway, Sweden, and Finland, Simma has produced or coproduced over two dozen short, mid-length, and feature-length films, which makes him among the most prolific of Arctic Indigenous filmmakers to date. In recent years, Simma's feature-length documentaries have emphasized contemporary lived Sámi realities. Topics and approaches include a poetic rumination on reindeer herding on the desolate Russian tundra in *Olga, To My Friends* (Finland, 2013); invasive tourism and ensuing ecological degradation in the Norwegian-Finnish cross-border region, and Sámi responses to these outsider "invasions," in *The Power of Yoik* (*Cumhachd a' Yoik,* Beatrix A. Wood and Paul-Anders Simma, Norway/UK, 2018); and the issues faced by a Sámi woman brought up in a Russian orphanage with the belief that her mother was dead, only to learn that she is in fact alive, and that there may be more to the story of why she was placed in the orphanage, in *My Dear Mother* (*Ráhkkis Eadni/Kära Mor,* Norway/Sweden, 2020). Simma's features include one fiction film: the irreverent geopolitical dramedy *The Minister of State* (*Sagojoga Minister,* Finland/Norway/Sweden, 1997), about the German Nazi invasion of Finland and Norway and the local population's attempts to negotiate and survive changing borders.

Simma's work constitutes an archive of Sámi interventionist historiography in its consistent emphasis on foregrounding critical perspectives on Norwegian, Swedish, Finnish, and Russian nation-state actions, environmental destruction, and the impact of geopolitics and globalization on Sámi culture and society. An example in this vein is the documentary *Legacy of the Tundra* (*Duoddara árbi,* Norway, 1994), which interweaves two story lines and combines several documentary approaches: embedded reporting, docudrama, staged scenes, and hidden cameras. The first plotline concerns whether the son of the Utsi family will assume responsibility for the family's reindeer herd; the second is the training of reindeer for the Sámi participation in the opening of the Lillehammer Olympic Games in 1994. Because of the global media embargo that gave only licensed operators access, Simma was not allowed to film during the ceremony. In an act of civil disobedience that effectively serves as a way to document Sámi first-person participation in the ceremony, young Utsi secretly mounted a video camera on his sleigh and pressed start as he set off with the reindeer caravan. Rather than being an outsider news media report, this act preserves the Sámi participation for posterity as an act of media sovereignty.

Give Us Our Skeletons! (*Antakaa Meille Luurankomme/Oaivveskaldjut,* Finland, 1999) is one of Simma's most influential documentaries. Internationally released through Icarus Films, this activist film has three interwoven plotlines that are echoed and reworked in subsequent Sámi interventionist historiography and media sovereignty production. The protagonist is Niillas Somby, one of the key protesters during the Áltá dam movement in 1979–80 (having escaped Norway when charged with arson, he lived in a Canadian Iroquois community for several years in the early 1980s). The film depicts Somby's struggle to learn the truth about the remains of his relatives Mons Somby and Aslak Hetta, beheaded leaders of the Kautokeino Uprising in 1852, only to find that their skulls have been preserved in the anatomy department of Oslo University and that the Scandinavian legacy of Sámi racial discrimination is alive and well among late twentieth-century academics. The film documents how the skulls eventually are returned to the community for burial.

The Dano-Norwegian state's brutal response to the Kautokeino Uprising in 1852—which included the execution of Somby and Hetta, as well as charges, imprisonment, and labor camps for thirty-three of the forty-seven Sámi involved—has been less emphasized in historical records than the fact that this is the only known violent uprising of Sámi against colonial settlers, with two Norwegians killed among those who were taken hostage. It also led to a bifurcation of religious allegiance among North Sámi, with one group following the breakaway Laestadian tradition that promoted temperance, communal forgiveness, and lay pastorship, versus the hierarchical Dano-Norwegian Lutheran denomination. A traumatizing and polarizing event in Sámi history, it has been told and retold in numerous forms (see Christensen 2012; Pearson 2016). A large share of inhabitants of Guovdageaidnu today are descendants of the uprisers, those who were held hostage, or those who put down the uprising, with families knowing their lineage often eight to ten generations back. The event remains divisive. The Kautokeino Uprising pitted different Sámi communities against one another for generations, largely depending on whether families tended to identify with those who had challenged state and religious authorities or not. Gaup's large-scale historical epic *The Kautokeino Rebellion* (*Kautokeino-opprøret,* Norway, 2008) retells the story from the perspective of a Sámi woman, Elen Skum (Anni-Kristiina Juuso), who follows Laestadian teaching and tries to keep the *siida* together as it faces pressures from commerce (and unrestricted alcohol sales), colonial edicts, the local pastor, and the visiting bishop of Christiania. In addition, *The Kautokeino Rebellion* is an example of media sovereignty and of truth and reconciliation work performed through filmmaking. Gaup is a direct descendant of one of the men executed. The film's producer, Nils Thomas Utsi, is a descendant of one of the uprising families. The film team worked closely with the community to arrive at the foundational aspects of the screenplay and characterization. This coordination functioned both as TRC work and as a means of interventionist historiography, to bring together various past and divergent family and social histories to work through trauma and conceptualize a new future in the process of fiction filmmaking.

As such, both *Give Us Our Skeletons!* and *The Kautokeino Rebellion* are examples of interventionist historiography, as these films tell the story of a traumatizing historical event and its long-reaching aftermaths, with an outcome of bringing both little-known facts to life and communities together. As interventionist historiography, *Give Us Our Skeletons!* is unapologetic. It interweaves Niillas Somby's personal story of political activism and challenges Lutheranism—one of the first scenes presents his statement that he "won't enter that building" (a church)—with one of the first exposés of phrenology, racialized genetics, sterilization, and forced hospitalization in documentary form. The film outlines the many and varying subjugation practices used against the Sámi from the 1840s onward, including ample photographic evidence. A particularly poignant sequence involves an older woman informant recognizing herself as a child in one of the photos collected by Sweden's State Institute of Racial Biology from the early 1920s onward. While

Norwegian, Finnish, and Swedish practices of Sámi subjugation varied, Simma emphasizes their similarities in mobilizing European theories, philosophies, and practices of racial biology during the 1920s to '40s (see Broberg and Roll-Hansen 1996). Simma's voiceover asserts that many of the women in Scandinavia forced into sterilization were Sámi (estimated to total some seventy thousand in Sweden, sixty thousand in Finland, and thirty thousand in Norway). Others were treated with electric shock or relegated to mental institutions against their will. Simma intercuts found footage from 1930s Germany racially profiling Jews to underscore his point. Key themes in the film include an emphasis on the Sámi as a group incessantly "researched"—as in measured, photographed, and recorded—by scientists as early as Linnaeus and through later generations of anthropologists and folklorists (see Broberg 1975). These themes are echoed by later Sámi filmmakers such as Katja Gauriloff, who documents the Skolt Sámi of the Fenno-Russian border region in *Kaisa's Enchanted Forest* (*Kuun metsän Kaisa,* Finland, 2016), and Amanda Kernell, whose *Sami Blood* (*Sameblod,* Sweden, 2016) is about the South Sámi in Sweden.

Give Us Our Skeletons! also provides an unequivocal criticism of museology and ethnographic collection practices, which not only removed thousands of artifacts from Sápmi to exhibit in the southern capitals of the Nordic countries, but also staged tableaux of "authentic" and "primitive" Sámi life—especially of reindeer husbandry—in museums, such as the well-known Lapp exhibit in the Nordic Museum of Sweden and the outdoor folk museum Skansen in Stockholm, which, for over a century, cemented a view of the Sámi as seemingly frozen in time in a remote mountain landscape (see Silvén 2014). This critique is especially strong in Simma's talking-head interviews with anatomy and pathology scholars at the Universities of Copenhagen and Oslo, one of whom laments the danger of "breaking up the collection" of Sámi preserved skeletons (parts of which had ended up in Copenhagen through a trade in artifacts, catalogued there as part of the "Natural History" collection), proclaiming that if the skulls of Somby and Hetta were to be returned to Sápmi, they would only be buried and thus destroyed. Simma shows that discourse about the Sámi as subhuman was unreflexively perpetuated by these spokespeople well into the 1990s, partly under the guise of benevolent paternalism rather than overt racism, which plays into the image of the Scandinavian welfare states as egalitarian, equitable, and transparent.

The implementation of racist and colonialist dogma and state policy varied in Scandinavia, with clear contrasts among nations. The "Norwegianization"

policies of forced assimilation during the period 1890–1950 included boarding school education in Norway and the prevention of Sámi property ownership, for example. Sweden and Finland adopted more segregationist practices
that sought, to a greater degree, to "preserve" the Sámi people's "premodern"
ways of life. This perspective deemed the Sámi primitive, unable to adapt to
the transformative changes brought about by industrialization, wage labor in
the large-scale resource extraction sector, and the social policies of universal
access to education, health care, and housing that were part of the *folkhemmet*
philosophy of the welfare state. This history would be a well-known backdrop
for Simma's Sámi audience.

Others interviewed in the film make reference to the contested political
status of Sápmi, Norway being the only Scandinavian state to have recognized the Sámi according to the UN's ILO-169 convention. The talking-head
interviews furthermore imply resistance among upper-echelon Norwegian
scientists to recognizing Sámi as the first inhabitants of the region, and an
interest in potentially using museology and the natural science collections to
obfuscate Sámi land and self-governance claims. These statements are contrasted with Niillas Somby, who, at the end, gives voice to the skulls, speaking for the elder, deceased Somby in the first person as he puts the cranium
into a wooden box for transportation to Guovdageaidnu. The documentary
emphasizes, perhaps for the first time in Scandinavian cinema, that Sámi
perspectives on their own history, told in Sámi languages, are part of a media
sovereignty ethos. The film makes clear that the status of the skulls and what
they represent to the Sámi and to the scientists are different things, but that
these human remains are inseparable from larger historical narratives. The
ways Sámi articulate their experiences of the past and the history of nation-
state colonization practices are contrasted with the nation-state's impetus to
preserve "all of history for everyone"—especially when supported by public
funding—as if subject positions and power dynamics were inconsequential
in that process.

Give Us Our Skeletons! is a film of interventionist historiography in that it
situates ongoing political and social debates with respect to Sámi relationships to the Scandinavian nation-states and the significance of the Kautokeino
Uprising and the Álttá dam protests as signature events of Norwegian Sámi
history. The documentary, however, posits an "unproblematic" narrative of a
Sámi history of grievance, trauma, and loss, without examining much of the
internal conflict, including gendered or contested and internally divisive
aspects of the Kautokeino Uprising. In Simma's documentary, political

aspects of Sámi history are foregrounded much more than those of cultural and everyday history. This includes outlining the historical context of the impact of World War II, when military invasion, combat, and changing state borders across Sápmi effectively closed off nomadism and expansive reindeer herding, in part linking these events to emergent land-claim debates and considerations of Indigenous self-governance and sovereignty. Largely absent in this film are themes of loss and revitalization processes of Sámi cultural identity and language, which are frequent themes in later Sámi films, especially by women (we discuss this more closely in chapter 5).

Give Us Our Skeletons! illuminates how the Kautokeino Uprising has been historicized. The original court case proposed that it was a primitive and barbarian reaction, a result of misguided religious fanaticism and the influence of Laestadianism, rather than about political rights. Twenty-first-century historiography, with Gaup's *The Kautokeino Rebellion* as a case in point, emphasizes Sámi cultural resilience, the region's multilingualism, inter- and cross-cultural engagement, and the long-established practice of sustainable environmental stewardship through reindeer husbandry. In addition, *The Kautokeino Rebellion* became an important media event in Norway upon its release, "provoking wide-ranging discussion about colonialism, state culpability for past wrongs, and the importance of recognizing and reversing the disenfranchisement of Sámi people" (Cocq and Dubois 2019, 137). It implicitly links colonialist commerce and settler-promoted alcoholism to ongoing challenges of Sámi cultural resilience, which includes those of mass tourism, globalization, and large-scale natural resource extraction.

At the same time, *The Kautokeino Rebellion* establishes an interventionist historiography of media sovereignty by opening with the scene of the execution, seen through the eyes of the child of the woman protagonist. In many other Arctic films—especially documentaries—we see the voyager arriving, then leaving at the end, but from the Sámi perspective in the film, these travelers did not breeze through on their journey of discovery. The film makes a statement on the oppression: this is what the explorers' and colonizers' supposedly transparent, scientific, cartographic, and geographic footprint looks like from other perspectives. Gaup does not sensationalize the uprising and the violence, but focuses on the underlying reasons for it; in doing so, however, the film also erases tensions within the community and glosses over how divided the Sámi community was over both Laestadianism and the uprising.

In contrast to *Pathfinder, Give Us Our Skeletons!,* or most other films made by Sámi filmmakers during the 1970s through 1990s, the focus in *The*

Kautokeino Rebellion is on a female protagonist. It is her point of view, mostly, that guides the historical representation, and her story, via voiceover, that frames the film's opening and closing sections. This strategy is an attempt at telling a story of a collective past that is constituted specifically from a woman's subject position, complementing the general approach of the film, which posits that Sámi, for the first time in the history of filmmaking, become "the *subjects* of their own history" (Cocq and Dubois 2019, 144). At the same time, Elen Skum here serves as an example of what Kuokkanen describes as "the myth of the strong Sámi woman" (2007, 2009)—as someone who holds the community together—but with her Indigenous and self-determination rights questioned by both settlers and Sámi men. *The Kautokeino Rebellion* aligns with depictions in contemporaneous Arctic Indigenous anticolonial films about women, especially the Arnait collective's *Before Tomorrow* (see chapter 3). The soundtrack by Mari Boine, a world-famous female yoik artist, supplements the narration. One of Europe's oldest vocal traditions, yoik was largely banned by both church and state during colonization in Norway. Being connected to shamanism and "primitivism," women who yoiked were especially challenged by authorities. Using yoik as a framing device, therefore, draws on traditional Sámi oral narrative strategies through a woman's voice while connecting to a long history of suppressed social justice, including the use of the Sámi language and environmental activism by Sámi women.

VISUAL ANTHROPOLOGY, AMBIVALENCE, AND SHAME IN *DREAMLAND*

The complicated representation of the Arctic is well known in visual anthropology, as Indigenous peoples all over the global circumpolar North have been subjects of the outsider's ethnographic gaze for centuries. In response to this history, Indigenous ethnography has grown in recent years as a retort and political tool. One of the most striking examples is Britt Kramvig and Rachel Andersen Gomez's documentary *Dreamland* (Norway, 2016), which mobilizes an insider's autoethnographic perspective to critique the Nordic nation-state, settler colonialism, and their epistemological and patriarchal foundations. What sets *Dreamland* apart from virtually any other Arctic documentary is Kramvig's embrace of her own self-identity as both Indigenous North Sámi from the Tromsø region and a formally trained anthropologist and professor at the University of Tromsø, whose prolific scholarship includes examinations of Indigenous knowledge practices, visual culture, Anthropocene studies, and history of science. *Dreamland* focuses on women, both in mode of production and in content. Kramvig, in collaboration with Australian philosopher Helen Verran, wrote the script, and Verran provides the English-language voiceover. As the protagonist of the film—akin to a road movie—Kramvig traverses the northern Norwegian coast from Kirkenes to Tromsø via car, interviewing friends, family, Sámi political activists, and members of the general public along the way, while also reading aloud reflections from her own journal and the film's script.

The film's title is derived from the poem "Dream-Land" (1844) by Edgar Allan Poe. As an armchair ethnographer, Poe was preoccupied, as were many Victorians, with the mythos of "an ultimate dim Thule—From a wild weird clime, that lieth, sublime, Out of SPACE—out of TIME" (on Victorian imaginaries of the Arctic, see Potter 2007; Craciun 2016). *Dreamland*'s appropriation of this poem conveys a multilayered approach to histories of exploration, anthropology, and Western imaginaries of the North. A monotonous, unidentified, male voice reads segments of Poe's poem multiple times during the film. *Dreamland* contrasts with many twenty-first-century Arctic Indigenous and feminist documentaries in being thematically less focused on cultural revivalism and identity politics, and more philosophical in its critique of colonialism. It is nonrealist, ambivalent, and ambiguous, seeking to make these relationships more complex. Fundamentally, the film meditates on the relationship between past and present—rarely, if ever, looking to the future. Kramvig, as the protagonist, is portrayed as consistently ambivalent about her own subject position—as are, at times, those with whom she interacts. *Dreamland* is part of an emerging body of Indigenous road movies that includes Chris Eyre's *Smoke Signals* (Canada/USA, 1998) and the Sámi film *Bázo* (Lars-Göran Pettersson, Norway, 2003; see Kääpä 2015). While often thought of as an American genre, road movies—which lend themselves to cultural, psychological, and geographical journeys—have become a global phenomenon since the late 1960s. As Wendy Everett notes, the road movie genre is "characterized by fluidity and open-endedness" and by a "self-conscious exploration of the relationship between the spatial and temporal displacement of journey and the discourse of film itself"; and it is a genre "whose inherent flexibility makes it ideally suited to the exploration of complex social tensions and concerns" (Everett 2009, 166). If there is a precursor to *Dreamland* in the history of road movies writ large, the most salient example is the characters wandering at the border of East and West Germany in

Wim Wenders's *Kings of the Road* (*Im Lauf der Zeit,* West Germany, 1976). *Dreamland,* like Wenders's film, deals in part with what one does when cultural and political memories are no longer readily accessible, and with how a journey can function to retrieve them on both personal and sociopolitical levels. The road movie, broadly defined, can be understood as a protagonist's journey either in flight or on self-discovery. The retrieval of cultural memory in road movies like *Dreamland,* then, is an instantiation of interventionist historiography. The retrieval of historical and cultural memories and bringing to light what has been elided from personal and cultural history allows a lost yet new history to emerge as an act of interventionist historiography.

As an Arctic road movie, *Dreamland* mixes inward-looking, self-reflexive, poetic rumination about the protagonist/filmmaker/anthropologist Kramvig with references to key historical events that have shaped Sámi culture. Unless one knows the territory through which she journeys, it is difficult to map *Dreamland* and understand where the characters are going. Indeed, *Dreamland* has few signposts and a nonlinear structure. An insider perspective facilitates the understanding of the film, whether one is an "insider anthropologist," "insider Sámi," or, as is the case with Kramvig, construed as both. In that regard, *Dreamland* is, in some ways, *for* insiders—for people who know the region and where the different locations are in relation to one another. At the same time, the route follows Northern Norway's officially designated "National Tourist Routes," a publicly funded initiative to support tourists traveling through Norway. The film offers viewers, then, a journey that is as much Kramvig's personal journey as it is a trip through a demarcated geography. One of the reasons for this is that the Sámi territory is a map that has maps of the Scandinavian nation-states imposed on top of it. The Sámi spatialization has undergone a form of partial erasure, and the traces of this cartography are rediscovered through Kramvig's journey. *Dreamland,* then, proposes a new Sámi cartography—psychological and material—to complement ongoing land claims and political self-governance efforts. In its experimental emphasis, *Dreamland* draws on the palimpsest. As if uncovering a palimpsest, Sámi cartography is another example of the film's interventionist historiography, where the faint traces of the past are drawn to the foreground, turning the traces of Sámi spaces into the film's map proper.

As an interventionist historiography, *Dreamland* addresses several traumatic events from anthropologist and autoethnographic perspectives. These include the Norwegian skeleton pillage from the Kautokeino Uprising in 1852, examined in Simma's *Give Us Our Skeletons!,* and the numerous museums and other institutions in Norway, Sweden, and Finland that continue to hold Sámi remains; the quelled Áltá dam protests against land rights infringement; and ongoing gender politics. Verran's voiceover describes Kramvig as she visits the Sámi parliament buildings in Kárášjohka: "Ambivalence and shame at being a member of the tribe of anthropologists, set alongside a feeling of having betrayed in some obscure way, in her simultaneously being a member of the tribe of Sami. Britt, cultivating that ambiguous life in reality, making moral passage in body and in text." Straddling two subject positions that may at first seem incommensurate (like Rasmussen, the anthropologist examiner and Indigenous subject in one), the film proposes that doing so is not only possible, but vital; one set of perspectives should inform the other. As the film transitions to a sequence in a university library, we see Kramvig the protagonist surrounded by books, including one with a historical map of "Ultima Thule," stating to the camera: "It was here I had to travel to find my path into the academic world in order to find a way to be at ease there that could heal the wounds I saw around me." The film then returns, again, to a male voiceover declaring lines from the first stanza of Poe's "Dream-Land." Lines from the poem recur throughout the film.

It is no doubt unexpected to frame feminist Indigenous history and Indigenous rights with reference to one of the most celebrated American writers of the nineteenth century, a writer whose works include "fake" ethnographic travel writings such as "The Balloon Hoax" (1844). Poe's reportage mimics the nature of nineteenth-century accounts of journeys undertaken by amateur explorers, ethnographers, and anthropologists. His first novel, *The Narrative of Arthur Gordon Pym of Nantucket* (1838), ends with a story of polar exploration. *Dreamland*'s continuous references to Poe are not inconsequential. Scholars such as Gabriele Rippl have analyzed Poe's poems and travel writings as a riposte to the colonial assumptions of anthropology in the way these usurp the idiom of narrative and discursive authority and tell plausible tales, even though they are fakes (Rippl 1996). Kramvig's use of Poe, and his Gothic aesthetic romanticizing of the Arctic as an end-of-the-world "Ultima Thule," devoid of Indigenous cultures and histories, speaks to the ways in which Poe's prose, poetry, and tall tales of fake journalism foreground the psychological, and not the material, components of the ethnographic journey. Using Poe as a touchstone also resonates with the genre crossing—ethnographic, documentary, autobiographical, psychological, the blurring between the real and the imagined—that is central to *Dreamland*.

Dreamland's voiceover furthermore intones: "For Edgar Allan Poe, Sublime is experience, and experience is Sublime." The line "out of space out of time" returns repeatedly in the film, almost as an incantation—functioning as "the fragmented and transient present" that Catherine Russell (1999, 8) outlines—yet it only appears once in Poe's poem. The sublime-as-experience turns conventional Kantian or Burkean understanding of "the sublime" around to foreground the lived experiences of Indigenous populations across Sápmi, just as it calls into question the nineteenth-century legacies of colonialism that have shaped the disciplines of anthropology and ethnography. The film nondidactically asserts its interest in challenging received understanding about the Arctic and polar regions, where the Victorian, Romanticist, Gothic conceptualization of the Far North is a blank slate for the explorer and colonizer to conquer and settle. Drawing on experimental, disassociative modes of documentary, anthropology, and Indigenous knowledge practices, Kramvig and Andersen Gomez reinforce, time and again, that Sápmi has a rich and complex history.

The Álttá dam protests were defining for Sámi self-governance movements, yet the significant role Sámi women leaders played in the protests has been underexplored. *Dreamland* seeks to partially redress that omission. Two sequences in the film are especially illustrative. In the first, Kramvig visits the activist Marry Ailonieida Somby's house, reflecting on her domestic art installation. Somby has xeroxed black-and-white photographs of her ancestors by some of the many "Lappologists" and visual anthropologists who documented the Sámi from the late nineteenth century onward—often for anthropometric reasons (see Lien 2018, and our discussion of *Sami Blood* below). Guiding the camera to focus on the visages in the frames on her living-room wall, Somby ruminates on how she grew up as one of the last generation of nomadic herders, mentioning that she has her own reindeer-herding mark. Those reindeer have been "lost," she explains, or rather stolen, as she chuckles. In the second sequence featuring Somby, Kramvig and the crew travel to visit the site of the Álttá dam and conduct an interview on site. The interview is intercut with historical news reporting the Sámi hunger strike and occupation of the Oslo parliament building in 1979. Although *Dreamland* does not address the protests by Sámi women who occupied prime minister Gro Harlem Brundtland's office, one knows the significance of these actions from Simma's *Give Us Our Skeletons!* The interview with Somby focuses on the hardship encountered by the Sámi and the scar the Álttá dam has left environmentally, culturally, and psychologically.

Dreamland addresses women's experiences first and foremost. The film includes two visits to the Steilneset Memorial in Vardø. Designed by artist

Louise Bourgeois and architect Peter Zumthor, Steilneset is a tribute to the seventy-seven women and fourteen men who were burned at the stake in the province of Finnmark during the seventeenth century, accused of witchcraft (the region had one of the highest per capita rates of witch hunting in all of Europe during the period; see Hagen 2006). Bourgeois's sculpture and contribution to the memorial—her last major work—is called *The Damned, the Possessed, and the Beloved*. Kramvig visits with a long-lost friend, and as the two walk through the memorial and Vardø Fort, where accused women were held in a subterranean "witches' hole," they do not specify that about a third of those accused of witchcraft were Sámi, including Sámi men accused of shamanism (Hagen 2006). Kramvig and her friend reflect on the different gendered aspects of life in Sápmi and Northern Norway: "As a woman, we had much more liberty to leave; the sons were expected to take over the fishing." These statements signal the keen awareness of gendered realities and how settler colonialism, the Dano-Norwegian state and the Lutheran church, welfare state policies, "Norwegianization," and a globalized market economy have influenced the region for centuries.

To conclude, we call attention to *Dreamland*'s interest in language. Like many other Sámi media works, *Dreamland* addresses the colonizer's severing of cultural and linguistic belonging. It includes some "informant" interviews, and Kramvig herself reflects on the intergenerational Sámi language trauma in conjunction with her participation in the Global Indigenous Preparatory Conference held in Áltá in 2013. While the film ruminates in nondidactic ways on language cultures and power dynamics, it also celebrates multicultural expressions that extend beyond the relationship between Norwegian and Sámi languages—featuring, for instance, an interview with a Thai woman who operates a takeout business in Eastern Norway. Moreover, the film moves between English, Sámi, and Norwegian languages without calling much attention to the practice. As Kramvig and Andersen Gomez emphasize in their essay "From Dreamland to Homeland" (2019), in its "aesthetics, interview questions, and voiceover," the film "queers the familiar sense of causality" (Kramvig and Andersen Gomez 2019, 323). They further describe their film as follows:

> *Dreamland* assembles as part of its aesthetics and theoretical underpinnings routes, trails, people, landscapes, ruins, books, maps, rituals, and memories. The film is not a linear road movie but instead moves in and out of the past and present, producing ontologically distinct spaces and times in its visual

and auditory composition. For instance, the figure of the Indigenous anthropologist investigates through interviews and voice-over reflection the possibilities of a "real" ontological turn to happen, where the colonial space of a *Dreamland* can be transformed into a decolonialized space of a Homeland for all the earthlings living in the region. . . . *Dreamland* produces a displaced colonized space, with the notion of Homeland as the becoming of a space of reconciliation. Importantly, the anthropologist character needs to come to terms with her participation in both of these stories. (2019, 323)

Dreamland thereby functions as a key example of contemporary interventionist autoethnography and historiography, reframing the role of the anthropologist through Kramvig functioning as someone both inside and outside Sámi culture. The film also reflects on the history of environmental political mobilization in Sápmi and on the roles that women played in its inception—and that they continue to play today in advancing these debates through multiple channels, whether in the Sámi parliament, as scholars, as artists, or as activists.

AMANDA KERNELL'S *SAMI BLOOD* AS INTERVENTIONIST HISTORIOGRAPHY

The discipline of anthropology and the status of visual evidence and photography connect *Dreamland* and *Give Us Our Skeletons!* with Kernell's fiction feature film *Sami Blood*. Critically acclaimed, it opened at the Venice International Film Festival, screened at Sundance, and sparked debate in Sweden throughout 2016 and beyond. *Sami Blood,* and the debates it ignited, addressed the country's long-suppressed colonialist treatment of the Sámi, especially in terms of 1930s race biology—extensively documented by photography—and the publicly accepted ideology of race relations in the country through the 1950s. This is called the "Lapp shall remain Lapp" principle, which infantilized and primitivized the Sámi during the welfare state's development, with policy seeking to protect this Indigenous group from a modernity they were assumed to be incapable of adapting to and thriving in. A related outcome of this policy was that it emphasized reindeer herding as an original and intrinsically worthwhile occupation for Sámi. As such, settler colonialism panned a narrow notion of Sámi culture's connection to the environment (figure 5.3).

FIGURE 5.3. Colonialism and Swedish eugenics in Amanda Kernell's *Sami Blood* (*Sameblod,* Sweden, 2016).

While *Sami Blood* is not a documentary, its main thrust provides an interventionist historiography of the twentieth century to counteract the one promulgated by mainstream Swedish sources. Furthermore, it provides ample, historically accurate renditions of social life, institutions, and practices in support of this mission. As the first fiction feature film made by an Indigenous woman filmmaker in Sweden, *Sami Blood* is part historical epic and heritage drama, part coming-of-age story. Kernell's film is also, in part, a reflection on contemporary Sweden's continuous suppression of the Sámi, including the marginalizing of Sámi languages and the impact of Swedish colonialism on Sámi lands and environments. This is the first film to be made with dialogue in South Sámi, a language with fewer than five hundred native speakers. The film spans a range of narrative and formal registers that have not previously been associated with Indigenous filmmaking in Sweden. *Sami Blood* has also emerged as a particularly significant work for the history of Sámi filmmaking in Sweden because of its production, funding, and circulation contexts, notably mobilizing the distinct—though at times not perfectly aligned—priorities of the ISFI and the Swedish Film Institute (SFI).

Sami Blood addresses a highly charged mechanism of Swedish colonial politics during the 1920s and '30s, dominated by systematic discrimination and racialization based on principles of eugenics (see also Sand 2022; Dancus 2022). It foregrounds a woman's perspective on colonialism, racism, discrimination, and the suppression of an Arctic Indigenous identity, which connects *Sami Blood* to the work of the Arnait collective in Nunavut—including through its emphasis on community practice, women's experiences of colonialism and settler contact, authentic language use, and environmental connections, especially the experience of being removed from the land. The film's aesthetic register challenges voyeuristic ethnographic filmmaking, "empiricist" visual anthropology, and "impartial" scientific photography dominant during the period (all of which reflect the underlying racialist endeavor), by focusing on the embodied experiences, not of the implied ethnographic viewer, but of the film's protagonist, a young Sámi woman named Elle-Marja (Lene Cecilia Sparrok).

The film begins in the present. Curmudgeonly, eighty-something Elle-Marja (Maj-Doris Rimpi)—now known as Christina—is being driven to her sister Njenna's (Mia Sparrok) funeral. Her son plays a yoik recording on the car stereo for her, which she decries as shrill. She has cut ties with her roots, including disavowing this central form of Sámi expression. Similarly, her son seeks to speak to her in a few phrases of Sámi, but she pretends not to understand. The film then shifts temporally to the past and her childhood, reflecting both auditory and visual cues of colonial suppression. Elle-Marja/Christina speaks South Sámi with her sister and mother but does well at the *nomadskola*—the Swedish form of forced schooling of Sámi children, away from their families—partially because she speaks Swedish better than her sister, and does so mostly voluntarily. When Uppsala University "officials" visit, she is chosen to recite a poem, in Swedish. This "honor" is short lived, however, for Elle-Marja is subsequently examined as part of the university's eugenics study. Despite this humiliation, and a brutal branding of her ear by young men (replicating springtime reindeer calf marking)—and indeed perhaps because of it—Elle-Marja decides to leave home to study in Uppsala. Her *nomadskola* teacher, who seems fond of her, nevertheless points out that the Sámi people are ignorant and unintelligent and that her people must stay in the North "or they will die out." Running away, Elle-Marja adopts the name Christina, speaks Swedish exclusively, and seeks to pass for Swedish, including as part of a romantic relationship with Niklas, a young man of the Uppsala establishment. She eventually assimilates. In the epilogue, she returns for the funeral and to apologize to her sister's corpse, while her son and granddaughter learn about reindeer herding and Sámi ways of life. The depiction of the past is framed by a present-day prologue and epilogue.

As interventionist historiography, this bifurcation in temporal and narrative structure replicates the plot's central conflict: the main character's split subject position as she grapples—first as a young woman reindeer herder and later as a retired school teacher—with maintaining or relinquishing a Sámi identity, constructing a Swedish persona, and then returning with family to a Sámi context. Switching between languages—South Sámi and Swedish—further supports this bifurcation. This return signals that personal, cultural, linguistic, and historical reconciliation may be possible.

We argue that while interventionist historiography has narrative, linguistic, and formal components, it also functions as a guiding ethos of the film institutes that funded *Sami Blood*. The ISFI, by funding pan-Sámi films in the Sámi languages, works to bring to light not only neglected aspects of Sámi culture in the present, but, with the production of films like *Sami Blood* and *7 Sámi Stories* (Sámi Film Lab, Norway, 2015), to place the present in dialectical tension with the past and to emphasize authentic language use as a historical and political tool. The SFI, through its funding of both Sámi films and films by women (made mostly in the Swedish language), gives historically marginalized people a voice in the public sphere and raises issues about unreflexive assumptions surrounding gender equality in Sweden and the other Nordic countries. *Sami Blood* thus functions as interventionist historiography and represents how media sovereignty reveals the unresolved tensions of colonialism and oppression in a historical past and their ongoing resonances in a politically contested multiethnic and multilingual present—in this particular case, in Sweden and Sápmi.

As an act of media sovereignty, *Sami Blood* contrasts with a long history of ethnic Swedish film perspectives about the Sámi. In Swedish film culture, a number of these productions were considered "big budget" in a small national cinema context, intended for international distribution while exoticizing and primitivizing the Sámi in the process. Examples include *Lappblod* (Ragnar Frisk, Sweden, 1948), Sweden's second color film and the country's most expensive production at the time; the European coproduction success *Laila* (Rolf Husberg, Sweden/West Germany, 1958), with cinematography by Sven Nykvist; and the big-budget, Nordic noir crime serial *Midnattsol* (*Midnight Sun,* Måns Mårlind and Björn Stein, Sweden/France, 2016), conceptualized in the vein of recent Scandinavian TV successes.

Sami Blood is Sámi and Indigenous: it is shot on location in Sápmi and with Sámi dialogue, by an Indigenous director, scriptwriter, producer, and cast. It is also a work of art that foregrounds distinctive northern environments throughout all scenes shot in Sápmi. The film's "sovereignty," moreover, reflects the rise of robust funding, production, and circulation support structures for Sámi film in Sápmi as part of the efforts of the ISFI, which awarded a substantial amount of coproduction funding for *Sami Blood*. The film can also be seen as a key outcome of recent SFI policies that allocate support for films that reflect a diverse Sweden and that align with the country's gender equality principles. These have been part of Filmavtalet, the Swedish film policy since 2000, with SFI CEO Anna Serner accelerating the ambitious agenda, promoting it as "50/50 by 2020" at the Cannes film festival each year since 2016. Kernell participated in an SFI mentorship program for women filmmakers, and *Sami Blood* was promoted from the beginning as a marquee film by the SFI in their international efforts.

For some, however, Filmavtalet has not been sufficiently effective. To call attention to the continuing gender disparity in film production, the association Women in Film and Television and the nonprofit social advocacy group Rättviseförmedlingen (The Equity Agency) launched a rating system in 2013 based on the Bechdel test to assess effective gender discrimination in films distributed in Sweden. The groups collaborated with four Swedish cinemas to launch the "A rating system," which received extensive media attention. Although not a part of SFI's official mandate, the A rating quickly became associated with Serner's ambitious agenda, which has been controversial. Some film critics, journalists, and producers see SFI's implementation of Filmavtalet as a counterproductive quota system (similar criticism has been leveraged elsewhere against various affirmative action initiatives), as numerous media reports in Sweden and beyond show. Reporting on the "A rating system" generated additional attention for these measures, and governmental support of gender equality in films, and in the film industry, is viewed favorably in the public sphere. Perhaps not surprisingly, Kernell's film reflects the priorities of both of these funding institutions: the production of Sámi-language films (ISFI) and those that signal gender equality and diversity of representation (SFI). In addition, the ISFI's impact on filmmaking and its active role in promoting Sámi film at festivals around the world and for broadcast, digital, and cinema circulation reflects trends in global Indigenous media (we discuss the ISFI in more detail in the next chapter). The involvement of these funding organizations, with Indigenous culture and European nation-states evolving side by side, points to *Sami Blood*'s intercultural nature.

Sami Blood *as Haptic and Intercultural*

Sami Blood is an example of Indigenous media sovereignty in which images from settler culture are destabilized in both the past and present while

Indigenous audiovisuality is foregrounded. Yet the film is also *intercultural,* in the context that Laura U. Marks elucidates the term. The combination of these two modes contributes to the film's status as interventionist historiography. Marks describes the concept of the intercultural—including in many examples from Indigenous and First Nations cinemas—as follows:

> "Intercultural" indicates a context that cannot be confined to a single culture. It also suggests movement between one culture and another, thus implying diachrony and the possibility of transformation. "Intercultural" means that a work is not the property of any single culture, but mediates in at least two directions. It accounts for the encounter between different cultural organizations of knowledge, which is one of the sources of intercultural cinema's synthesis of new forms of expression and new kinds of knowledge. . . . [T]he term avoids the problem of positing dominant culture as the invisible ground against which cultural minorities appear in relief. Instead, it implies a dynamic relationship between a dominant "host" culture and a minority culture. (Marks 2000, 6–7)

Addressing intercultural cinema, Marks notes that it "moves backward and forward in time, inventing histories and memories in order to posit an alternative to the overwhelming erasures, silences, and lies of official histories" (2000, 24). *Sami Blood* further engages a diachrony (a plot spanning two time periods), thematizes "the possibility of transformation," mediates culture in multiple directions, and engages a complex, often unequal, but always "dynamic relationship between a dominant" and a "minority culture." *Sami Blood* mobilizes a relationship between the intercultural and the haptic. Marks notes: "It may be . . . obvious why intercultural works deploy haptic visuality. The apprehension of being seen, categorized, and killed into knowledge informs many of the works that speak from a place between cultures, given the ethnographic (in the broad sense) tendency to fix its object. . . . The critique of visual mastery in such works speaks from an awareness about the destructive and literally imperialist potential of vision" (2000, 24). The haptic, then, functions in *Sami Blood* as a means to destabilize the process of "empirical" categorization that the "ethnographic" eugenics photography takes as "killing into knowledge." In addition, *Sami Blood*'s emphasis on Sámi language use—and its recognition of centuries of suppression and even eradication of linguistic communities—furthers Marks's arguments to demonstrate that haptic intercultural cinema is also auditory.

The Contested Past: Eugenics and the Image of the Sámi in Visual Culture

Sami Blood achieves its interventionist status by engaging in a self-reflexive manner and by emphasizing gendered realities against the backdrop of a long history of Swedish visual culture that has shaped images of Sámi. Photography, film, and museum exhibits have been especially influential in this regard. *Sami Blood* challenges this tradition, both aesthetically and thematically, as part of an international movement by which Indigenous filmmaking in the twenty-first century takes issue with practices of institutionalized film production and image making more generally. Kernell exercises what Fatimah Tobing Rony calls "a sustained critique of the pervasive form of objectification of indigenous peoples" that constitutes the "label Ethnographic" (1996, 5). Photographic and cinematic cultures during the 1920s and '30s infamously arose as cultures of spectacle through world's fairs exhibiting nonwhite peoples or through open-air museums such as Skansen in Stockholm—just as the "scientific" practice of collecting Indigenous images conjoined with moviemaking. *Sami Blood* integrates within its narrative and aesthetic registers a profound critique of these traditions and assumptions.

Indeed, in the film we see interventionist historiography at work: *Sami Blood* rewrites the history of the function of eugenics photographs as, to use Walter Benjamin's term (1999, 473–75), the *dialectical image.* In the 1930s, the photos stood as "empirical proof" of the "biological inferiority" of the Sámi. In Kernell's reframing of the images, they now signal a different kind of empirical proof: that of the history of Swedish colonialism. Addressing institutionalized image making, *Sami Blood* criticizes a long history of ethnographic film as a phenomenon of colonialism. *Sami Blood* promotes what Christopher Pinney calls a "destabilization of the relationship between anthropology and photography" and anthropology's distrust of verbal "native testimony" in favor of the visual, construed as direct transmission of impartial and empirical observation through photography (Pinney 2011, 12). A central scene in *Sami Blood* is the photo-documentation of Elle-Marja and other Sámi students visiting Uppsala University scientists. This scene, in which Elle-Marja is forced to disrobe in front of the camera, speaks to the long-standing, racialized codification of Indigenous peoples, based on observation and the disinterest of "many nineteenth-century anthropologists . . . in 'culture'"—they were interested in "comparative anatomy" (Pinney 2011, 15). The *Sami Blood* scene is brutal and uncomfortable, with the light and sound of the camera's flash going off foregrounded, as if it were a gun-shot. Elle-Marja's naked body is not shown to the viewer, who instead sees the faces of the schoolchildren observing her denigration (waiting in line themselves to be photographed), the proud teacher, and, through a gap in the curtain, a leering boy from a neighboring farm—incidentally the one who has been tormenting her (for a complementary reading of these scenes, see Tavares 2021).

The photography scene is visually rich and ethically complex, as Elle-Marja attempts to shield herself from a range of spectators who occupy power positions with respect to her body and person. It also silences her. No words of protest are uttered either in Swedish or in South Sámi. The scene thereby also provides a commentary on the privileging of visual anthropology over the spoken word, the authentic utterance of the "subject" under considera-tion. The shame at being exposed in front of her school friends through the visual register of the science of physical anthropology is palpable; for Elle-Marja, this is especially problematic, as she is construed as the most Swede- and Sweden-friendly of the school group. Here, the haptic comes into play, as the eugenic images created are not framed within the ethnographic, voyeur-istic gaze of spectacle and instead focus on the physical impact of the image-making process—Elle-Marja's body twitching as the flash bursts. As Marks notes, "haptic images may raise ontological questions about the truth of pho-tographic representation [and] encourage a more embodied and multisensory relationship to the image" (2000, 172). *Sami Blood* therefore turns the proc-ess of objectification into an embodied, subjective experience, making inter-ventions in several different colonial histories. In doing so, *Sami Blood*'s use of haptic visuality "give[s] the impression of seeing for the first time, gradu-ally discovering what is in the image rather than coming to the image already knowing what it is" (2000, 178). This process reframes eugenics documenta-tion, allowing the viewer to see the images anew outside their received func-tion. No spoken justification is provided or challenge uttered—the power of the colonizer's visual techniques of silencing Indigenous voices is clearly conveyed.

The scientist photographer in *Sami Blood* is a fictitious version of the well-known eugenics and race biologist Herman Lundborg of Uppsala University. The Swedish Institute for Race Biology (SIRB) was founded in 1922 by unanimous parliamentary approval; the institute's dogma of visual documen-tation and physical anthropology was instrumental in the ongoing racializa-tion of Sámi during the following decades. Lundborg's theories built on Carl

Linnaeus's classification system that categorized *Homo sapiens* into different sub-races. Lundborg's practice was also consistent with mainstream methodologies of international anthropology at the time, which depended substantially on visual categorization and measurements of body parts. "Scientific" photography of Sámi was a key part of this practice. More than twelve thousand photographs are held at the SIRB and Lundborg archive at Uppsala University. Lundborg's extensive photo-documentation was done in the service of establishing white, Nordic races as biologically—and, by extension, culturally and socially—superior to all others, the Sámi included. As an intra-filmic reference to this context, the *nomadskola* teacher has a book with photographs of "Lappar" on display and removes it when Elle-Marja starts looking at the images, as if to hide that part of problematic history from her student, and also, presumably, to spare herself any confrontation over it. This brief sequence is yet another example of how *Sami Blood* turns the perspective around—racialized photo-documentation now serving the purposes of interventionist historiography.

The assumption guiding Lundborg's work is Sweden's policy of ethnic segregation during the first part of the twentieth century, which complements the promulgation of discrimination based on physical characteristics. That policy was based on the notion that the Sámi were not only culturally inferior to Swedes, but also racially unequipped to adapt to life in an industrializing, urbanizing welfare state guided by ideologies of uniform democratic progress. This policy stipulated that "a Lapp shall be a Lapp," or "lapp skall vara lapp" (see Lundmark 1998; Lantto 2012), and provides the context for Elle-Marja's teacher's remark that the Sámi tribe must remain as nomadic reindeer herders in the North, at the risk of perishing elsewhere, and the admission that the *nomadskola* does not provide the same level of education as would be provided to Swedes (see Taveras 2021). The photography of eugenics classification practices dovetails with government policy of the time, just as the political establishment—regardless of party affiliation—supported Lundborg's work, at least during the early 1920s.

Photographing and documenting Sámi peoples and practices was also foundational to a related set of visual cultures: those of museology (in exhibits and through the collecting of artifacts). As Rony notes, "until the 1920s, anthropology in Europe and the United States was still based in the museum," and the ethnographic gaze of spectacle influenced images depicting "non-Western" cultural practices (1996, 63). For instance, the ethnographer Ernst Manker—a curator at Nordiska Museet in Stockholm from

1939—documented the Sámi, not as part of the eugenics movement, but through the related vein of salvage ethnography, to capture the quickly receding past as a fixed moment in time. His extensive collecting of materials and artifacts in Sápmi—near the region where *Sami Blood* is shot—shares many similarities with the principles guiding Lundborg's work in physical anthropology, though to slightly different ends.

In 1934, Manker argued that modernization of Swedish society was inevitable and, indeed, that it was desirable for the Sámi, as it "would allow development among the Sámi and offer them the same possibilities of a comfortable life in modern society like other citizens. By leaving artefacts no longer needed to the museum, the Sámi could, he said, with pride build a monument to their traditional culture—and move on, since he saw modernization as natural, inevitable, [and] righteous" (Silvén 2014, 59–60). Manker started his ethnographic work in Sápmi in the 1920s, avidly photographing and collecting artifacts. There are two scenes in *Sami Blood* that suggest his presence: one in which two of the Uppsala students describe their excitement over a summer field-site visit to the North as part of their studies in anthropology and ethnography; and another in which Elle-Marja/Christina asks her mother to sell her father's silver belt, presumably to a museum collector such as Manker. Indeed, Manker was the main curator for Nordiska Museet's long-standing exhibit *Lappar* ("Lapps," 1947–77), which construed reindeer herding and nomadism as quintessentially and irrevocably "Sámi." There were thus many movements coming into alignment that constructed a reductive and reactive image of Sáminess. Three paradigms were instrumental for how visual cultures of the time represented Sámi life: (a) salvage ethnography; (b) social policy ("A Lapp shall be a Lapp"); and (c) physical anthropology (through racialized measurements). *Sami Blood* directly challenges these dominant racializing paradigms in the critical scenes documenting Elle-Marja's subjugation and her subjective, haptic experience at the *nomadskola*. She is interpellated by "science" (as eugenics), by "ethnography" (in the visitors' comments about the skilled Sámi handicraft), and by "social policy" (her teacher orchestrating the events). The film acknowledges the forceful amalgamation of eugenics, ethnography, and policy in visual culture. This critical scene exemplifies *Sami Blood*'s mode of interventionist historiography, destabilizing supposedly normative modes of empiricism and reframing them as part of a racist, colonial endeavor.

Modernization—what scholars such as Lundborg and Manker, as well as Swedish politicians at the time, saw as detrimental to "authentic" Sámi culture—is also often dismissed by characters in the film. The striking exception is the depiction of Elle-Marja/Christina herself, who is presented as committed to her own process of modernization: escaping by train, burning her *gákti* (a traditional Sámi garment), verbally suggesting and initiating sex, applying makeup, cherishing the creamy confections from an Uppsala bakery, and speaking only Swedish, thereby leaving behind what she sees as the "backwardness" of Sámi culture. Her various verbal attacks (in South Sámi) on her sister and her mother are a pastiche of both forms of colonialism: these altercations express that while Elle-Marja feels that other Sámi can never change (and therefore will always be inferior), she is equally committed to the notion that she can change herself. *Sami Blood* does not shy away from such complexity in its analysis of the often bifurcated nature of subjectivity, and the way language politics plays into this. The film can thereby be understood as interventionist historiography as well as through Marks's terminology: it "avoids the problem of positing dominant culture as the invisible ground against which cultural minorities appear in relief" (2000, 7).

In addition, the film's central scene of "scientific" photo-documentation of Elle-Marja also speaks to Kernell's interest as a filmmaker and visual storyteller in presenting and challenging the internalized shame that she—and many others—argue have been part of Sámi culture for generations. As Kurt W. Fischer and June Price Tangney write, "people are ashamed . . . because they assume that someone (self and/or other) is making a negative judgment about some activity or characteristic of theirs" (1995, 4). Psychologists describe shame as one of the social and self-conscious emotions, "inextricably embedded in cultural contexts" (1995, 34). *Sami Blood* addresses some of these issues both visually and auditorily through its characterization and mixed language use, thereby "uncover[ing] the social grounding of emotions" (1995, 6). This is evident in *Sami Blood,* most forcefully in Elle-Marja's rejection of Sáminess and her own mother tongue, both in the past and in the present. The character's actions signal a belief that if she were to become "Swedish," the shame (the sight and sound) of Sáminess would also be washed away. At the same time, *Sami Blood* speaks to other Arctic Indigenous TRC films, with Kernell emphasizing in an interview that this is "not an educational film. It is a film about healing. I wanted to explore shame and the colonization of the mind" (cited in Cocq and Dubois 2019, 183). These aspects of *Sami Blood* connect the work of Kernell to other contemporary works of interventionist media sovereignty across the global circumpolar north, and to TRC movements in Nunavut and Greenland specifically.

Integrating aspects of both Indigenous media sovereignty and haptic audio-visual cues of intercultural cinema, *Sami Blood* also has narrative, thematic, and formal characteristics that mark it as "Swedish art cinema." Because of this code mixing, the film makes it possible to tell a story of colonialism, suppression, and racism in ways that are understandable to multiple audiences and that echo familiar cinematic conventions. For instance, the film emphasizes the insider-outsider plot common in Swedish cinema, with a protagonist revolting against what she perceives to be several overbearing collectives—whether colonialist, gendered, or Indigenous. This tradition of rebellious coming-of-age stories ranges from the series about Astrid Lindgren's Pippi Longstocking character, known from children's books, films, and TV adaptations, to more recent coming-of-age films, including Lasse Hallström's historical drama *My Life as a Dog* (*Mitt liv som hund,* Sweden, 1985), Lukas Moodysson's queer classic *Show Me Love* (*Fucking Åmål,* Sweden, 1998), and Tomas Alfredson's vampire film *Let the Right One In* (*Låt den rätte komma in,* Sweden, 2008). *Sami Blood* also evokes Swedish art cinema tropes, following an open-ended narrative path and leaving many questions about Elle-Marja/Christina's actions unresolved.

Similarly, the film's cinematographic attention to characteristic Northern environments—to expansive landscapes, and to the gradations of natural light in late spring through summer and early fall—resonates with viewers who cherish these distinctive aspects of the Swedish film tradition. Specifically, the film's evocative use of light, skillful manipulation of depth of field, emphasis on long sunsets, traveling shots in the wilderness, and comparatively brightly lit scenes in Uppsala call to mind the cinematography of Sven Nykvist. The northern environments extensively pictured in the film are deeply familiar to both Sámi and Swedes, though for at least partially different reasons. This difference is also addressed in the dialogue, with Elle-Marja and Njenna remarking on the profound personal connection they have to characteristic aspects of that environment, specifically Northern Great Mountain. One of the most profound scenes involves Elle-Marja and Njenna yoiking the mountain, sharing a moment of connection that spans culture, environment, language, and practice. A number of recent examples of contemporary cinema by young women directors in the Nordic region address "erasures" and "silences" that echo those in *Sami Blood*. These include feature films about refugee assimilation, migration, and diasporas in present-day Sweden, such as Gabriela Pichler's *Eat Sleep Die* (*Äta sova dö,* Sweden, 2012), Sophie Vuković's *Shapeshifters* (Sweden, 2017), and Rojda Sekersöz's quasi-*banlieue* film *Beyond Dreams* (*Dröm vidare,* Sweden, 2017).

Sami Blood is an example of this growing intercultural phenomenon of conceptualizing "Sáminess" and "Swedishness" as an ongoing contact zone of identity and its performance; related approaches are evident in both Jarl's *The Threat* and Kramvig and Andersen Gomez's *Dreamland.* One of the distinct characteristics of *Sami Blood* is its focus on racialized identity as an act of (linguistic) performativity, and how that performance might play out. Elle-Marja/Christina decides that if she could fully "perform Swedishness," then her problems—having to face daily racism, being denied a "proper" education—would cease, and she would "become" Swedish. Within Swedish cultural and racialized discourses, this manifests as a possibility, in part, because the Sámi and Swedes can look similar. *Sami Blood* therefore questions how and why racialized identity is performed and whether this performance can transform a person. Indeed, Elle-Marja/Christina behaves as if the performance of one identity equates to embodiment of it. This speaks to a contradiction constructed by the film, which it cannot resolve. Moreover, the project of assimilation undertaken by the Swedish school system is one that is colonial on two fronts: valorizing Swedishness and the Swedish language above Sámi language and culture on the one hand, while on the other upholding the notion that the Sámi, within the colonial narrative, can only be interpellated by Sweden but not "become" Swedish.

This process of becoming also speaks to the heart of the colonial endeavor and the ideologies (manifested in this film through eugenics) that underpin it. While both Elle-Marja/Christina and the Swedish state may want the Sámi to "be" Swedish to certain and contradictory degrees, the inverse is not present as a possibility: no one in this film, nor in Swedish society at the time, would talk about the possibility of "becoming" Sámi. But how does one become a Swede? No one is born into an essentialized Swedishness any more than one is born into an essentialized Sáminess. Becoming Swedish—as with any other ethnically, culturally, and linguistically defined national or ethnic identity—is a process of social construction. For instance, Kernell notes that "we"—and one presumes that this *we* is the bicultural plural: Sámi and Swedish)—"learn nothing about this history. I was also taught only one sentence about the Sámi in school" (Agebro 2016, 7). This elision of Sámi culture, then, is a constituent part of "becoming Swedish," though it is also a part of "becoming Sámi," as this process of interpellation "allows" the Sámi to know their place in the Swedish imaginary. This elision of the Sámi plays itself out in other ways in the film. As Elle-Marja/Christina sits in a bar in the hotel she goes to—instead of staying with her family, and still pretending not to "be" Sámi—she talks to an ethnically coded Swedish woman (she is blond and speaks Stockholm-inflected Swedish) who wants "peace and quiet" during her vacation in the nature preserve on the Indigenous homelands of the Sámi people. The elision of the Sámi and their languages is a central aspect of "becoming Swedish," and despite the national imagination that the country is governed by principles of radical equality, this elision, the film points out, is still at the heart of Swedish national consciousness.

Sami Blood represents a contested colonial history to foreground that Sáminess and Swedishness are not, and have never been, separate—and that power dynamics between them have not been equal. The production circumstances around language use offers a case in point. The film adheres to the ISFI's mandate that about half of the dialogue in its productions must be in a Sámi language. The film itself becomes a way to strengthen South Sámi and present Sámi culture as multivalent and pluralistic. As Cocq and Dubois argue, "Kernell inserts South Sámi into the filmic soundscape of Nordic film, to stand alongside the now relatively familiar North Sámi of previous [films by Gaup and others]" (2019, 187). In addition, speaking Sámi or Swedish is a point of contention in the film, as language-switching signals a shift in outsider/insider status, with Elle-Marja/Christina using Swedish to set herself apart from her family and community, and Sámi characters speaking Sámi to her to pull her back into that cultural context. Swedish-speaking characters remark on her dialect or assume that her name must be something "exotic." Therefore, *Sami Blood* is both Sámi and Swedish, made in the Swedish language and in South Sámi, just as its narrative engages various cultural assumptions of what it means to be Swedish and Sámi, in the past and in the present. The film takes seriously the dual and multiple subjectivities of those who have experienced similar points of separation or erasure of communal and personal history, emphasizing that film can represent such stories in a complex manner that generates agency and not only victimization, with the narrative structure of the film replicating this duality as it moves between the present, the past, the present and points toward future continuations. This movement between temporalities, which aligns with interventionist historiography's desire to rupture hegemonic histories, is one of the key aesthetic and political achievements of *Sami Blood*.

Challenging the portrayal of normative Swedish and Sámi gender performativity is also an aspect of *Sami Blood*'s interventionist historiography, one that aligns with the SFI's and ISFI's focus on women directors, producers, and scriptwriters to produce new, counter-hegemonic works and modes of production. Sámi women documentary filmmakers have reconfigured the role of the active woman participant in the telling of Sámi history from women's perspectives, often with an explicitly gender-conscious approach, acknowledging that women's points of view in Sámi filmmaking have been largely ignored (we discuss this further in chapter 6). *Sami Blood* integrates this set of priorities into its project of interventionist historiography, destabilizing national, gendered, and colonial hierarchies. On the one hand, Elle-Marja/Christina is portrayed throughout most of the film as strong, independent, and resilient. Yet her independence is directed toward assimilating, if not becoming, the Swedish colonizer. In one way, her sister and mother seem far less empowered, but they nevertheless demonstrate their own form of resilience by staying within, and therefore preserving, the Sámi community. *Sami Blood,* then, through its three female Sámi characters, avoids totalizing Sámi women as only being able to resist in one manner, challenging myths created both by Swedes and by Sámi.

Sami Blood ends on a note of reconciliation, pointing positively toward the future, as Elle-Marja/Christina is walking at daybreak through her community's reindeer-herding grounds, toward her immediate and extended family. "Estranged" Sámi (her son and granddaughter) are integrated into significant and symbolic cultural practices—wearing a *gákti,* practicing phrases of South Sámi, and participating in reindeer-marking on the mountain, for instance. While such reintegration or self-acceptance of a multiplicity of identities and intercultural subject positions may not be possible for everyone, *Sami Blood* demonstrates that film can portray such complex and contested processes in a manner that resonates across generations, audiences, and ethnicities. While the coming-of-age historical drama is the center of the film, it is the split subjectivity of the protagonist that is the overarching frame. While Elle-Marja/Christina has lived through and with this split, in part by her own volition, the reconciliation is left to the next generation to overcome. What the conclusion of the film points to is that while reconciliation for the present generation is possible, the scars of the past continue to reverberate. While racism may be presented in a more genteel fashion in an ethno-Swedish normative consensus culture that is guided by principles of egalitarianism, it is certainly still present. The film nevertheless affirms agency and autonomy on behalf of both individual characters and the Sámi collective.

This chapter has examined a number of Sámi interventionist works from the late 1970s to the present. It is important to note that in the Scandinavian context, these films have effectively served as interventionist historiography for both Sámi and non-Sámi audiences. Films such as *Give Us Our Skeletons, The Threat, Kautokeino Rebellion, Dreamland,* and *Sami Blood* represent events actively neglected and suppressed as part of the Scandinavian welfare-state mythology of democracy, class equality, gender equity, and environmental stewardship. These works also excavate the histories of Nordic colonialism, rarely represented in film and video work, and most often unacknowledged in the Scandinavian public consciousness. As a body of work, the examples also span multiple narrative and formal strategies, representing a breadth and range of voices and perspectives. These are examples of Indigenous interventionist historiography that continue to shape public debate and political mobilization. In the next chapter, we turn to recent experimental documentary works by Sámi women to further an examination of Indigenous self-determination and environmental politics in Scandinavia.

SIX

Sámi Feminist First-Person Documentary and Women's Activism

SCANDINAVIAN INDIGENOUS FILMMAKING has grown in the past decade, often supported by the programs of the International Sámi Film Institute (ISFI), the Sámi Parliament, and nation-state funding agencies. Many of these films are made by women and almost all engage the transnational, cross-regional Sámi Indigenous heritage that spans northernmost Scandinavia, supporting Sámi rights to self-determination in the form of audiovisual self-representation that counters centuries of assimilation and colonial, nation-state, and settler politics. They are examples of the media sovereignty we've explored in previous chapters. These works, we argue, construct an interventionist historiography that engages the significance of cultural and linguistic revitalization movements for cinematic expression; the political mobilization that either is reflected in or has ensued from this body of work; the ways in which these interventionist historiographies are gendered and often feminist; and how these works relate to ongoing debates about the form, function, and implementation of Indigenous self-determination in Scandinavia. Examples we discuss in this chapter have had international distribution at film festivals and in cinemas and include documentaries by Liselotte Wajstedt, Katja Gauriloff, and Elle-Máijá Tailfeathers.

The works discussed span a range of documentary cinematic forms in ways that illustrate the breadth and depth of twenty-first-century media practice in Sápmi, including the essay film, self-reflexive video art, and first-person testimonials, along the lines of what Alisa Lebow terms the "cinema of me" (2012), while integrating techniques of found footage, animation, and temporal disjunctures, to name a few. The cinema of me is characterized by what Lebow calls "first-person films." These films place the documentarian herself as a constituent part of the pro-filmic world and narrative: "The designation 'first-person films' is foremost about a mode of address: these films 'speak' from the articulated point of view of the filmmaker who readily acknowledges her subjective position" and who, when making "a film with herself as a subject, . . . is already divided as both the subject matter of the film and the subject making the film" (Lebow 2012, 1, 4). Lebow uses the term *first-person film,* in contrast to *autobiographical documentary* or *self-reflexive film* (2012, 2, 6), in order to accommodate a multiplicity of selfhood, acknowledging that "the 'I' is always social, always already in relation, and when it speaks, as these filmmakers do, in the first person, it may appear to be in the first person singular 'I' but ontologically speaking, it is always in effect, the first person plural 'we'" (2012, 3). This means that first-person films also express "our commonality, our plurality, our interrelatedness with a group, a mass, a sociality, if not a society" (2012, 3).

The topics, themes, and narrative strategies of these works are also highly varied. These include a focus on first-person identity discovery, as Wajstedt uncovers her family's Sámi heritage in *Sámi Daughter Yoik* (*Sámi Nieida Jojk,* Sweden, 2007). Other first-person documentaries thematize journeys and trans-Indigenous commonalities to address transnational connections between Canadian First Nations and Norwegian Sámi. Tailfeathers's *Rebel* (*Bihttoš,* Canada/Norway, 2014) and Gauriloff's *Kaisa's Enchanted Forest* (*Kuun metsän Kaisa,* Finland, 2016) are examples of interventionist historiography that engage settler and Indigenous politics. In Wajstedt's *Kiruna: Space Road* (*Kiruna—Rymdvägen,* Sweden, 2013) and other Kiruna films, as well as *The Silence in Sápmi* (*Tystnaden i Sápmi,* Sweden, 2022) (see chapter 11), the impact of resource extractivism in the Scandinavian North and the silenced histories of violence against women within the Sámi community take precedence. All these works implicitly or explicitly engage matters of repatriation, Sámi cultural and political recognition, and gendered realities in Sápmi and beyond.

The experimental documentary cinematic forms are especially important in interventionist historiography because of their centrality in documenting, archiving, collecting, and providing evidence of the lives of Indigenous communities in the face of settler colonialism (see also Wilson 2015). The role of Indigenous filmmaking collectives has been especially important in this regard, as we showed in examining the Arnait and Isuma collectives in Nunavut in chapters 3 and 4. While the filmmakers addressed in this chapter do not formally constitute a collective, their contributions are part of a contemporary media production movement in Sápmi, wherein women's documentary filmmaking about gendered histories mobilizes a range of approaches, including capacity building, cultural capital, and support catalyzed by the ISFI.

Political scientist and Indigenous studies scholar Rauna Kuokkanen argues that many if not all Arctic Indigenous self-determination processes replicate the patriarchal foundations of colonial and settler paradigms, which are also dominant within Indigenous contexts. The ways in which these processes have been and continue to be gendered have been largely excluded from debates both within Indigenous communities and in settler and colonial contexts. In her landmark book *Restructuring Relations,* Kuokkanen does not posit *"whether* gender regimes exist in Indigenous self-government institutions, but [asks] *how* and *what kinds* of gender regimes dominate in those institutions" (Kuokkanen 2019, 175; see also Kuokkanen 2007, 2011). Kuokkanen confronts some of the strongest gender equality taboos in Sámi cultural politics since the 1970s cultural revivalism movements, which were first most strongly articulated in Norway, most centrally during the protests against the Álttá (Alta) dam in 1979–81. Kuokkanen notes how the creation of political and governmental systems is looked upon as essential and masculine work of self-determination, while the work of community is considered the secondary work of women (2019, 169). Kuokkanen furthermore argues that the Sámi political, intellectual, and civic society has demonstrated an ambivalence toward so-called soft-policy questions, such as language enhancement and strengthening cultural and arts practices, in relation to hard-policy questions such as political autonomy, enforceable land claim rights, or Indigenous independence within Sápmi. These soft-policy questions are not awarded the significance they deserve, she argues, noting "a problem when major issues are cast aside as apolitical and less urgent, such as transmitting language, land-based knowledge, and skills to the next generation" (2019, 169). Representing the significance of these practices has emerged as an important function of Indigenous media production in Sápmi, especially by women and feminist practitioners.

In the context of Sápmi, then, the impetus of political self-determination and land rights is deployed when addressing the larger global community of Indigenous movements, while within Scandinavia, so as not to upset or cause concern in the Nordic countries, the "soft" issues of culture and language are mobilized. For these reasons, Kuokkanen laments the fact that culture is the main form of self-determination on offer, which she concludes may not be a form a self-government at all: "The construction of the Sámi rights and Sámi self-government in cultural terms is a prime example of such culturalization and turning Sámi rights into minority rights" (2019, 89–90). The term *informal channels* (2019, 158) is highly significant with respect to twenty-first-century media production in Sápmi. There are two aspects we want to highlight. First, Indigenous women's political participation may be more commonly seen "in support and auxiliary roles in Indigenous nation-building and self-determination struggles" (2019, 158) than in elected positions of official power—though Sámi women are also represented in these positions. More importantly, informal channels are important because political self-determination has been weak. Documentary films, and the systems that support media production (especially in Indigenous languages), effectively examine, critique, and provide new models to address Kuokkanen's concerns about the sidelining of language, cultural, and social issues, promoting the notion that these cultural systems are among the most significant for equitable Indigenous self-determination frameworks.

THE INTERNATIONAL SÁMI FILM INSTITUTE, INFORMAL CHANNELS, AND THE POLITICS OF INDIGENOUS CINEMAS

The ISFI is one of the most salient examples of the informal channels concept of political power processes. Led by women, a large portion of filmmakers supported by the initiatives are also women, and many productions address topics central to women's Indigenous experiences, including the gendered regimes of self-representation discussed by Kuokkanen above. Norwegian North Sámi Anne-Lajla Utsi, as managing director and supported by film commissioner Liisa Elisabet Holmberg, launched the ISFI as the International Sámi Film Center in 2007; it became an "institute" in 2015. The ISFI functions as a key example of what Faye Ginsburg calls institutions supporting Indigenous media sovereignty, namely the "practices through which people exercise the right and develop the capacity to control their own images and words, including how these circulate" (2016, 583; see chapter 3).

The primary aim of the ISFI is to support filmmaking in the Sámi language, with a stipulation that funded films have at least 50 percent Sámi dialogue. It does so by fostering Sámi film production, training, distribution, and circulation from across Sápmi—through, among other ventures, its Film Lab for script development, its sound and post-production studio Duottar, and its presence at Indigenous and major-market film festivals around the world. In addition, the ISFI promotes the exchange of cinephile culture beyond Sápmi by supporting collaboration within the Arctic Indigenous Film Fund (AIFF); by launching events at international film and art festivals such as Sundance, the Berlinale, the Venice Biennale, and imagineNATIVE; and through "cooperation between Indigenous film workers at a global level" (ISFI website). The ISFI plays an important international advocacy role for Sámi and Indigenous cultural knowledge in the global entertainment sector, as exemplified by the close collaboration between Sámi representatives and Disney on Sámi cultural representation in *Frozen II* (Chris Buck and Jennifer Lee, USA, 2019). This collaboration is heralded by international Indigenous advocacy groups as a model to mitigate negative effects of cultural appropriation. As Utsi says, "I think it's a good example for every other . . . film [company] in the world who want[s] to be inspired by Indigenous culture. If you want to do it, you have to collaborate" (Last 2019). Building on experience from the Disney collaboration, the ISFI developed the document "Ofelaš— The Pathfinder: Guidelines for Responsible Filmmaking with Sámi Culture and People" to serve "as a practical guide for filmmakers who want to include Sámi themes, topics, and persons in their movies" and as an ethics manifesto "to point out what values we Sámi treasure and provide guidance for avoiding misrepresentation and appropriation" (ISFI 2022, 2).

The ISFI has also provided film packages of shorts for an array of international film festivals, promoting a form of "branding" that would be lacking from a single short being submitted to a film festival. Prominent examples include the "ÁRRAN 360" installation of Sámi film and digital media at the Venice Biennale in 2022 and the "7 Sámi Stories" program launched at the Berlin International Film Festival in 2015, accompanied by training workshops and network opportunities and subsequently traveling to imagineNATIVE, among other festivals. The styles of "ÁRRAN 360" and "7 Sámi Stories" films are heterogeneous and don't simply subscribe to an ISFI "look." This is important because it signals the ISFI's interest in supporting a range of aesthetics and approaches, with directors working across multiple genres.

The ISFI is owned and operated by the municipality of Guovdageaidnu (Kautokeino), with additional funding from the Norwegian Sámi Parliament and the Norwegian Ministry of Culture (extended collaboration with the national and Sámi governments of Finland and Sweden as well as with the Sámi Parliamentary Council are, at present, pending). The designation "institute" puts the ISFI on par with the Swedish, Norwegian, and Danish film institutes and the Finnish Film Foundation as representatives of a nation-state, and makes the ISFI the first Sámi transnational cultural entity to operate with a similar rhetorical and cultural authority as that of a nation. To that end, the ISFI has adopted many of the highly successful models of the Nordic national film institutes for the shaping of film policy, for the marketing of films, and for enhancing international reach and exposure. By changing its name from a "center" to an "institute," the ISFI recognized the weight and impact of the Nordic national film institutes, thereby adopting a nation-state "sovereignty" model to better support the production and circulation of Indigenous media.

The ISFI's approach can also be seen as distinctly Scandinavian; in general, the public perception of government in Scandinavia is positive, and state-supported cultural initiatives are construed as benevolent and not necessarily as counterproductive to Indigenous self-determination. The ISFI's approach of using established nation-state structures to further the reach of Indigenous voices effectively signals Sápmi as an autonomous region on par with that of a self-determining country. For small Nordic nations, as well as small nations in many other parts of the world, film productions (especially art cinema and documentary) have been important tools of nation-building and nation-branding processes for the past five decades (see Hjort 2005).

Through its support of the entire production and circulation cycle of Sámi-language films across Sápmi—and with increased collaboration between Indigenous filmmakers there and around the world—the ISFI mitigates Pamela Wilson and Michelle Stewart's argument that "Indigenous peoples have long had an ambivalent relationship with the mass media. Even though Indigenous groups and artists have produced their own expressive media for generations, the industrialized, mass-produced messages and images—and accompanying technologies—in most cases have represented the perspectives, values, and institutional structures of empire" (Wilson and Stewart 2008, 3; see also Ginsburg 1991). In this way, the ISFI's efforts under Utsi's leadership are strongly oriented toward influencing public policy and enacting transnational cultural diplomacy through film production.

Many ISFI-funded or collaborative fiction film productions prioritize interventionist Sámi historiography, as is certainly the case with Amanda Kernell's fiction feature *Sami Blood* (*Sameblod*, Sweden, 2016; discussed in detail in chapter 5), *Frozen II*, or Katja Gauriloff's *Je'vida* (Finland, currently in production). On the other hand, for Sámi filmmakers who are not tightly aligned with the ISFI, or who are critical of aspects of Sámi culture and governance or of the Nordic nation-states providing funding, the ISFI's language production requirements, strong branding, and extensive networks can pose limitations. In addition, the women filmmakers supported by the ISFI have diverse sets of aesthetics and politics. Nevertheless, their work can be seen as complementary to current Sámi feminist movements gaining prominence in the North. The work of the ISFI operates across a range of registers, modes, and political frameworks, effectively serving as one of Sápmi's most influential informal channels to advance discourses on self-determination and Indigenous sovereignty.

THE FEMINIST POLITICS OF SÁMI CULTURAL MEMORY IN CONTEMPORARY EXPERIMENTAL DOCUMENTARIES

There are several significant trends in recent Sámi film production by women. One trend foregrounds ongoing cultural revitalization processes in terms of Sámi language, culture, and agency, emphasizing that film production and cinematic representation are political acts. Another trend focuses on new approaches to interventionist historiography—especially about topics that address settler colonialism, racism, resource extractivism, and the role of the Nordic nation-states in relation to Indigenous populations. These films construct an alternative historiography of Sámi culture, not only in relation to the Nordic nation-states' imposed and often deliberate misrepresentation of both Sámi agency and the intricacies, complexity, and multivariate state of Sámi history, but against the idea that these groups "had no history." In Sweden, this assumption became codified during the early twentieth century as the nation-state attempted to "protect" the Sámi against modernity and implemented a far-reaching social transformation that was part of the Swedish welfare state experiment—what was officially called the "Lapp shall be Lapp" policy, or, in Norway, the priority of Norwegianization in language and culture. The works also support the notion that in the twenty-first century, Sámi and Sápmi are international entities and globally connected, not least as the Arctic region becomes even more important geopolitically and environmentally, which is reflected in the *The Sámi Arctic Strategy,* a report from the Saami Council (2019), a voluntary, transnational organization advocating Sámi rights and interests.

A third trend we will emphasize is how women filmmakers bring together these discursive, societal, and historiographical strands, where language, cultural identity, self-determination, and personal histories are unequivocally posited as political and historical, not relegated to a "domestic" sphere, as if that sphere were an apolitical one. We examine a prevailing trend over the past decade of documentary and essay films about Indigenous "identity suppression" and, relatedly, "identity rediscovery" that effectively integrate gender into the historiography to convey stories silenced, forgotten, or downplayed by mainstream Nordic media. These works address women's discovery of their Sámi identity and the mechanisms that contributed to the suppression and silencing of that identity; they effectively function as an alternative form of historiography—one that, we argue, is both interventionist and political. The works of these women span the autobiographical aspects of Sámi cultural revivalism and question the often masculinist assumptions that underlie Indigenous self-determination movements, as argued by Kuokkanen.

For most of the twentieth century, films about Sámi by non-Sámi filmmakers carried traces of ethnic, racial, and gendered Othering, exoticizing cultures and complex histories in service of colonial interests on behalf of the Scandinavian nation-states. The roles women played in Nordic fiction feature films about the Sámi produced by outsiders mythologized Sámi women, infantilizing them, portraying them as "pagan" or non-Christian, and placing them outside of modernity. Examples include the lethal shape-shifting temptress in *The White Reindeer* (*Valkoinen peura,* Erik Blomberg, Finland, 1952)—reminiscent of Jacques Tourneur's RKO horror classic *Cat People* (USA, 1942)—and the fears of miscegenation that are thematically central to all three versions of the melodrama *Laila* (two made by George Schnéevoigt in Norway in 1929 and 1937, and one by Rolf Husberg in Sweden in 1958). These films feature stereotypical images of Sámi women, drawing heavily on the melodramatic tradition (on the history of Sámi representation on screen, see Cocq and Dubois 2019; Mecsei 2015, 2019; Sand 2022). In some fiction feature films by male Sámi filmmakers, the roles of Sámi women have been more varied, including the female protagonist Elin in Nils Gaup's historical epic *The Kautokeino Rebellion* (*Kautokeino-opprøret,* Norway, 2008). Yet most recent genre films by male Sámi directors represent women characters as largely passive and outnumbered, following conventional Western narrative structures, characterization, and dramaturgy, in films such as Tommy Wirkola's Tarantino parody *Kill Buljo* (Norway, 2007), Gaup's "Northern" *Pathfinder* (*Ofelaš/Veiviseren,* Norway, 1987) and police procedural *Glass Dolls* (*Glassdukkene,* Norway, 2014), or Simma's comedy *The Minister of State* (*Stol paa ministerien,* Sweden, 1997).

There is also a limited realist feminist or queer documentary tradition that depicts Sámi women's lives in ways that address identity politics, cultural history, and reindeer herding as a contested practice. These include works such as *Daughters of the Midnight Sun* (*Ovanlandet,* Ylva Floreman and Peter Östlund, Sweden, 1985; coproduced with Swedish public broadcaster SVT); *Lesbian in Kautokeino* (*Lesbisk i Kautakeino,* Nils John Porsanger, Norway, 2001); *Herdswoman* (*Hjordeliv,* Kine Boman, Sweden, 2008); *Suddenly Sami* (*Min mors hemmelighet,* Ellen-Astri Lundby, Norway, 2009); and Suvi West's *Me and My Little Sister* (*Sparrooabbán,* Finland, 2016) and *Our Silent Struggle* (*Eatnameamet: Hiljainen taistelumme,* Finland, 2021). These realist documentaries provide a robust historical background of Sámi filmmaking against which the emergent experimental tradition of Sámi documentary should be positioned, especially the ways in which experimental practices allow for a new interventionist historiography to emerge.

Within the burgeoning field of documentary film studies, the dominant mode of interventionist documentary filmmaking has been realist and activist, often connected to social justice, but the "expressive or aesthetic function has been consistently undervalued" (Renov 2013, 347). Michael Renov further notes that

> there has been an explosion of recent work in which film- and video-makers have explored the historical world from diverse perspectives, employing a range of methods and approaches which, through their innovations, have drawn the documentary film ever closer to the realm of contemporary art. The work of these artist-practitioners is drawn to the world 'out there' . . . but shaped and informed by the world 'in here', by personal experience, cultural and sexual identities, political and aesthetic engagements. (2013, 349)

The works of Wajstedt, Tailfeathers, and Gauriloff are part of an emergent and growing global practice of artistic, self-reflexive, and experimental Indigenous documentaries, made in both Indigenous and settler languages, reflecting that the world "in here" may be multilingual or unilingual in ways that don't always match with expectations of the world "out there." This set of priorities was also part of a major Sámi art and video installation exhibition at Sven-Harry's Art Museum in Stockholm in 2021, called *Bakom hörnet vindens jojk* ("Around the corner the yoik of the wind"). The show featured several video installations, by artists such as Marja Helander and Joar Nango, that bridge environmental concerns in Sápmi with the personal. Several contemporary directors mobilize their own autobiographies and positions as hybridized subjects in terms of national, Indigenous, and linguistic identities, generating a Sápmi interventionist historiography that is also a statement of media sovereignty. It is especially important to note that these investigations are formally complex. For example, *Sámi Daughter Yoik, Rebel,* and *Kaisa's Enchanted Forest* deploy experimental techniques that challenge a realist tradition of Sámi documentary to emphasize, through cinematic experimentation, the contradictions, tensions, and variegated set of experiences that constitute Sámi culture. These three films employ formal experimentation that includes superimposition, time-lapse photography, and animation techniques such as stop motion, pixilation, and cell animation, along with non-diegetic sound and music. The works of these Sámi women directors problematize questions of identity, inclusion, and exclusion, and the hybrid nature of ethnic and national identities. In all three films, the dialectics of the past and present, tradition and modernity, are expressed through sophisticated aesthetic experimentation. To this end, the films need to be contextualized within a broader set of considerations about gender, cultural revivalism, and aesthetics that are shaping Indigenous screen cultures in the Nordic region and beyond.

CONTESTED HISTORIES, MULTIPLE IDENTITIES: *SÁMI DAUGHTER YOIK*

Liselotte Wajstedt is a Sámi filmmaker who engages in many different styles in the service of her feminist interventionist politics and aesthetics. The shift she represents in terms of Sámi self-imaging is part of a generational shift that is more broadly defined by the challenges posed by feminist and queer movements, and in partial response to the implicit patriarchal structures of Indigenous 1970s and 1980s self-determination movements. Many of Wajstedt's films engage explicitly or implicitly with multiple and hybrid subject positions, illustrating Renov's argument about twenty-first-century documentary, wherein "the subject in documentary has, to a surprising degree, become the subject *of* documentary" (Renov 2004, xxiv). Wajstedt's films have screened at film festivals and in art exhibits or installations in Europe and North America, and have been broadcast on public television in Scandinavia. Her work reflects feminist Indigenous work that aligns closely with the ISFI's interest in supporting multiple perspectives on Sámi culture and history, but they are not always in the Sámi language. However, in the recent experimental hybrid documentary *Sire and the Last Summer* (*Sire och den sista sommaren,* Sweden, 2022), based on the death during childbirth of Wajstedt's great-grandmother, Wajstedt uses the Sámi language nearly exclusively. Moreover, questions of mobility come into play, with the director having trained in Stockholm and living parts of her adult life in Giron (Kiruna). Many of her films address rediscovering Sámi heritage (including learning to speak the language); for others, her work may not be seen as quintessentially Sámi because of her locale, mobility choices, and non-native-speaker status. As a director, Wajstedt has repeatedly stressed that she wishes for her works to be considered as "films" in and of themselves, not necessarily to have them exclusively grouped under a "Sámi Cinema" rubric (Stenport 2013, 2019). Furthermore, her films speak explicitly to the hybridized nature of her experience, placing her both inside and outside Sámi culture and the dominant strains of contemporary Sámi filmmaking, while also providing evidence that these subject positions are constantly evolving. In the essay "Frispel 2020," for example, Wajstedt reflects on her interventions in Sámi art and culture, how these might have been received as challenging, and her insistence on an individualistic approach (2021, 132).

Sámi Daughter Yoik's experimental documentary aesthetic employs animation, superimposition, and digital media in ways that break with the norms of realist documentaries (figure 6.1). These strategies challenge prescriptive frameworks and emphasize that a story of personal and cultural discovery is never straightforward and needs to be punctuated by self-referential aesthetic strategies that signal an awareness of the performative and storytelling project under way in the film. The film is rooted in transnational and international artistic practices, whereby works are created to travel across borders while maintaining their foundation in a specific first-person experience.

Sámi Daughter Yoik foregrounds travel—across, to, and from Sápmi—and journeying, both emotional and psychological. *Sámi Daughter Yoik* is also a road movie whose content extends beyond a narrow autobiographical context. Wajstedt, as the urban protagonist, returns from Stockholm to Giron with the explicit purpose of learning more about Sámi culture, heritage, and language as well as about her own family history. The film self-reflexively visualizes its road movie aesthetic as highways connect various villages through animation and superimposition, both as red arteries and as embroideries found on a *kolt* (Wajstedt uses the Swedish word for the traditional Sámi garment also known as *gákti*). This animated map connects the

FIGURE 6.1. Experimental mapping in Liselotte Wajstedt's *Sami Daughter Yoik* (*Sámi Nieida Jojk,* Sweden, 2007).

transnational space of Sápmi as a conglomeration of cultural, linguistic, and identity practices in a formal strategy that equates the map legend with life-giving blood.

Wajstedt narrates her experiences in a first-person voiceover, interspersed with dialogue, usually consisting of the director's recorded conversations with others—sometimes in Sámi, sometimes in Swedish—as she herself learns the Sámi language, which was denied her as a child. As a first-person road movie, the film's content is self-reflexive, questioning whether one can learn to *become* Sámi, while simultaneously representing and performing the act of learning to *be* Sámi through the filmmaking process. Rendered on film, her experience becomes not only a personal journey but a collective one, reflecting the loss of language by Wajstedt's generation. Sámi parents of the time chose Swedish enculturation for their children, in the process denying them one of their mother tongues and their connection to—or even knowledge of—their Sámi identity. *Sámi Daughter Yoik* is, in many ways, a political act of interventionist historiography. The film is not simply about "recovering" lost identity through language acquisition; Wajstedt "learning"

Sámi stands as a sign of what has been lost in terms of inculturation. The film's political intervention is its illustration of how lost identity can only ever be partly recovered, whether by Wajstedt learning Sámi or learning about her family's elided history. *Sámi Daughter Yoik,* then, is not so much a work of cultural revivalism as it is a work that, as an interventionist historiography, delineates what has been lost as a means of documenting that loss as a significant component of Indigenous media sovereignty.

The silenced—unspoken—history engaged in *Sámi Daughter Yoik* also connects to the centrality of the oral tradition in Sámi and Indigenous cultures, which documentary filmmaking is exceptionally poised to record and regenerate in new forms, thereby moving it from the private to the public sphere. This strategy relates to what Coppélie Cocq and Thomas Dubois (2019) term "Sámi Indigenous agency" and what Faye Ginsburg (2016, 583) calls Indigenous media sovereignty. Wajstedt herself states that she "became" Sámi only as an adult, in part through the documentary process, and in part by engaging in rites of inculturation: taking language classes, visiting spiritual locations, sewing a *gákti*, attending a cultural festival, reconnecting with family and friends from her youth, and participating in a yoik workshop. At the same time, Wajstedt questions the point of her journey, asking in a voiceover, when she puts on the *gákti*, whether she is "pretending to be a Sámi," admitting that she wants to wear it "to make me feel more Sámi," while noting that she "can't decide if I am dressed up or dressed." *Sámi Daughter Yoik* uses statements like these in multiple instances to call attention to the ways in which self-reflexive documentary strategies and linguistic utterances—such as where the director is also the protagonist in the "cinema of me" context—question realism and the seemingly coherent and unchanging knowledge structures that underpin it. The fragmentation offered by the self-reflexive documentary parallels the fragmented and hybridized subject position of Wajstedt-as-protagonist. The political act of acknowledging fragmented, multiple, and hybridized identities, and where these come from through family and cultural history, is what makes *Sámi Daughter Yoik* an interventionist historiography.

Sámi Daughter Yoik presents the director's hybrid identity as a continuously evolving process, just as Sámi cultures are multiple and dynamic. The Sámi musical form yoik is presented as a key example in this regard, especially in the ways it is combined with Western rock and folk music performed, for instance, by Sámi DJ Ande. This hybridization is both acknowledged by DJ Ande and ridiculed by a Sámi woman on stage at the festival, who states that the Sámi get together and "dance once a year to some bad rock band doing yoik covers." Wajstedt and her grandmother address the settler colonial repression of yoik and the Sámi language as an important part of Sámi history; as her grandmother affirms, once Sámi were Christianized (in some cases forcefully), yoiking was considered a sin: "Only drunks yoik," she states. The stereotype of drunk Sámi yoiking masks a trauma, however. Only when drunk could a core aspect of Sámi culture, history, and identity be expressed, in one's native tongue. When sober, that cultural form of expression needed to be suppressed. For Wajstedt, the yoik is a central and multimodal trope of her own political, and psychological, journey. She states in a voiceover that "my film is a yoik"—it conveys and contains knowledge, culture, tradition, and interconnectedness. The film-as-yoik becomes a political act that turns the oral tradition of yoiking into a documentary expression that speaks both to the power of Indigenous media sovereignty and to the power of documentary as interventionist historiography.

At one point, Wajstedt debates a racist ethnic Swede at a music festival in Kiruna. His face is obscured by an animated talking head with only his lips showing, stridently questioning why she would ever want or need to "rediscover her identity"—given that he, as an ethnic Swede, can fully attest that there has never been any discrimination of Sámi in Swedish society. He insists that Wajstedt's Sámi inculturation process—documented on film—is pernicious: "You want to segregate from Swedish society." She defends her right to embrace her Indigenous heritage and language, which he refuses to comprehend, by asserting: "You have your own identity today." What the Swede inadvertently identifies is the mixed and hybrid position of *Sámi Daughter Yoik*'s director—the very topic of her experimental and political approach. On the one hand, the film depicts the search for a "Sámi" identity, while at the same time relaying many other strong identity positions occupied by the director: as an educated, urban, practicing artist, a Swedish-speaking Sámi-Swede, Wajstedt can ask questions and challenge norms that other people in the film and the communities she depicts cannot, because their identity positions are construed as fixed and normative, whether "Sámi" or "Swede."

In a few instances, *Sámi Daughter Yoik* calls attention to self-determination movements in Sápmi, situating the personal journey of Wajstedt through the road movie format as part of ongoing calls for political action and mobilization. For instance, the film documents a political rally as well as a televised debate about Indigenous and minority rights in Sweden. The topic of these political events is the issue of ratifying ILO-169, a binding international convention that led to the formulation of the UN Declaration on the Rights of Indigenous Peoples. While Norway has ratified ILO-169, Sweden has not. The ratification is contested both among the Swedish political establishment—because of its potential granting of land rights to Sámi governance in some of the country's most active resource-extraction areas (mining, logging, and hydro-projects, some by state-owned entities such as LKAB in Kiruna)—as well as among Sámi. In a filmed debate, *Sámi Daughter Yoik* brings to the fore potential inequities of an ILO-169 implementation, which reflects the contested nature of these conversations in Sápmi. The fact that Wajstedt joins the Sámi protest rally in Stockholm, and includes reference to televised political debates, speaks to another part of "becoming" Sámi: that of dawning political awareness. While reclaiming language, culture, and personal identity is no doubt part of this process, becoming a public political agent is also a key part in the trajectory. Here, the artificial boundaries between "public" and "private" politics merge and intersect, as Wajstedt engages with a new frame of political reference for her, as a member of a colonized minority that, dominant political rhetoric aside, is excluded from Sweden's so-called egalitarian and consensus-based political culture. While Wajstedt, to the extent that she identifies as an ethnic Swede, may have believed in the Swedish political imaginary of egalitarianism and consensus, as a Sámi, she becomes politically aware that the ways in which the Sámi are imagined into the Swedish political imaginary does not reflect the dynamic, multiplicitous, and hybrid experiences of being Sámi. The emergent consciousness, on the part of Wajstedt-as-protagonist, of this schism between state narratives and lived realities of being Sámi—as excluded from this imaginary—makes *Sámi Daughter Yoik* interventionist historiography. As such, Wajstedt's political agency as a Sámi woman is significant. The "informal channels"—using Kuokkanen's term (2019, 158)—mobilized in, by, and through her documentary practice bring gendered aspects of Sámi culture to the forefront, asserting that lived experiences, family history, community relationships, language, and domestic spaces and practices *are* political acts and actions, especially when leveraged as components of Indigenous media sovereignty.

The experimental aesthetic strategies employed in *Sámi Daughter Yoik* illustrate Renov's argument about contemporary art and self-reflexive documentary (2013, 349). Another self-reflexive recent work that also thematizes the tensions between urban life and Sámi connections to the land and northern environments is Wajstedt's short film *On the Run* (*Moai Báhtaretne*, Sweden, 2021). Shot during the first COVID-19 lockdowns, the film chronicles a train journey the director and her daughter took from Stockholm to Kiruna to see Wajstedt's mother and father (see also Wajstedt 2021). Because of the infection risk, the reunion is aborted and the mother and daughter instead quarantine in the family's isolated cabin on the land. The film juxtaposes footage of the Stockholm concrete jungle—a highway overpass, a deserted central station—with expansive views of snow-covered mountains and a frozen lake. The director's voiceover—alternating between Swedish and Sámi, touching on the fear of contagion and a longing to be reunited with the land and with family—is interspersed with interviews with her daughter, Meja. Animation scenes include the insertion of ptarmigans dancing on the snow; Wajstedt is also shown embroidering similar motifs on face masks. As a self-reflexive work of art of the pandemic era, *On the Run* is a statement of the fragility of human connection and the precariousness of Indigenous identity when separated from community and family connections. At the same time, it is also a work of Indigenous media sovereignty and is one of the first of Wajstedt's films to receive ISFI funding. As such, it is an important counterpart to *Sámi Daughter Yoik*'s radical interventionism, while also maintaining a strong historical grounding as part of Sámi documentary film traditions that include both cultural revivalist films and solidarity feminist filmmaking. However, by consistently challenging a realistic mode of documentary storytelling through the use of experimental techniques, the tension necessary for complex and interventionist historiography becomes central to the documentary form and the impact of Indigenous media sovereignty.

FIRST NATIONS AND SÁMI SOLIDARITY AS AUTOETHNOGRAPHY: TRANS-INDIGENOUS SOLIDARITY IN *REBEL*

Canadian Blackfoot and Sámi artist and filmmaker Elle-Máijá Tailfeathers made *Rebel* as a commission for the 2014 Embargo Collective II project for imagineNATIVE. The Embargo Collective is "inspired by Lars von Trier's [and Jørgen Leth's] *The Five Obstructions*"; its members "create creative and technical rules and challenges for each other to employ whilst making their films" (imagineNATIVE 2014, 96). Each member devised rules for another member and the "creative and technical rules and challenges," including adopting each other's film styles and challenges of personal expression. *Rebel* has since screened at numerous festivals, both as a stand-alone work and as part of the Embargo Collective II program, to substantial acclaim. *Rebel* is also one of the very few films—whether fiction, experimental, or documentary—funded and produced outside Sápmi (though it does have some ISFI funding) that addresses Sámi history and the struggle for their rights. Moreover, it is an outlier in terms of Indigenous production writ large, as most Indigenous works are made on their territory within a nation-state context. In the First Nations Canadian context, *Rebel* is part of a growing body of work made by Indigenous women filmmakers supported by private and public funding packages from APTN, the NFB/ONF, imagineNATIVE, and Canadian cable networks such as Bravo. It is thus part of a larger body of First Nations and Inuit filmmaking by women, including Embargo Collective II directors such as Tailfeathers, Caroline Monnet, Lisa Jackson, and Alethea Arnaquq-Baril.

Rebel addresses the relationship between Tailfeathers's father, a Sámi rights activist and a key figure in the first wave of the Sámi liberation movement in the late 1970s, and her mother, a Canadian Blackfoot Indigenous activist, and the impact of that complex relationship. *Rebel* is trans-Indigenous, linking self-determination and settler-colonialism protest movements that connect Sápmi with Canada and the United States. Like *Sámi Daughter Yoik,* Tailfeathers's first-person film traces a journey of rediscovery of silenced aspects of her Sámi heritage and their political implications. Tailfeathers's film is darker, however, pointing to a history of family and social trauma, as exemplified by her father's personal history and the ways in which it echoes a suppressed history of Sámi colonization.

Rebel is a first-person, self-reflexive short in which the "expressive" and "aesthetic" (Renov 2013, xxx) modes of documentary are mobilized to address both personal and public history and how these are intertwined. Experimental techniques stitch these two realms together, with Tailfeathers using cutout and painted animation, stop motion animation, computer animation of family photographs, photo-journalistic montage, and the sounds of 35mm slide and 16mm projectors to mediate the various histories unfolding in the tripartite narrative. The first of the three sections is nearly exclusively cutout and stop motion, telling what the filmmaker's voiceover calls the "legendary" story of the early romance between Tailfeathers's parents, a story intertwined with Canadian First Nations self-governance movements and the intersection with global Indigenous political activism. In some sequences, the stop motion animation is used to comic effect, echoing the notion of animation as both a child's form and a self-reflexive alienation device, to foreground the mythical aspects of the story. This combination infuses the sequence with both the parents' personal history and the public understanding of the romance. This story is told in Sámi, voiced by a young girl who is a stand-in for Tailfeathers.

Obliquely, the first section of the film also addresses the global coalitional politics of the Sámi during the 1970s and the movement's "internationalist" wing, which supported Indigenous self-governance movements in Oceania and North and South America and also wanted to use the World Conference of Indigenous Peoples (WCIP) "as a weapon in the struggle to promote the Sami cause," noting that "WCIP was a natural extension of the organizational history of the Sami" (Minde 2003, 85). Sámi internationalist political engagement—including at the WCIP in Canberra in 1981, where Tailfeathers's parents met—led to the global Indigenous acceptance of Sámi, in the parlance of the time, as "White Indians," despite the perception that their (uncomfortable) integration within, and cultural assimilation to, democratic Scandinavian welfare states would set them apart from the struggles of many third- and fourth-world WCIP participants, as Henry Minde notes (2003, 84–86). For "Nordic camp" and "anti-Internationalist" Sámi political activists, WCIP membership was controversial, because some activists thought it might lead "to reprisals on the part of the Nordic states" (2003, 86). *Rebel* does not address this historical context directly; rather, the child's voiceover in the first part of the film situates this political context as a part of Tailfeathers's personal history, in parallel with certain formal choices, like the partly comedic cutout animation and the symbolic use of a child's view-finder toy, through which the construction of these events gets funneled. What is important, however, is Tailfeathers's insistence on telling this part of Sámi coalitional and internationalist politics from the perspective of a young Sámi-Blackfoot child and how such political engagement shaped her own upbringing and, arguably, her political commitment.

In the film's middle section, the filmmaker reflects, in voiceover English, on her childhood in Sápmi and a subsequent move to North Dakota. Techniques deployed include animated photographs, along with restaged dramatized material to tell the story of the father's mental anguish and the dissolution of the marriage. Experimental techniques in this section include actors playing her mother and father against a slide show of family photographs, and a deliberate use of a spotlight as sole illumination in restaged scenes. The third section, told in the present with Tailfeathers as an on-screen presence, moves fully into the register of realism, with a combination of restaged scenes and documentary footage, and a reduction in the use of processed still images. This trajectory allows for the narrative to move from the realm of memory and myth to the personal and political histories that underpin the family and their political action. The animated sections work to foreground mythmaking, gradually stripping away falsities and moving toward a newfound realism about the family's history, and therefore uncovering how elided Sámi histories motivated the personal and political positions and motivations of Tailfeathers's father. As interventionist historiography, *Rebel* uses the bringing to (partial) light of the family's past to unveil the histories of colonialism and silence left out, until recently, in the history of the Sámi.

In the third section—essentially a short road movie—Tailfeathers uses handheld cameras in a documentary mode, coupled with panoramas of Sápmi landscapes. In restaged scenes, she uses her own voice to reflect on the reconstruction of childhood memories to gain an understanding of the public narrative and myths surrounding her own parents as Indigenous political activists. In the final documentary realist scenes, in which Tailfeathers engages with her father on screen for the first time, her voice and agency convey a political message, which bridges personal history with public knowledge. Like Kernell's *Sami Blood* (see chapter 5), *Rebel* addresses aspects of Scandinavian settler colonialism suppressed in the grand narratives of the nation-state by referencing her father's traumatic experience in boarding school. Tailfeathers states: "I heard things I was not prepared to hear" and "I had completely underestimated the damage that the Sámi boarding school system had on my father and our family." These personal reflections illustrate the lasting impact of the forced Norwegianization process in place in Norway from the late nineteenth to the mid-twentieth century, and against which activists such as Tailfeathers's father fought, which included governmental and church procedures serving to assimilate Sámi into the ideology and structure of a homogeneous ethnic Norwegian nation-state. The filmmaker situates this history in a trans-Indigenous political context about self-determination, which is an interventionist strategy. Sámi scholar Troy Storfjell accentuates the importance of these trans-Indigenous contexts: "But juxtaposing Sámi and First Nations experiences of colonial schooling also serves to help Sámi viewers make more sense of their own familial traumas. By highlighting similarities between what happened in Norway and what happened in Canada, *Bihttoš* [*Rebel*] helps Sámi viewers on the Norwegian side put their own histories into a larger trans-Indigenous context. What

FIGURE 6.2. Elle-Máijá Tailfeathers—as a child and as an adult—in *Rebel* (*Bihttoš,* Elle-Máijá Tailfeathers, Canada, 2014).

happened to Indigenous people in Canada also happened to Indigenous people in Norway" (Storfjell 2019, 284).

Rebel thereby aligns little-known connections of political history with those of intimate, psychological, and family experiences. Trans-Indigenous knowledge allows both the Sámi and the Canadian First Nations peoples to know their own histories and each other's in ways that reflect Chadwick Allen's call for "*purposeful* Indigenous juxtapositions" that engender interpretations "*across* various categories" (2012, xviii), foregrounding the notion that experiences of one individual, family, group, or people are not isolated from global Indigeneity.

Two images strikingly connect the second and third parts of *Rebel:* in the first, we see a snapshot of a young Elle-Máijá seated on a beach, holding a stone in front of her eye; in the other, the adult Tailfeathers is crouching in what appears to be the same spot, holding a stone to her eye as she had done about twenty years earlier (figure 6.2). Tailfeathers's repeated use of the photographic scene, with a nod to obscured or partial vision, is illustrative of the personal and political connections the film seeks to make. The return to this image, from child to adult across decades, thematizes the complexity involved in historical hindsight and the challenges of telling stories of a traumatic personal and political legacy in the present. At different points in the film, Tailfeathers tells the stories of her parents, both as a couple and as individuals, and the concluding section synthesizes these stories as her own, as stories that tell as much about how she became who she is as they tell about who her parents were. The use of her own voice, in both English and Sámi, at the film's conclusion functions to unite the different, trans-Indigenous aspects of her history. The obscure and partial vision of her past, as a woman, family member, and Indigenous person, comes more fully into focus through her use of realism. As the fragments of her past come into clearer focus, her own agency as an Indigenous woman becomes the subject of the film—her reclaiming of her personal and political, previously obscured, past. *Rebel,* like *Sámi Daughter Yoik,* addresses both hybridization and fragmentation—the fragmentation of identities and of families and how the next generation needs to piece together the remnants of the past to get a more complete picture. This need to put the pieces back together is reflected in the film's aesthetic movement from experimental animation to realism at its conclusion.

GEOPOLITICS AND INTERVENTIONIST HISTORIOGRAPHY IN *KAISA'S ENCHANTED FOREST*

Katja Gauriloff's *Kaisa's Enchanted Forest* is a significant example of trends emerging around the interventionist historiography performed by Indigenous documentary. Gauriloff has made several films since 2002, including the Skolt Sámi short *A Shout into the Wind* (*Huuto tuuleen,* Finland, 2007) and the documentary *Canned Dreams* (*Säilöttyjä unelmia,* Finland, 2012), and she received funding from the Finnish Film Foundation to begin work in 2022 on a feature film in the Skolt Sámi language called *Je'vida. Kaisa's Enchanted Forest* earned several awards in Finland and circulated internationally through the Toronto International Film Festival, Sundance, imagineNATIVE, and other festivals and distributors. The documentary interweaves the director's own Skolt Sámi lineage—Gauriloff is the granddaughter of Kaisa, the storyteller featured at the center of the film—with family recollections, archival material, animation, and found footage. Having grown up hearing her grandmother Kaisa's legends, she knew that they had been collected by Russian-Swiss author and ethnographer Robert Crottet, with amateur photography by Enrique Méndez, Crottet's life partner and collaborator. Crottet and Méndez traveled extensively in Sápmi on multiple occasions in the 1930s through the 1960s, staying as embedded observers with Kaisa Gauriloff and other Skolt Sámi, including working closely with ethnographers, anthropologists, and politicians who were involved in the numerous relocation and resettlement processes of this Sámi community. This historical background is central to the documentary; indeed, the Skolt Sámi are anecdotally known as the most researched Indigenous group in the world. Their relative proximity to European capitals and universities, in combination with the growth of the disciplines of ethnography, anthropology, and "Lappology" during the first half of the twentieth century, as their communities faced multiple attempts at "cultural salvaging," provided a self-sustaining cycle.

The Skolt Sámi are known in the Nordic context as the only Indigenous group that has been forced to repeatedly move, relocate, and resettle into near-reservations—a practice common in other parts of the world—reflecting what Jukka Nyyssönen describes as "an odd mixture of integration, discrimination, and . . . segregation policy" on behalf of Finland during the twentieth century (2009, 45). In this way, the Finnish settler colonialist approach contains elements of both the segregationist policy of Sweden that sought to preserve cultural "authenticity" (the "Lapp shall be Lapp" policy) and the Norwegianization process, which emphasized integration and suppression of Sámi cultures, languages, and identities. The Finnish categorization of the Skolt Sámi, moreover, fell along two additional lines, both of which are echoed in *Kaisa's Enchanted Forest*. As Sámi scholar Veli-Pekka Lehtola argues, the "semi-nomadic life of the Suenjeli (Suonikylä) district," near present-day Murmansk, "was looked on as authentic and unspoilt" and deserving of preservation, while the Skolts who lived side-by-side with Finns in Paatsjoki and Petsamo were despised as degenerate" and their impoverished lifestyle—too closely associated with modernity—was seen as one that would "inevitably die out, and steps were even taken at times to encourage this" (V.-P. Lehtola 2000, 210). Lehtola's many insightful writings about the Skolt Sámi repeatedly emphasize the role of ethnographers, anthropologists, and scholars in cementing this bifurcated, essentialized, and static view of the Skolt Sámi. Finnish officials, administrators, and the general public, moreover, tended to perceive the Skolt Sámi even more negatively. Lehtola writes:

"With their 'Russian' ethnic characteristics, which included Eastern Orthodox religious beliefs, the Skolts were for many Finns 'racially, behaviourally, linguistically and religiously obnoxious products of Russian culture'" (2000, 209). The Skolt community was relocated after the Petsamo, Suenjel, and Paatsjoki regions' annexation into Finland from the Soviet Union as part of the 1920 Tartu agreement; other relocation processes occurred during and after World War II, as the borders changed during the German invasion and between the Soviet Union and Finland. The environmental degradation of the Kola Peninsula and the geopolitics of changing borders that tore families and communities apart, with negative impacts on Skolt Sámi language and culture, are also engaged with in the documentary *Great Grandmother's Hat* (*Šaamšiǩ*, Harry Johansen and Anstein Mikkelsen, Norway, 2022).

In the interwar period, when Crottet first visited Kaisa and her family, the Skolt community of Suenjeli would have been at the forefront of Finland's preservation project for the region, which essentially amounted to the creation of a reservation (see also Lehtola 2005; 2017, 90; 2018). With the 1920 annexation, the Skolt Sámi had to assume Finnish citizenship, acquiesce to the limitations of new nation-state borders that disrupted reindeer movement and access to fishing and hunting grounds, and give up access to their winter village and pastures, which were now inaccessible in the Soviet Union. When the larger Petsamo regions were reclaimed by the Soviet Union in 1945, Skolt Sámi faced difficult decisions on whether to remain Finnish citizens, thus being forced yet again to resettle. In 1947, relocated Skolt Sámi in Finland were granted permission to settle and rebuild their community around Lake Sevettijärvi, with support from the Finnish government. This is the community depicted by Méndez's photography and film in *Kaisa's Enchanted Forest*. Although the identification of Sevettijärvi and Ijjävri (Lake Inari) was made with Skolt Sámi input, the new geography and numerous traumatic migrations and relocations negatively affected the characteristics of Skolt communal life. These changes included technologies of modernity, inequalities of wage labor, and near-eradication of the Skolt language, as well as the loss of "possibilities to meet together, talk about the past, and tell stories," so that "conditions for such a natural exchange of stories and narratives were disrupted" (Mazzullo 2017, 46).

This traumatic history of cultural, community, discursive, and familial disruption underlies the premise of *Kaisa's Enchanted Forest,* the director's first film, *A Shout into the Wind* (Finland, 2007), and the forthcoming *Je'vida*. While *A Shout into the Wind* is a conventional and realist first-person, low-budget documentary, featuring interviews with Sevettijärvi community members and family about the plight of the Skolt Sámi and their efforts to maintain their language and customs, the emphasis on recovering a lost history and recreating one that is interventionist for the twenty-first century unites both films. As an example of interventionist historiography, *Kaisa's Enchanted Forest* takes a different approach to media sovereignty in its storytelling about the Skolt Sámi and Gauriloff's own family. Although several essay books and collected legends were published by Crottet and Crottet-Méndez, and translated into multiple languages, these works were largely forgotten during the past half century. Gauriloff and her team found Méndez by chance in 2013, realizing that he had in his possession multiple film reels, sound recordings, and still photographs of Kaisa and the community, as well as field notes and letters (for more on the production context, see Cocq and Dubois 2019, 167–69). This little-known personal archive of text, images, and recorded voices became the core of the film.

In contrast to the deliberately self-reflexive first-person documentaries of Wajstedt and Tailfeathers, Gauriloff's presence and voice as director-protagonist is nearly completely absent in *Kaisa's Enchanted Forest.* Her only presence comes just prior to the end credits, in the form of a still photograph of Kaisa holding her toddler granddaughter Katja in her arms. The film interweaves several strands and formal strategies. One component is the director's use of found footage and faux silent-era intertitles to present Crottet's personal biography, the little-known history of the Skolt Sámi plight during multiple relocations and resettlements, and the trauma brought about by World War II and its aftermath. In the first instance, the use of found footage provides a backdrop of geopolitical events in Europe at the time, from the Russian Revolution, Finland's Winter War, and London during the Blitz to Crottet's extensive public information and philanthropy campaign in the United Kingdom after the war to raise relief funds for the Skolt Sámi. At times, canonical films take on a new register when transformed into images that convey, in part, the personal narrative of Crottet. Found footage from Sergei Eisenstein's *October: Ten Days That Shook the World* (*Oktyabr': Desyat' dney kotorye potryasli mir,* USSR, 1927) and Esfir Shub's *The Fall of the Romanov Dynasty* (*Padenie dinastii Romanovykh,* USSR, 1927)—the latter itself a compilation film—are reedited to represent the Russian Revolution, while images culled from the General Post Office Film Unit, such as *London Can Take It!* (Humphrey Jennings and Harry Watt, UK, 1940) and *Fires*

Were Started (Humphrey Jennings, UK, 1943), delineate the effects of the Blitz. These works, all of which were produced as various forms of propaganda, take on a new valance when used as documentary evidence. Their use here—and their recognizable nature—demonstrates the ways in which the experiences of the Skolt Sámi are left outside the grand narratives of the representation of European history, which *Kaisa's Enchanted Forest* seeks to rectify.

In another formal register, a Skolt Sámi origin myth of the Northern Lights, as told by Kaisa to Crottet, is recreated using black-and-white charcoal and stop motion animation. This use of black-and-white animation contrasts with the use of black-and-white found footage to retell the events of the Soviet revolution and the Blitz. Both formats connote the past: one mythic and spiritual, the other lived and material. Another register in the film is the different modalities of (auto)ethnography, including black-and-white and color still and 16mm footage of Kaisa and the Skolt community shot by Méndez between the 1930s and 1960s; Crottet's taped field recordings and notes from his conversations with Kaisa; and Crottet's essays and memoirs, sometimes narrated verbatim by a male actor through voiceover. Crottet's prose is rich in mid-twentieth-century racist and ethnocentric interpretations of the "primitive nature folk" of the Sámi. Although Gauriloff states in an interview that she "was very critical about it" (interview with E. Nina Rothe, 2017, cited in Cocq and Dubois 2019, 170), the film does not directly take issue with Crottet's infantilization of Kaisa and primitivization of the Skolt Sámi community. Méndez's films—which also function as archival films—offer another layer of reality, intertwined with animated myths and geopolitical found footage. As such, the film combines a range of formal practice to place the history of the Skolt Sámi, Kaisa's spiritual and cultural legends, and the legacy of Sámi folkloristics and ethnography into the transnational context of the devastation wrought throughout Europe during World War II—from which the Skolt Sámi were not spared. The film conveys how lost histories are reconstructed through the remediation of numerous sources to tell a multilayered history that spans the personal and the political.

A number of Sámi activists, historians, curators, librarians, and scholars have been actively shaping the repatriation and archives debates for decades, using displaced artifacts, documents, and stories to shape a new Sámi history. As Faye Ginsburg notes, media sovereignty includes not only Indigenous production and circulation but also archival remediation:

These are activities linked to indigenous efforts to assert their rights to self-representation, governance, and cultural autonomy after centuries of assimilationist policies by surrounding states, part of a spectrum of practices of self-conscious mediation and cultural mobilization more generally that began to take on particular shape and velocity in the late 20th century. . . . From small-scale video and local radio to digital projects, archival websites, and mobile phone films, to national indigenous television stations and feature films, indigenous media makers have found opportunities for all kinds of cultural creativity, increasingly on their own terms. (2016, 582–83)

Importantly, repatriation and reclamation allows for what V.-P. Lehtola calls "an archaeology of knowledge in reverse" to promote expanded access, new interpretation of existing sources, increased influence of Sámi thought and practices, and the rediscovery of ancestral voices to make these a constituent part of the Sámi present (2005, 84–87). Lehtola confirms: "The right to one's own history means criticism and re-evaluation of conventional readings, and discovery of one's own voice and context" (2005, 87). In this way, Sámi historiography becomes newly constituted, less based on the traditional disciplines that underpin "Lappology" (ethnology and linguistics), whose "'facts' abstracted from their context have often resulted in biased and utterly erroneous conclusions" (2005, 87). The work of Indigenous documentary filmmakers, such as Gauriloff, who address neglected aspects of Sámi history—by reconfiguring material from ethnologists who collected and documented Sámi practices and narratives (e.g., Crottet and Méndez)—are thus part of a repatriation and reclamation process that exemplifies Lehtola's call for an archaeology of knowledge in reverse. On the one hand, Cocq and Dubois argue, *Kaisa's Enchanted Forest* "seems at first to break with broader patterns of Sámi ethnopolitical filmmaking" by avoiding "a pan-Sámi perspective," yet, on the other hand, "Gauriloff's film makes a place for the Skolt story within the broader narrative of Sámi colonial history" (2019, 172, 173). The director's emphasis on repatriation of audiovisual material engages in a Sámi interventionist historiography that goes beyond national and linguistic borders for one specific time.

The status of repatriated, reclaimed, and repurposed audiovisual material in *Kaisa's Enchanted Forest* is ambiguous; it is an experimental documentary that gains its aesthetic and thematic force by juxtaposing form and content. The film integrates material that can be understood as ethnographic, including Crottet's voice recordings that formed the basis for his publications about the Skolt Sámi and Méndez's 16mm films documenting Sámi life. Yet the

Crottet-Méndez audiovisual material was never used to make an ethnographic film, and Méndez's material often contained images of his partner Crottet, so these images could easily be seen as home movies taken on family voyages. And Gauriloff doesn't demarcate the material as ethnographic, but celebrates it as a trove of happily retrieved connections to her family's past—not only lost, but heretofore not known to exist. This also speaks to interventionist historiography, as the film's reframing of materials that could be construed to be ethnographic—home movies and tourist photography, or even both—gives them a new status as reverse archaeology. Indeed, Gauriloff's insertion of her own family photography at the end of the film (a faded color print of her grandmother holding her as a baby) functions as a nodal point for the personal, private, and public archives and narratives that are woven through the film.

What Gauriloff's film allows for, then, is the repatriation and repurposing of Méndez's images by reframing them within a Sámi Indigenous interventionist historiography, building on V.-P. Lehtola's reverse archaeology of knowledge. Yet, by the very nature of moving images as a form that circulates, Gauriloff's work also reaches beyond Sámi communities to other Indigenous groups and contexts. Her film, then, does not represent a static or primordial past of lost and displaced Indigenous pasts, which may have been the ethnologist's priority, but functions as part of a dynamic evolution of the documentary and cinematic forms. In that way, such interventionist historiography allows for new modes and new opportunities for Sámi influence—a key priority for Lehtola—to engage with audiences around the world, including digital and festival modes of circulation that reach both Indigenous and non-Indigenous publics, thereby enabling reciprocal approaches to be shared as method and told as stories. In this way, *Kaisa's Enchanted Forest* is a central artifact of Skolt Sámi interventionist history.

An Indigenous Essay Film?

By drawing together these various public, personal, and private strands, *Kaisa's Enchanted Forest* can be described as an Indigenous essay film. While it hews closely to the content of Crottet's essays, it also raises salient questions about the definitions of the documentary essay film. Timothy Corrigan describes essay films as often being "anti-aesthetic" works, "combining the 'personal' with 'actuality'" and mobilizing the "tendency of the essay to respond to and depend on other cultural activities that precede them"

(2011, 4). To Corrigan, the distinguishing feature of the essay film lies in the "encounter between the self and the public domain" (2011, 5). These characteristics are clear in the ways in which *Kaisa's Enchanted Forest* combines personal and public histories through its multimodal aesthetics and the interwoven relationship between many different pasts and the present. *Kaisa's Enchanted Forest* incorporates many modes of fictional and nonfictional cinematic representation, from media reportage to experimental autobiography. As such, Gauriloff's Indigenous essay film engages in "practices that undo and redo film form, visual perspectives, public geographies, temporal organizations, and notions of truth and judgment within the complexity of experience" (Corrigan 2011, 4). With respect to *Kaisa's Enchanted Forest* hewing so closely to Crottet and Méndez's material, the essay film's close association with the European Enlightenment and with the literary heritage of the essay (*l'essai*) is important.

Specifically, *Kaisa's Enchanted Forest* foregrounds "a practice that renegotiates assumptions of documentary objectivity, narrative epistemology, and authorial expressivity" to examine "key issues in the historically varied and multidimensional relationship between film and literature" (Corrigan 2011, 6). More importantly, however, Gauriloff's documentary expands the essay film framework when put into the context of Sámi and international Indigenous storytelling about settler colonialism, forced relocation, loss of culture and language, and collaboration among Western ethnographers and Indigenous populations in creating alternative historiographies to those of the grand narratives of the nation-state. Documentary, as Pamela Wilson has argued, is central to documenting, archiving, collecting, and providing evidence of the lives of Indigenous communities in the face of settler colonialism (Wilson 2015). The essayistic film and Indigenous storytelling in documentary form are not interchangeable; oral and performative practices are predominant in Indigenous cultures, lending themselves to audiovisual representation without a literary counterpoint. Similarly, the counter-public sphere in which films such as *Kaisa's Enchanted Forest* circulate—and to which they respond and which they shape—is one of suppression, marginalization, and even erasure by the dominant public sphere. *Kaisa's Enchanted Forest* repatriates Crottet's Western European essays and incorporates them into an Indigenous essay film as an act of interventionist historiography.

Kaisa's Enchanted Forest has a special relationship to the essay film and to the gendered, politicized, historiographical intervention that is part of contemporary women's filmmaking in Sápmi. While explicitly engaged with a public discourse of social events that has tended to suppress or erase Sámi "history"—assuming that because it was not written or integrated within the nation-state's grand narrative, it did not exist (Lehtola 2005, 88)—the film partially removes Kaisa's coauthorship by following Crottet's written words so closely. The partial erasure of the Indigenous voice is curious. It prompts the question: Whose self and whose public history does the film seek to portray? In addition, the film's presentation and reinterpretation of cultural and political events, past and present, is filtered through Crottet's biography. *Kaisa's Enchanted Forest* is thus closely related to textual precursors, especially the European tradition of the literary essay as written predominantly by educated male authors of means, shaped by an educational system that encouraged their attempts at essays from an early age. *Kaisa's Enchanted Forest* demonstrates a range of formal characteristics that link it to the emergence of an experimental, interventionist tradition of Sámi documentary made by women directors, yet its reliance on the male "expert" ethnographer voice sets it apart from other contemporary interventionist works by women directors.

The works of Tailfeathers, Wajstedt, and Gauriloff speak to an emergent practice, wherein photographic recontextualization functions as a process of aesthetic and political defamiliarization, allowing for the elided in the photos to dialectically reappear. Their work can thus be seen as central to interventionist historiography. Their films are not simply documentaries of the past; they are meditations on the influence of the traumatic, elided past's continuing echos in the present. Moreover, they speak to the ways in which understanding one's family history and one's own past is both inherently personal and political. Indeed, the works of these filmmakers postulate that to be able to engage in the political, one must first self-reflexively examine one's own past, which is also inherently political.

EXPERIMENTAL FORM AND CIVIC PROTEST:
LISELOTTE WAJSTEDT'S GIRON (KIRUNA) FILMS

Wajstedt's recent experimental works have taken a more directly political turn in their interventionist historiography, centering on the multiple and conflicting discourses caused by the partial razing and relocation of her hometown of Giron in northern Sweden. A century of underground mining has destabilized the bedrock to the extent that the structural integrity of buildings and infrastructure has been compromised. The discourses surrounding the move of Kiruna include issues of land rights, community engagement, social engineering, municipal and state policy, environmental degradation, settler colonialism, and cultural and personal belonging, with Wajstedt integrating her personal reflections and subjective experiences as key components of her Giron films.

Kiruna: Space Road and other early Giron shorts by Wajstedt such as *A Soul of a City* (Sweden, 2012) and *Giron* (Sweden, 2011), and later shorts such as *Bromsgatan* (Sweden, 2020), *The Drift Block* (*Ortdrivaren,* Sweden, 2018), and *The Girl Kiruna* (2020), are among the very few films that present alternative viewpoints to the official and corporate approaches to the city's transformation, including the dominant perspectives presented in a major exhibition about Giron at Stockholm's Arkitekturmuseet in 2020, where some of Wajstedt's films screened (see Wajstedt 2021). Her films emphasize women's and children's perspectives. They also include references both to Wajstedt's Sámi background and to multifaceted aspects of Sámi history, presence, and political engagement in Giron that paint a rich picture of the complexity of Indigenous, Swedish, nation-state, and Sámi land-claim priorities, demonstrating the ways in which these are intertwined, nonbinary, and constantly evolving. Employing experimental and animation techniques, auto/biography, and an environmental politics, these films are among the most significant and aesthetically interesting Sámi land-claim films.

The basic premise of *Kiruna: Space Road* is that Wajstedt returns north from Stockholm precisely because she feels she must document—from her own and other denizens' perspectives—the partial razing and moving of the city. The interplay between activities underground, in the bowels of a beautiful valley marked by a century of iron ore mining operations and the tracing of memories seemingly lost in the dark recesses of the filmmaker's mind, provides a compelling juxtaposition of vantage points. Both the material nature of the mine and the psychological nature of the mind are shown to be split and bifurcated. Moreover, the Giron films posit that extraction, abandonment, razing, and relocation are not only land-based and historical processes but also personal, aesthetic, and formal ones. In this way, *Kiruna: Space Road,* especially read in the context of Wajstedt's other Giron films, functions as an interventionist historiography in ways related to other films discussed earlier in this chapter, with an obvious point of connection to Gauriloff's uncovering, in *Kaisa's Enchanted Forest,* of the Skolt Sámi's multiple forced relocations in service to competing needs of the nation-state.

The Kiruna Representational Concept and the Technological Mega-system of Northern Sweden

Before delving further into Wajstedt's mode of interventionist historiography, which includes oblique references to her own Sámi heritage and Giron's central location in Sápmi, it is important to briefly sketch what we can term the established, colonial, and nation-state-promoted representational concept of Kiruna, which has developed through multiple iterations over a hundred-plus years. The new city plan and relocation of Giron reflects a long history of internal colonization and massive planning efforts by the state. A useful concept is the industrial, technological, and infrastructural "mega-system" in northern Sweden (see Forsell 2015; Hansson 1998; Sörlin 1988). This is the principle upon which Sweden's industrialization in the late 1800s, its economic growth, and its ongoing development into a welfare state and an export economy depended throughout the 1900s: mass logging and pulp production, iron ore mining and smelt refinement, hydropower, railways and ports, and military defense and space exploration installations—all on lands inhabited by Sámi.

With the LKAB mine established and a 50 percent ownership by the state implemented in 1907, Kiruna was planned from Stockholm on the basis of Camillo Sitte's climate, landscape, and urban planning principles. Kiruna was to be a model company town. Later planning imported from the outside includes architect Ralph Erskine's neighborhood the Drift Block (1966), based on the social engineering and modernist large-scale design principles of Sweden's publicly commissioned high-rise *miljonprogram* (1965–75), aimed at generating modern housing for a rapidly expanding postwar population. Along these lines, the Erskine complex in Kiruna was planned to efficiently house miners, with the design replicating the inverse of a mining shaft. As Håkan Forsell argues, Kiruna was created "to support resource extraction" as a nodal point for "national economic development" that was organized to meet "an originally international export-driven exploitation logic" (2015, 201; see also Carrasco 2020, 31–38). It is worth noting that these settler colonial priorities were intended from the start to make Kiruna an Arctic hub in a global network of commodities trade. Similar sentiments guided the design for LKAB's main office by architect Hakon Ahlberg, specifically to support this notion of Kiruna as a global hub. The building replicates the UN's headquarters in New York City; when completed in 1960, it was the tallest building in northern Sweden. Similar priorities are echoed in the promotional video and corporate exhibits shown to tourists who pay a visit to the LKAB mine, which promote the environmentally friendly processes of the iron ore extraction process in Kiruna, the efficiency of operations, the egalitarian corporate culture and safe working conditions, and the modern and seamless global dissemination of Kiruna iron ore all over the world. These planning priorities, and their relentless implementation in Kiruna, make visible in the landscape how "Sámi rights had gradually been undermined . . . during the last decades of the nineteenth century" and continuously into the twenty-first century (Forsell 2015, 200). This aspect of Swedish history has been consistently overlooked in the dominant discourse about Kiruna and the role of LKAB. Wajstedt's works are important counterpoints to these elisions.

The philosophy of the mega-system implicitly and explicitly undergirds the planning and design of "New Kiruna." After an international competition, the Stockholm-based firm White Arkitekter, in conjunction with Ghilardi + Hellsten Arkitekter, won the bid to relocate Kiruna two miles to the east, with the move planned to be fully completed by 2033. In a 2014 press release, White Arkitekter states: "White's vision for the transformation of Kiruna will take place in phases with the aim to create a sustainable model city, a city with a diverse economy that is less dependent on the world market for iron ore and a city which better serves it diverse and thriving community" (2014). The firm then launched a "consultation" with the people of Kiruna, along with a promotional campaign about the efficiency and benefits of the relocation. To this end, they produced a four-minute promotional video about the move, called *Kiruna 4-ever* (White Arkitekter, Sweden, 2016). This video features realist environments, architectural art, and "expert" talking heads extolling how attractive, functional, and well designed the new town will be. Along with nature shots of the city and its environs, the video also features slow-motion shots of Stockholm-based Swedish lead architect Mikael Stenqvist, who stops moving at times to express the brilliance of White's plan. Time is also given to Swedish social anthropologist and consultant Viktoria Walldin, a woman of color, who advises on social sustainability, while connoting Swedish multicultural diversity. These interviews are intercut with futuristic hyperreal images of the new city, with the global fast-fashion retailer H&M featured prominently on the new Main Street.

Kiruna 4-ever promotes the ingenuity and social conscience of urban Stockholm, but it says little or nothing about the specificity of Giron and nothing about Sámi inhabitants. The only local who appears is Eva Ekelund, Giron's land and development manager. As a promotional video, the work foregrounds how the moving of the city will let it flourish for another hundred years, conveniently sidelining any question of whether perhaps the city should *not* be moved, or whether the iron ore extraction should be ceased. Indeed, in a lecture entitled "How We Moved a City" by White's architect and partner Krister Lindstedt at the reSITE architectural conference in Prague in 2016, little time is spent on resistance to the move, or to the fact that the town and the mine are on Indigenous lands. Curiously, the Sámi are represented in White's materials solely, and briefly, by a herd of reindeer migrating through town, reflecting traditional herding routes still in existence; there is nothing in the design elements or cultural context that recognizes Sámi history or presence in Giron. A twenty-first-century notion of mainstream Swedish ideals are instead foregrounded in White's materials, including attention to Giron as a town of foreign national and multicultural migrants, as globally connected, and as inhabited by residents who enjoy the outdoors (skiing, hiking) and cultural events (museums, performances) and who will benefit from enhanced public transit, relinquishing their reliance on cars.

Swedish ingenuity, societal norms, and a long history of social engineering of urban environments are central to White's presentation of the "New Kiruna"; the presentation also engages in "greenwashing," taking for granted the necessity of perpetuating mining while simultaneously playing up notions about Swedish conscientious concern for the environment. Overall, the presentation functions to elide any local resistance to the move, "understanding" why older people may not want this change but presenting it as a necessity and putting all agency on Giron by stating that "the town decided to move." Lindstedt equates the extraction of iron ore with climate change, making the false equivalence that in both cases humans need to adapt to the damage they have wrought, instead of changing their practices to produce different and more environmentally sustainable outcomes. This kind of greenwashing reflects ongoing settler colonial approaches to Sápmi and the downplaying, in mainstream Scandinavian media, of Sámi opposition to implementations of the technological "mega-system" philosophy, as recently apparent in the Gállok mining protests and the concerns voiced over the Arctic railway through Finland and new hydropower projects proposed in Norway.

Wajstedt's Giron *and* A Soul of a City

Two of Wajstedt's experimental shorts, *Giron* and *A Soul of a City,* offer critiques of the mega-system philosophy as well as prescient counternarratives to White Arkitekter's *Kiruna 4-ever* and Krister Lindstedt's presentation at reSITE in 2016, foregrounding both the destructive environmental impact of a century of mining in Kiruna and the impact this has had on the social fabric of the city. *Giron*'s experimental montage is divided into three parts. The first section contains various levels of superimposition, including a line drawing of a Giron house slated for relocation or demolition, briefly shown photographs of Sámi, cutout animation of ore trains, and animated birds, all set against a wintry mountain landscape. The second section directly engages the Sámi language to challenge the normative history of Giron. Wajstedt superimposes Sámi words for "mountain," "holiness," "Rock Ptarmigan," and other significant local concepts onto the static line-drawn images of two children, with one claymation child playing as animated birds fly by. The Sámi words for "not inhabiting" (*il orrut* in Sámi; *inte bo* in Swedish) are on the screen. After a break of video static, the third section consists of animation depicting urban life in Kiruna. This practice signals interest in making the claim that Sámi language, and its traces mapped onto the city, are important land-claim-argument issues. Language, then, is mobilized as part of Giron's physical landscape to affirm that whether urban or mountainous, these *are* part of Sámi territory. Wajstedt denotes this in Sámi terms that bridge the environmental (as in *varri*, "mountain"), spiritual (*bassi*, "holy"), and emotional/psychological (*il orrut*) to signal that history and future, such as experiences of the impending move of Giron and the planned demolition of housing, cannot be separated from one another, just as Sámi experiences of land, location, lived experience, and spirit are intimately connected. Indeed, as Jarno Valkonen, Sanna Valkonen, and Tim Ingold argue, "a landscape might seem to outsiders to be uninhabited, untouched and in natural state," but such juxtapositions, they write, "encounters reality, or in fact two realities: Western reality, predicated on a sharp distinction between nature and culture, meets Sámi reality, where such a distinction is not conventionally made" (2019, 4). These aspects of lived environmental experience are central to Wajstedt's Giron works.

In regard to both experimental shorts, *Giron* and *A Soul of a City,* it is of central significance to note Wajstedt's priority in signaling that the name *Kiruna* is a Swedified version of the Sámi word for Rock Ptarmigan, *Giron;* the mountains Kiirunavaara and Luossavaara, known by Sámi to contain iron ore for centuries, are also known in popular parlance as the silhouettes of two nesting Rock Ptarmigans against the horizon. Rock Ptarmigan line drawings and animation by Maja Fjällbäck figure prominently in both shorts. The connections between Sámi experience, history, language, location, and land rights are thus intertwined in these experimental works, even if the words in Sámi are not uttered—instead, the experimental soundtrack does the talking.

Part of the interventionist historiography and Wajstedt's experimental practice in these shorts lies in the lack of a voice-of-god explanatory narrative. Like *Giron*, *A Soul of a City* operates on non-denotative registers, an experimental amelodic soundtrack—by Mathias Josefson in *Giron* and by Maja Ratkje in *A Soul of a City*—signaling disjuncture, dissonance, and the juxtaposition of the technological, mechanistic, and industrial with Indigenous lands. As a founder of the Oslo Industrial Ensemble and a composer whose works expand the chromatic twelve-tone scale, Ratkje presents a dystopian sound-and-vision-scape. *A Soul of a City* includes mostly landscape shots of mountains around Giron, with an emphasis on the LKAB site as one of monumental scar-like ruins, with close-ups of the rusting and disintegrating infrastructure—remnants of surface operations before the mining went subterranean. These shots challenge any conventional portrayal of the northern mountainous landscape as pristine, wild, or "beautiful," but instead starkly showcase the disastrous effects of massive industrial intervention mobilized by the technological mega-system. In both *Giron* and *A Soul of a City*, Wajstedt includes various forms of animation—cutouts, superimpositions, computer-generated images, line drawing—to underscore an environmentalist critique through formal juxtaposition between sound, image experimentation, and landscape cinematography.

In addition to the environmentalist critique, *A Soul of a City*'s interventionist historiography includes two references to Sámi presence and habitation. About a minute into the film, there is a brief superimposition of a newsprint photograph of an unnamed traveling Sámi group—likely a family or a *siida*—resting on the mountain, with some of the members of the group looking directly into the camera (a similar if not identical image is also included in *Giron*'s first section). There is no comment on this picture and no obvious historical reference; as the superimposition dissolves into blowing snow, the message is clear: this was once Sámi land, the land rights are dissolved, and at no point has this fact made it into "official history"—Sámi habitation of these locations remains anonymous, fleeting, and seemingly evaporated into thin air, and can therefore be unaccounted for in the grand narratives of the nation-state as well as those of LKAB and White Arkitekter.

In *A Soul of a City*, line-drawn Rock Ptarmigans emerge from the landscape; they fly out of the mining infrastructure, emphasizing the missing, disregarded, and eroded aspects of Sámi history. By mechanically reinserting the Rock Ptarmigan into the industrial landscape, Wajstedt effectively repopulates the territory with Sámi through land-rights politics, insisting both on the historical and contemporary presence of Indigenous populations in Kiruna and on the roles these have played in the LKAB mine. Both experimental techniques—the superimposition and dissolve of newsprint photography without any commentary and the recurrent motif of Rock Ptarmigan animation—remediate the landscape to tell histories buried within the realist image through an act of interventionist historiography.

A Soul of a City also remediates European, modernist, and utopian mega-system ideologies. For instance, between 1964 and 1966, British architect Ron Herron, as part of the Archigram collective, devised a series of massive "walking cities" as postapocalyptic "survival pods." Archigram is described in Herron's obituary as "the radical architectural powerhouse which incited young architects around the world to question fiercely the practice of Establishment Modernism" (Lyall 1994). Wajstedt takes one of these walk-ing-city models and, through cutout animation that resembles Terry Gilliam's, has its round body and millipede-like legs rampage through Giron's industrial minescapes. Here, the "survival pod" becomes one of the causes of the destruction of the landscape. In Krister Lindstedt's 2016 talk, he describes Herron's work, using an image very similar to the one Wajstedt animates. Martin Johnson of White Arkitekter makes an even more striking proposi-tion, asserting that the relocation of "Kiruna will be more a walking milli-pede than Ron Herron's eight-legged procession of urbanity in the Archigram project 'Walking City'" (Johnson 2014). While Lindstedt insists that Herron's "walking city" is the kind of model, in fact, that White Arkitekter do not wish to follow (Lindstedt 2016), *A Soul of a City,* made four years earlier, stipulates that the walking industrial city has already done its damage to Giron. *A Soul of a City* also functions as a riposte to the modernity cele-brated in the "city films" of the 1930s and '40s, many of which were experi-mental in nature and celebrated massive technological systems and industri-alization as the epitome of progress.

Kiruna: Space Road—*Politics and the Animated Documentary*

Wajstedt's feature on Giron, *Kiruna: Space Road,* as well as the shorts *The Drift Block, Bromsgatan,* and *The Fire* (Sweden, 2017), are as much works of memory and memorialization—of Kiruna and the director's family, friends,

FIGURE 6.3. Experimental animation in Liselotte Wajstedt's *Kiruna: Space Road* (*Kiruna—Rymdvägen*, Sweden, 2013).

and childhood—as they are experimental documentaries and interventionist historiographies about the events surrounding the town and the mine (for more on these films through the frames of eco-justice and resource extractivism, see Fish 2018, 2019). While the town is collapsing, these films are as concerned with the risks of social disintegration of the community—especially in terms of the effects on and perspectives of women and children—as about the shifts taking place, in terms not of what will be saved in the city, but of what will be lost. The experimental aesthetics and antirealist formal choices in *Kiruna: Space Road* and *Bromsgatan* reflect these concerns. Some of the most significant techniques include cell animation, superimposition, computer animation, stop motion, and claymation, which juxtapose personal memories against official and LKAB narratives.

Kiruna: Space Road is, in part, an "animated documentary," an often unrecognized hybrid form stretching back to the beginnings of cinema that parallels the evolution of the realist tradition of documentary and is often used to present counternarratives (figure 6.3). A quick survey of the two forms points to the fact that not only do animation and documentary have a long, entangled history, but the emergence of the animated documentary precedes the supposed arrival of the documentary itself (see MacKenzie 2008; Roe 2013, 1–16). These animation techniques function in several interweaving ways, all of which reflect the bifurcated nature of the director's assessment of her hometown and her own subjectivity. *Kiruna: Space Road,* like *A Soul of a City* and *Giron,* features superimposed images of Rock

Ptarmigans drawn by Fjällbäck that connote the suppressed Sámi history of the area. By contrast, stories of Wajstedt's childhood are told through use of the children's toy Play-Doh in claymation, to foreground the fact that these memories are reconstructed and mediated ones. In addition, computer animation of the cosmos, which reflects on the smallness she feels, are juxtaposed with superimposed images of the street where Wajstedt grew up, Rymdvägen, which means "Space Road." This computer-animated journey signals Giron's central role in the Cold War space age. The Esrange Space Center—built in 1964, forty kilometers outside the city, by the agency now known as the European Space Agency—became part of the city's imaginary and continues to actively launch rockets for satellites. For Wajstedt, the connection is personal: she addresses how infinitely small she felt in an ever-expanding universe and how the Cold War tension frightened her and her friends. Incidentally, Giron would have been on the rocket trajectory between the Soviet Union and the United States, as reflected in media coverage of the Arctic's central role as a heavily militarized and geopolitically vulnerable region during the Cold War (see Stenport 2015). Significantly, these animation strategies allow for a process of defamiliarization of the realist aesthetic so often dominant in documentary cinema and reinforce the importance of subjectivity in the documentary.

Aboveground and Belowground: Sámi Kiruna Politics

The aesthetics of splitting, superimposition, juxtaposition, and fragmentation support the politics of *Kiruna: Space Road* in several ways. The documentary opens with panoramic views of, and the director's voiceover explanation of, the Sámi significance of the Kiirunavaara and Luossavaara mountains as a pair of Rock Ptarmigans, reinforced by bird animation along the lines of the Sámi land-claim representations in *Giron* and *A Soul of a City*. She remarks on the fact that the "New Kiruna" will be moved to a location colloquially known as Death Valley, both because it is so cold and because it will be positioned around the cemetery. This is contextualized with her reference to Sámi beliefs that those who are dead do not die as long as they are remembered by others. There are only a few other explicit references to Sámi presence in the film, including the mention of Sámi working in the mines and a remark attributed to a friend of the director, to the effect that it was not only the Sámi who were teased during their childhood, but also those from families who raised cows and farmed, rather than working in the mine.

Wajstedt's work addresses a number of different and related political issues, and the director's subject position is dynamic and shifting: she is a woman, child, Sámi, insider, and outsider, all at the same time. The film complements earlier works such as *Sámi Daughter Yoik* and *The Girl Kiruna* by emphasizing that the recollection of childhood and family memories, and the stories of Giron past, present, and future, are personal and part of the city's collective identity, whether from the position of a Sámi woman, an ethnic Swede, or both. In this way, the film complements and implicitly critiques, as part of its interventionist historiography, the masculinist and colonialist myths around the technological mega-system. These myths are propagated in media material by the LKAB mine, the municipality's public relations affirmation that the city must be relocated, and videos such as *Kiruna 4-ever*. However, the traces of Sámi culture and history are mostly absent in *Kiruna: Space Road*, which reflects their absence also in the White Arkitekter promotional video.

Such a personal-political dialectic marks the inherently conflicted status of Sámi Indigenous rights in northern Sweden. Moreover, the moving of the city also leads to the eradication of Wajstedt's personal history. One of the ways in which *Kiruna: Space Road* differs from Wajstedt's other Giron films is the first-person voiceover and the direct mobilization, in both experimental and realist terms, of the director's own childhood and memories as drivers of the film. To recall Renov's formulation, Wajstedt's directorial presence in *Kiruna: Space Road* can be construed as "the subject *of* the documentary" (2013, xxiv). One of the ways in which this connection is made most explicit is through shots of crumbling and cracking walls, empty buildings, and discarded living spaces about to be razed. Through these shots, a conflicted past emerges: these splits in the bricks of the houses metaphorically let us see through to the childhood of Wajstedt, while guided by her reminiscing voiceover. The integration of public history with personal experience and memory is reflected in other accounts offered in the film. Indeed, one interviewee addresses the difficulty of conceptualizing and acknowledging that the active parts of the mine are directly under her feet, some thirteen hundred meters below the building she works in every day and the house in which she lives. Right after this sequence, Wajstedt expresses, through voiceover, that she cannot reconcile her own personal memories. This statement leads into a section of the film focused on Wajstedt's teenage diary, which, like the building's wall, is split open to reveal her past. These accounts include both personal reminiscences of Wajstedt's junior high period—replete with underage drinking, allusions to anorexia and potential sexual abuse, the presence of the mine in their lives, heartbreak, and Queen Bees/Mean Girls scenarios—and her revisiting of people and locations depicted in the diary in the present day. These accounts further emphasize the multiple levels of connectivity within the city's past and present, as well as the acknowledgment that the personal is political. These connections echo through many layers, foregrounded by the images of splitting that undergird the ongoing suppression of Sámi agency and history in present-day Giron.

Material and landscape metaphors, including those of buildings, govern the figurative register of the film. Wajstedt, in this context, is like one of the split buildings in the film: her own foundation and that of the town's identity are undercut and striated by opposing and sometimes contradictory assumptions, like the mining shafts that run under Kiruna and cause its demise. This figurative register is further pronounced in the abstract experimental shorts *Bromsgatan* and *Welcome to Kiruna* (Sweden, 2016), in which Wajstedt features multigenerational actresses who, dressed in all white, perform a ghost-like figurative dance constructed through dissolves, made to embody the director's memories and fraught emotions over the pending demolition of the apartment area where she grew up. These shorts do not employ dialogue or voiceover but operate on the associative plane; their political impetus is to integrate the perspectives of children and women into the narrative of Kiruna's "transformation." The figures are both ghostlike and dreamlike in their movements and through the dissolves, conjuring up personal memories and the haunting traces of those memories in the built environment.

Wajtstedt's many depictions of the relocation of Giron are not explicitly situated in a Sámi land-rights context. Centrally, *Kiruna: Space Road* is a work about precisely how conflicted and controversial the Sámi claims over land are within the normative Swedish nation-state and for many Sámi communities—and how conflicted they are for the filmmaker, too. In the film, family members and friends remark about the director that she has become like a Stockholmer, that she is "posh" and different from what she used to be. Indeed, the different facades of identity—ethnic, class, generational, personal, family, social—pervade the public and private debates in *Kiruna: Space Road*. Suppression of Sámi land claims or governance politics in the film is thus partly a reflection of Wajstedt occupying a normative Swedish position within it, but also—and this is critical—Sámi were always also employed by the LKAB mine, participants in and citizens of the social-democratic welfare state, and dependent on it for their livelihood as modern, urbanized, industrialized Swedes, which is also part of what it means to be Indigenous within this particular context. *Kiruna: Space Road* is a political film, but a tentative one about the questions it raises and the stances it takes, paying heed to the complex narratives taking place above ground: many locals, Sámi included, are supportive of the relocation, because the mine is a key means of financial support for women and men alike.

Instead, *Kiruna: Space Road* is an attempt to argue for a new kind of politics, one that includes "informal channels" (Kuokkanen 2019, 158) and that takes both daily lived experience, outside of official political rhetoric, and gendered experience as starting points. On the one hand, previous work by Wajstedt, such as *Sámi Daughter Yoik,* exemplifies a "typical" Sámi political stance; that film does not question a discursive constitution of the Sámi Indigenous subject as, first and foremost, based in language and identity politics rather than in legal, political, or policy positions that can be argued and represented against a dominant or colonizing nation-state, or through the paradigm of gender disparity. In this way, the film highlights that Sámi self-determination has been weak in terms of anti-extractivist political mobilization. *Sámi Daughter Yoik* and *Kiruna: Space Road* contrast with the centrality of the environmental aspects for land-claim activism as a historical foundation for political mobilization, especially in Norway around the Áltta dam conflict.

In *Kiruna: Space Road,* the nation-state is represented through the mining corporation LKAB, an incorporated, for-profit entity fully owned by the Swedish government, whose priorities, the film contends, are supported by the Giron municipal government. The spokespeople or representatives of these legal, political, and official entities are not represented in the film. Instead, the interviews are mostly with local individuals who offer accounts of their feelings about the town and the mine; these are juxtaposed with larger issues about capital and environmental concerns. Instead of giving voice to the politicians and industry representatives, Wajstedt critically remediates these points of view through her experimental techniques. She does not directly question the foundations of the mega-system. Instead, she deploys a range of voices: personal, political, Sámi, ethnic Swede, female, male, young, and old—all in the service of building a new, nonbinary manner of exploring personal and political history.

The representational concept of Giron that we trace above is also beginning to be challenged by filmmakers other than Wajstedt. For example, Greta Stocklassová's realist and observational documentary *Kiruna: A Brand*

New World (Czech Republic, 2019) features a large number of shots of Kiruna as well as the surrounding landscape, foregrounding both the mining operations and city life, while interweaving stories of three Giron inhabitants. Only one of the three protagonists, Timo, a teacher at the local high school, is seen as actively engaged in reflecting on and documenting the city's move and what it means to the community. Maja, a teenager about to complete high school, is actively rediscovering her Sámi heritage, which includes scenes of her participation in a youth network's political mobilization meeting, interacting with her Sámi grandparents, and learning the Sámi language. The third character is an orphaned, Arabic-speaking Yemeni teenager named Abdalrahman, noted in the documentary as the first refugee from the Syrian war and migration crisis to arrive in Giron. The film documents his adaptation to life on his own in an asylum-seekers' dormitory—learning Swedish, visiting the LKAB mine, awaiting permanent residency decisions. These three characters represent new and old Giron, reflecting the changing demographics of the town and the presence of multiple languages, ethnicities, and cultures, including Sámi. This approach supports the representation in *Kiruna 4-ever* and other media by White Arkitekter about Giron as a globally oriented multicultural city, but also contrasts with it by including Sámi perspectives, though no Sámi openly criticizes the move or the LKAB mine in Stocklassová's film. The subtitle *A Brand New World* must be read as ironic, given how the film leverages recurring shots of the expansive and "stable" landscape versus the dynamic and "transforming" city. Through its references to environmental and social engineering, Fordism and technological oppression, and striated class structures, the film seems to critique the "megasystem" philosophy in ways that strongly recall Aldous Huxley's dystopian 1932 novel *Brave New World*.

Like *Kiruna: Space Road*, *Kiruna: A Brand New World* focuses on loss, though in a far different manner than the approach taken by Wajstedt. If *Kiruna: Space Road* has Wajstedt's interior experiences and emotions splitting open through her use of animation, *Kiruna: A Brand New World*, through the mode of observational documentary, foregrounds the characters' external experience of decay and loss—whether this is centered on the city itself, as is the case with Timo; on the recovery of Sámi identity lost by her parents' generation, as is the case with Maja; or on the trauma of adapting to an interstitial space between home and new world felt by the refugee Abdalrahman. In the tradition of Frederick Wiseman's seminal observational documentaries *High School* (USA, 1968) and *Juvenile Court* (USA,

1973), *Kiruna: A Brand New World* allows the viewers to draw their own conclusions on the move of the city and its inhabitants without the need of a voiceover to tell them how to think about the various kinds of loss being felt in the city.

In this chapter, we have examined a range of different documentary films by women filmmakers, foregrounding the rich development of Sámi cinematic expression in just under fifteen years. Postulating that ethnic and national identities are performative and shifting, and not ahistorical and fixed, the films by women discussed in this chapter demonstrate the personal and autobiographical aspects of performing one's culture, personal, and public histories, gendered identities, and memories as political statements. All their works intersect through the expressive and aesthetic component of a self-reflexive tradition of first-person filmmaking as an antidote to voice-of-authority realism, and in so doing function as interventionist historiographies, affirming documentary as a constituent part of Indigenous media sovereignty. Moving away from documentaries that posit a need to "revive" traditional modes of Sámi culture, these Sámi women filmmakers problematize questions of identity, inclusion, exclusion, and the hybrid nature of ethnic and national identities. The dialectics of the past and present are very much present in this body of work, oscillating between tradition and modernity, through sophisticated aesthetic experimentation that mobilizes both environmentalist and feminist practices and discourse. Mobilizing the impacts of cultural memory, Sámi identity, language, cultural revitalization, discourses about self-governance and land rights, the characteristics of the essay film, and the social and political power of Indigenous media, these films construct remediated interventionist historiographies. These filmmakers' self-reflexive, often first-person, experimental documentary practices have greatly contributed to enhancing knowledge of Sámi culture and practices in Europe and globally.

SEVEN

Global Greenland and Postcolonial Cinema

LOCALLY PRODUCED GREENLANDIC CINEMA is very much a twenty-first-century phenomenon. Films have been made in Greenland since the early 1900s, but local production—shot in Greenlandic, made with Greenland-based practitioners and funding, and marketed as Greenlandic cinema—is a recent development. Nonetheless, emergent Greenlandic film production has increased in recent years, garnering festival screenings worldwide. For instance, the Greenland Eyes International Film Festival was founded by Danish-Greenlandic filmmaker Ivalo Frank in Berlin in 2012. Beginning in 2014, the festival toured Greenland and other Nordic countries, including Denmark, the Faroe Islands, Finland, Iceland, Norway, and Sweden. The following year, the tour program was screened at the Smithsonian National Museum of Natural History in Washington, DC. The festival's programs combined the work of emergent Greenlandic and Danish-Greenlandic filmmakers, including Frank, Aka Hansen, Inuk Silis Høegh, and Ulannaq Ingemann, with older films that were shot in Greenland, about Greenland, or based on material related to Greenland. Examples of those films are *SOS Iceberg* (*SOS Eisberg,* Arnold Fanck, Germany/USA, 1933), *And the Authorities Said Stop* (*Da myndighederne sagde stop,* Per Kirkeby and Aqqaluk Lynge, Denmark, 1972), *Before Tomorrow* (*Le jour avant le lendemain,* Marie-Hélène Cousineau and Madeline Piujuq Ivalu, Canada, 2008; discussed in chapter 4), and *Blok P* (Rikke Diemer and Peter Jensen, Greenland, 2014).

Both domestically and abroad, Greenland was not thought of as a filmmaking country or as one that had any sort of film history. The Greenland Eyes festival disabused many of that notion, screening a rich body of emerging work (see Frank 2019, 339–44). Moreover, the strategy of launching a touring festival functioned as a means of branding and drawing attention to a body of work, in a context where individual films might not get any focused attention. This was also the strategy of FILM.GL, a nonprofit organization that develops talent, promotes production, strengthens capacity, and facilitates runaway productions interested in shooting Greenland's spectacular landscapes. In a country of sixty thousand people, the burgeoning film industry is necessarily small, with around fifty active practitioners. Against this backdrop, of both branding works as Greenlandic cinema and branding Greenland as a filmmaking destination, numerous production companies have emerged, including Inuk Silis Høegh and Emile Hertling Péronard's Ánorâk Film; Malik Kleist and Aka Hansen's Tumit Production; Inuk Jørgensen's NukkiNukki Productions; Pipaluk Kreutzmann Jørgensen's Karitas Productions; Péronard's and Jørgensen's Polarama Greenland; and Nina Panninguaq Skydskjerg Krisiansen's PaniNoir (see also Grønlund 2021c). The production companies engage in a range of industry-building collaborations. Unlike IBC and APTN in Nunavut, or the Sámi-language and Sámi-run broadcast television in Finland, Norway, and Sweden, the Greenlandic Broadcasting Corporation (Kalaallit Nunaata Radio, KNR) has remained, since home rule in 1979, a Danish television transplant in terms of leadership, staffing, and adopting similar forms of TV broadcast journalism and documentary. However, since 2009, broadcasts have increasingly been in Greenlandic. The training continuum from broadcast television and radio that has been important for developing Indigenous and film and cinema talent in Nunavut and other parts of the Arctic was never especially strong in Greenland. The recent proliferation of film, video, and other forms of moving-image production and circulation is thus quite remarkable (see Montgomery-Andersen 2021; Grønlund 2021a). Taken together, the past decade has seen the rise of a robust set of Greenlandic media sovereignty practices, which correlates with the political mobilization that achieved self-government in 2009, a watershed for the Greenlandic independence movement. Self-government has a range of implications for the country, not only in terms of politics, but also in terms of the role of cultural production for Greenlandic representations on a global media stage.

CINEMATIC CONTEXTS OF THE SELF-GOVERNMENT ACT AND GREENLANDIC INDEPENDENCE

The preamble to the Act on Greenland Self-Government (hereafter "Self-Government Act") "expressly recognizes the Greenlanders as a people under international law with the right to self-determination," and a section of the act states "that the decision on independence is made by the Greenlandic people" (Mortensen and Barten 2017, 115). While the Danish Parliament has to endorse independence, the act's language assumes that a good-faith agreement should be reached when Greenland seeks to request it. Meanwhile, Greenland continues to rely financially on Denmark through an annual block grant. Greenland's current inability to maintain a welfare state and stable economic conditions on its own is the biggest obstacle to its independence, and the extraction and export of natural resources—especially rare-earth minerals, aluminum, and uranium—are seen as critical to the independence potential (Lindroth 2019).

The Danish colonization policies, usually called "modernization policies," had long-term effects. They effectively continued until 1979, even though, during the period 1953–79, Greenland was not an official "colony" but rather a "county" in the Realm of Denmark. These policies included forced relocation from small coastal settlements; discriminatory and unequal education, labor, and employment policies between Danes and Greenlanders; prioritizing mono-industries such as large-scale fisheries and fish processing; a monopoly on export and import through the Royal Greenlandic Trading Company; and the privileging of the Danish language, Danish perspectives, and Danish government priorities, developed and implemented with little substantive input from Greenlanders. These policies were outlined in ten-year plans (called G-50 and G-60) that envisioned Greenland "as a modern welfare country" (Rud 2017, 123), with G-60 implementing large-scale relocation efforts. These plans led to the effect of "Danes being supervisors and guides and the Greenlanders were to be supervised and guided," Jens Heinrich argues, such that Greenlandic perspectives were "disregarded without any real influence on the events. Apathy, drinking and rootlessness became commonplace" (Heinrich, cited in and translated by Rud 2017, 124). In addition, high rates of abuse, violence, and suicide have been strongly associated with the modernization project, and this has led to the "publicly accepted" understanding that the hasty process had profound sociological and psychological effects (Flora 2019, 18). At the same time, for decades, "Denmark has legitimized its domination of Greenland as a benevolent guiding hand on the path towards modernity" (Andersen 2019, 216). This perspective can be seen as an aspect of Scandinavian exceptionalism, which continues to suppress debate around—and policy changes that address—centuries of internal colonialism.

The modernization policies posited a divide between Greenlandic "tradition" and Danish "modernity" and coded this divide in terms of private and public, Indigenous and Westernized. Greenlandic scholar Karla Jessen Williamson stressed in a public lecture in 2016 "that to a large degree Inuit culture had been extinguished or relegated to the private sphere of individuals with no place in the official public domain" (Andersen 2019, 241). As part of Denmark's protectionist stance during the first two-thirds of the twentieth century, Greenland remained largely isolated from other Inuit and Indigenous peoples. This changed in the early 1970s, with the rise of Arctic circumpolar pan-Inuit and Internationalist pan-Indigenous movements during that decade providing momentum for decolonization, self-determination processes, and pan-Arctic environmental protection and climate change mitigation (see Rud 2017, 125; Shadian 2018, 331–47; Sejersen 2015, 16–19).

The use of the term *Indigenous* in the Greenland context raises specific issues about the country's history and future. Significantly, the Self-Government Act makes no reference to Indigeneity. This signals ambivalence with respect to pan-Inuit and Indigenous contexts and reflects the legacy of colonial suppression of Inuit cultures, practices, and values supposed to be inherently inferior. Attached to the term *Indigenous* are "connotations of underdevelopment and suppression—and anti-modernity" (Thisted 2013, 235). There are also important legal issues about the term's usage, because an *independent nation-state* has a greater degree of autonomy than an *Indigenous people,* which is considered a population subset of a state, granted special rights, according to definitions in the UN Declaration on the Rights of Indigenous Peoples, adopted in 2007. At the same time, the absence of language in the Self-Government Act pertaining to Indigeneity also aligns with a long-standing colonial practice by Denmark, which "treats the Inuit in Greenland as a closed group. This stands in contrast to the Inuit themselves [as represented by the ICC clauses] as a single people or a single, indigenous people across all boundaries in the Arctic" (Inuit Circumpolar Council 2009, Art. I [3] and I [4], 2009).

Language politics are significant in this context. Notably, Greenland's parliament passed an act on language policy in 2010 (hereafter Greenlandic Language Act) that "aims at preserving knowledge of Greenlandic in the future as the language serves as a bearer of the traditional Greenlandic culture" (Mortensen and Barten 2017, 113). Danish rule during the 1950s prioritized Danish-language use for everyone, meaning that native Greenlandic-speakers became inherently disenfranchised, forced to learn Danish at school while losing formal instruction in their native language. Couched as Danish benevolence, this doctrine became one of the most forcefully contested during the 1970s, with the burgeoning independence movement arguing for Greenlandic to be the language of Greenland, for *all* Greenlanders, ethnic Danes included. Kirsten Thisted writes: "The quest for a Greenlandic identity and political self-determination began as a quest to strengthen the Greenlandic language, and the Home Rule Act of 1979 specified Greenlandic as the primary language of the country. The Greenlandic language became the marker of Greenlandic identity" (2017a, 284). Language preservation also led, for a time, to a rejection of all aspects of modernity in Greenlandic independence discourse. Thisted continues: "Suddenly, the modern Greenlandic culture was seen as suspect because modernity and urbanity had become synonymous with 'Danish'. Paradoxically, the independence movement thus returned to the rhetoric of the colonial administration, praising and idealising the cultural heritage of hunting traditions as genuinely 'Greenlandic' while condemning modernity as something foreign and incompatible with Greenlandic society and tradition" (2017a, 284). Others have argued that this political belief shifted in the years leading up to self-government. As Ulrik Pram Gad noted in 2009, the future-oriented construction of Greenlandic identity was fluid in the ways in which Greenland attempted to preserve its tradition, especially in the use of language, while at the same time arguing for and upholding aspects of the modern welfare nation-state, while trying not to have that model seem Danish (Gad 2009, 144). This allowed for the possibility of other discourses of modernity being de-linked from the specific Danish context. A specific form of modernity has indeed exploded in the post-2009 context: film and media production.

The years following the momentous passage of the Self-Government Act in 2009 were characterized by a great amount of energy in and around artistic, media, and film production by Greenlanders, in both Greenlandic and Danish. In this wealth of production, two distinctive discourses stand out. In the first, we see a futurity-oriented set of priorities, imagining Greenland as the hub for spokes of a new wheel. Greenland is conceptualized as a central geopolitical actor, thanks to global interest in the environmental effects of climate change that paradoxically generate political and economic opportunities for expanded Arctic resource extraction and the postulation of faster shipping routes across a polar ice cap traversable by commercial traffic. In a similar vein, Greenland's connections and contributions to global artistic, cultural, and political phenomena become vehicles to bypass Copenhagen and Denmark to brand Greenland as a distinct, autonomous cultural entity. Often seen as a core tool in the nation-building process of the twentieth century, film production is not insignificant in this context, and we have seen the International Sámi Film Institute, Isuma, and Arnait engage in similar priorities as an aspect of self-determination processes in other Arctic regions. We also note the emergence of ice and climate crisis imaginaries and how Greenlandic perspectives over recent decades, especially those of women filmmakers, have consistently reassessed both the futurity and globalization discourses so dominant around the time of independence.

The second major cinematic trend is the interest in reassessing Greenland's relationship to its colonizer, Denmark, and to Greenland's own assessment of itself as a diverse community in which Greenlandic-Danish discourses on the past, present, and future are complex and intertwined, as exemplified by the Greenlandic Reconciliation Commission (GRC) that reported to the Government of Greenland from 2013 to 2017. We will examine some examples of GRC-related film production more fully in chapter 8. Most of the post-self-government films that we discuss in this chapter and the next exemplify the ambivalence that characterizes this second trend; none engage directly with the issues as part of the film's diegesis, but most offer opportunities to identify the complexity and ambivalence as part of post-2009 Greenlandic identity.

THE FUTURE IS NOW, AND OURS: GREENLANDIC FUTURITY AND GLOBAL CONNECTIVITY

The first generation of Greenlandic fiction feature films were directed and produced by young people and featured young main characters, focusing on the perspectives of youth and on global youth culture in the Greenlandic language. The first feature film produced by a Greenlander, in Greenlandic, is Ujarneq Fleischer's 2008 low-budget comedy *Fore Finger, Middle Finger, Ring Finger, Little Finger (Tikeq, Qiterleq, Mikileraq, Eqeqqoq)*. Set in the small town of Sisimiut in western Greenland, the film was shot on low-grade video with a budget of roughly US$100) and presents a day in the life of a group of friends. The film is resolutely contemporary and situates Greenland as an important part of global youth culture. The conventions of Hollywood youth and high school films get reworked, parodied, and ironized in ways that also reflect local priorities. In the film's presentation of Greenland on the cusp of self-government, the future is one of youthful, albeit male-centered, optimism. That optimism infuses the production itself: getting the film made can be seen as an act of exuberant agency on behalf of the director and his buddies, and in circulation their work can contribute a different imagery of Greenland than was found in previous colonial or Danish depictions.

Similar in its male-centered depiction of contemporary teenage experience and featuring a goofy high schooler in quest of a life purpose and romance, Angayo Lennert-Sandgreen's *Hinrik's Dream* (*Hinnarik Sinnattunilu*, Greenland, 2009) is another significant example of the reimagining of global Hollywood tropes for the self-government generation. It is the first exclusively Greenlandic-language film, produced by Aka Hansen and Tumit Production, and correlates with the Greenlandic Language Act in committing to the language as a vehicle of artistic originality, community building, and the promotion of a broader range of public discourse. Hansen contextualizes *Hinrik's Dream* as significant for the self-government generation because it establishes a counter-discourse and, we add, exemplifies what media sovereignty looks and sounds like. Hansen states: "A lot of the things that are shown about Greenland are from a Danish perspective. And they really like to portray us either in the idyllic—natural context—or as a country riddled with abuse and alcohol. I really want to show how I see Greenland and not through that binary" (Henderson 2019). An important component in the role of these films as intervention is that Greenlandic becomes a part of the public discourse through cinematic release and circulation in the public sphere. For many Greenlanders, Hansen continues, seeing *Hinrik's Dream* "was the first time they could understand a film from start to finish. That was a big eye-opener for me, because I have spoken both Danish and Greenlandic since I was young and I did not know this kind of representation could have so much meaning" (Henderson 2019). Hansen's priorities as a filmmaker have continued to be shaped by these experiences.

Hansen produced another early Greenlandic fiction feature in the same vein of integrating global youth culture into Greenlandic contexts, in the country's own language: the teenage horror film *Qaqqat Alanngui* (Malik Kleist, Greenland, 2011). This film, too, tells a kind of story familiar from Hollywood: a group of young friends take a trip to a desolate summer cabin, where the environment is haunted by a threatening supernatural power—in this case, a *qivittoq*. As Janne Flora notes, the figure of the *qivittoq* is multivalent and very much a part of contemporary Greenlandic belief systems; it is someone who, because of an "extreme situation . . . has crossed the threshold and is firmly outside, not as a liminal character but as a not-even-human out-sider . . . an archetype of permanent loneliness from which there is no return" (Flora 2019, 64). Although the state of *qivittoq* is not necessarily related to suicide (as if leaving one's community alone for the wilderness would imply death), a correlation was articulated as the modernization project of Greenland accelerated: "Suicide came to be seen as motivated by the same sort of problem facing the young man who would have chosen to become a *qivittoq*—namely, that there was no room for him in society" (2019, 72).

By contrast, what these three early fiction features share is their unequivo-cal commitment to representing the youth of Greenland—especially young men—as part of and contributing to an emergent Greenlandic global popu-lar media context, which they are self-confidently modifying to reflect, one could argue, the optimism and future-oriented discourse in the Self-Government Act. These films also propose alternative on-screen representa-tions of Greenlandic youth to television narratives and media reports that emphasize a suicide epidemic affecting primarily male youth. One example among many is the first season of *Borgen* (Adam Price, Denmark, 2010), the internationally known Danish television series about a female prime minis-ter, whose first visit to Greenland focuses on suicide mitigation. Yet, as Flora emphasizes, the suicide discourse must be put into a broader context, given Greenland's small population—which skews country-by-country compari-sons (but makes good media and television fodder)—and the fact that when suicide happens, it happens in small communities, where "it affects almost everyone" (2019, 123). Like that of Michael Kral in Nunavut (see chapter 4), Flora's work foregrounds the many competing discourses surrounding suicide—including a pervasive "tradition-modernity" argument that is also gendered (2019, 126)—while also noting the profound emotional, psychologi-cal, and social challenges of talking about suicide in small communities. The growing media agency of youth in recent film production can be seen as a way to proclaim a different kind of youth activism. The fourth season of *Borgen* (Adam Price, Denmark, 2022) focuses entirely on Greenland, Arctic geopoli-tics, and the environmental and political aspects of resource extraction. The dozen years that have passed between the two seasons shed light on the evolv-ing colonial relationship between Denmark and Greenland.

An important film in the context of the burgeoning Greenlandic film industry is *Nuummioq* (*A Man from Nuuk,* Otto Rosing and Torben Bech, Greenland, 2009), which is the first fiction feature film with international reach produced entirely in Greenland, nearly exclusively in Greenlandic, with domestic funding and a cast and crew of Greenlanders. The film received positive reviews at Sundance and was Greenland's nomination to the Oscars for the year's best foreign-language film—the country's first such nomination, signaling its interest in mobilizing film and the Greenlandic language as part of a global hierarchy of symbolic value as awarded by Hollywood nobility. Released the year that the Self-Government Act was passed, the film makes several political and symbolic gestures. Producer Mikisoq H. Lynge describes in the documentary *Faith, Hope and Greenland* (Ivalo Frank, Greenland, 2009) how the film should be understood as a deliberate statement about Greenland and Greenlanders, helping them move past the years of colonial stigma constructed and propagated, at least in part, by Danish and other ethnographic films. He also describes the team's efforts to *Variety* "as a historic project" and as "wanting to get in the game," to go beyond outsider representations: "Usually you see these cliches about Greenland as if they're going through a checklist: a kayak, hunting in the ice or showing the indigenous as alcoholics" (Jaafar 2008). The work, then, functioned to question and undermine unreflective and racist stereotypes (see also Grønlund 2021b).

As a bromance, *Nuummioq* foregrounds the mundane experiences of thirty-something construction worker Malik, an "everyman" in contemporary Greenland (the film's title translates to "A man from Nuuk"). The film alludes to inequities between the country and its former colonizer—for example, high-quality health care for needed cancer treatments cannot be accessed on the island. It also signals Greenland's increasing role in a global community shaped by climate change, as Malik and friends plan to export to East Asia melting glacial ice, tens of thousands of years old, to meet the appetite of Japanese whiskey drinkers looking for the ever more exotic. While Malik is partly portrayed as a young man without purpose, the film makes a poignant reflection on contemporary politics in Greenland. The export of melting glaciers can be seen as Greenlanders taking control of their own resource extraction and positions Greenland as globally interconnected through export, potentially imagining a world of trade beyond the European Union (its largest trading partner) and Denmark. *Nuummioq* thereby also turns the tables on the many decades of Danish film that reinforced tropes that Greenlanders—somehow lost in a sea of modernity—should envision themselves as agents in a geopolitical and globally interconnected context, and slyly making a point that not all political films about Greenland are necessarily earnest documentaries.

By launching at Sundance, *Nuummioq* was also making another kind of statement about global interconnectedness. This international exposure places the film as the country's first "independent" film for the world community. This global outlook also calls into being a discourse or imagining of futurity: by being the first, one is implicitly stating that one is not the last. Beyond the story the film tells, this sense of productive momentum and more to come is one of the greatest aspects of *Nuummioq*'s legacy. The burst of films that followed speaks to a form of cultural branding, implicitly stating Greenland's independence—in filmmaking and as an interesting subject on its own terms—even if its status as a fully fledged nation-state is still in flux. There is a related ambivalence in the film: it projects futurity while also having a sense of nostalgia.

It is notable how many of the early fiction features produced in Greenland have male-centered narratives. This trend continues in Marc Fussing Rosbach's two science fiction features to date, *Among Us: In the Land of Our Shadows* (*Akornatsinniitut—Tarratta Nunaanni,* Marc Fussing Rosbach, Greenland, 2017) and *Among Us: The Masked Men* (*Akornatsinniittut—Kiinappalik,* Greenland, 2021). These films mobilize the tropes of global Hollywood science fiction to parodic and self-aware ends, incorporating Greenlandic myths and legends, albeit with teenage male leads. In these and most other Greenlandic fiction feature films, male characters have agency while female characters have little. This is not surprising, for that is the dogma of Hollywood and global film cultures; it is also a convention familiar to some other Arctic Indigenous productions that are engaging discourses of media sovereignty and self-determination (see, e.g., discussions of Zacharias Kunuk and Nils Gaup in previous chapters).

Reacting to a masculinist context embedded in colonialism and imperialism, Greenlandic cinema has recently explicitly challenged heteronormative frameworks. LGBTQ2S documentary *Eskimo Diva* (Lene Stæhr, Denmark/Greenland, 2015) focuses on trans drag queen Nuka Bisgaard from Nuuk. Also an activist, Nuka founded Gay Pride in Nuuk in 2010. In the film, Nuka and her straight friend Aqqalu Engell, who goes by the name Lu, tour remote towns in Greenland with their drag-electronica show, in the face of homophobic slurs and threats from older Greenlandic men. Their goal, as Nuka states in the film, is to be at the forefront of "a new, open-minded and more progressive Greenland." To that end, *Eskimo Diva* resonates with what Joanne Barker describes as the purpose of critical Indigenous gender, sexuality, and feminist studies, namely, to "grapple with the demands of asserting a sovereign, self-determining Indigenous subject without reifying racialized essentialisms and authenticities" and to "grapple with the demands of denormalizing gender and sexuality against the exceptionalist grains of a fetishized woman-centered or queer difference" (Barker 2017, 7). *Eskimo Diva* does not turn its eye away from the kind of struggles this goal necessitates. The authors of a study conducted in Nuuk and published in the *International Journal of Circumpolar Health* write: "Community informants told us that being LGBTQ2S was not socially acceptable so most LGBTQ2S individuals left Greenland for Copenhagen as soon as they were old enough. Our own observations suggested that this was true, that a hidden LGBTQ2S population would be very small" (Gesink, Mulvad, and Koch 2010, 36). Nuka's decision to be out and proud in Greenland ought to be read against this backdrop, along with the consequences that ensue from this action. For instance, the constellation between homophobia and suicide is addressed in *Eskimo Diva;* a friend of Nuka has committed suicide, and Nuka himself tried to do so through an overdose of paracetamol. Yet the duo's friendship and dark humor—in the face of not only homophobia, but Nuka's cancer crisis—foreground the youthful aspirational element of reimagining Greenland and postulating a new form of inclusive futurity.

Eskimo Diva is part of an emergent LGBTQ2S global Arctic group of films made by both Inuit and settler allies. The realist and activist documentary *Two Soft Things, Two Hard Things* (Canada, 2016), by queer allies Mark Kenneth Woods and Michael Yerxa, offers a historical account of the suppression of plural marriage and the experience of being a queer Inuit youth in a contemporary context. Woods and Yerxa's work considers how the history of Christian colonization still reverberates in Nunavut—for example, in the debate over sexual orientation in the proposed Nunavut constitution in 1999. Alethea Arnaquq-Baril's *Aviliaq: Entwined* (Canada, 2014; see also chapter 4) is the first Inuit lesbian-themed film, a piece of fictional reenactment made by a straight Inuk woman that exposes how Western codes of sexual conduct upended Inuit approaches to sexuality.

Recent Greenlandic works by women directors have explored queer identities and experiences further. The short film *Tuullik* ("Great Northern Diver/ Loon," Berda Larsen, Greenland, 2021) is shot in both Danish and Greenlandic and blends a realist mode with a symbolic one. The first part of the film presents a group of university students in Nuuk and their social interactions as they plan a Pride Parade; the last part focuses on the intimacy between two women lovers as they take a moment to themselves, away from the constraints

FIGURE 7.1. Anori (Nukâka Coster-Waldau) on the ice in *Anori* (Pipaluk Jørgensen, Greenland, 2018).

of extended family. The ways in which this intimacy is portrayed are significant: the lovers paint each other's bodies with black ink, drawing abstract forms from the legend of the Raven and the Loon directly on their skin. The film ends with an affirmation of their intimacy and leaves untold how they will navigate the constraints of external pressures to conform, whether from immediate family or society at large. *You'll Be Okay* (*Ajornavianngilatit,* Aka Hansen, Greenland, 2021) tells, in a realist fictional mode, the story of a single Greenlandic working mother who goes out for a night on the town and begins a relationship with a woman for what the audience takes to be the first time. Intimacy is again center stage, though in a less experimental vein, more akin to the tropes of a journey-to-discovery narrative. All these works postulate an LGBTQ2S Inuit futurity that transcends both colonial legacies and homophobia within Inuit communities themselves. These works also point toward various related, but at times oppositional, forms of solidarity—whether they intersect with Inuit communities of the global circumpolar North or with transnational Arctic LGBTQ2S communities.

More recently, new discourses about gender and a globalizing, diverse Greenland have emerged. Pipaluk Jørgensen's *Anori* (Greenland, 2018) was the first woman-directed and woman-centered fiction film to be released in Greenland (figure 7.1). Sections of the film, while in concept development, also screened at multiple international festivals, including at imagineNATIVE and the Berlin Native section in 2015. Produced by Jørgensen's own

Nuuk-based production company, Karitas Films, in collaboration with the Government of Greenland and the International Sámi Film Institute (ISFI; see chapters 5 and 6), *Anori* represents Greenlandic media sovereignty, while also expanding this concept to include an Arctic pan-Indigenous context in its collaboration with ISFI. The plot centers on the main character Anori's (Nukâka Coster-Waldau) experience of her partner Inuk's (Angunnguaq Larsen) hospitalization in New York City, following a seafaring accident with the Greenlandic Coast Guard, where he is saved by Malik (Ujarneq Fleischer).

The film is interspersed with several antirealist sequences that integrate Greenlandic myth. In contrast to Nuuk and New York City, where much of the film takes place, *Anori* also integrates a mythical and psychological register set in a mountainous landscape of ice and snow. Notably, there is no formal mention of Denmark in the film; the only indicator of the Danish presence is the Danish Defense coat of arms on the life vests of Inuk and Malik, who both work for the Joint Arctic Command. The geopolitics are instead related to the Arctic as a contested space, where multiple countries seek dominance, while also reflecting the fact that Denmark retains formal control over Greenland's foreign affairs (this theme is also explored in the fourth season of *Borgen*).

To receive the care needed after his accident, Inuk is transported not to Copenhagen but to New York City, suggesting that this location is faster to reach via medevac while symbolically commenting on Greenland's cosmopolitanism and global interconnectedness. When Anori travels to meet him, we see her stepping on board a plane in Nuuk and exiting at LaGuardia Airport. Thus, the fact that traveling to most locations from Nuuk requires going through Denmark is excised. In this way, the film posits Greenland as North America–oriented, reflecting some of the arguments Kirsten Thisted (2011) makes about USA-oriented youth culture and futurity in Greenland. At the same time, Anori's experiences in New York City are deeply traumatic: Malik, the supposed friend, rapes and nearly strangles her in his jealousy over her relationship with Inuk (indeed, the film is open-ended as to whether Malik was the cause of "the accident" to begin with). When she pleads with doctors that she must bring Inuk "home," we are to understand that he—and she—needs the connection to Greenland to survive. She cannot, alas, and Inuk dies at the film's conclusion. The plot also includes the suicide of Malik's former girlfriend, performances of Anori's rock band (Coster-Waldau is a vocalist and performer), and a celebration of traditional Greenlandic mask making and culture, thus providing a multifaceted view of contemporary Greenland. Anori maintains narrative presence throughout the film, making this the first Greenlandic film with a woman protagonist—with an emphasis on her experience of events, including those that are traumatic—and that raises questions about the culture of silence around questions of rape, abuse, and sexualized violence by making them visible.

ACTIVISM AND ICE IMAGINARIES IN GREENLANDIC VIDEO ART AND ART CINEMA

The rapid expansion of media production in Greenland has not been limited to the development of the film industry. A concurrent vein of production can be seen in video art and digital media, in which Indigenous women have played a key role. Filmmaker Pipaluk Jørgensen and two Greenlandic artists, Pia Arke and Jessie Kleemann, are of note in this regard. Their works intersect with questions of gender and ice as a constituent part of the Arctic imaginary. Arke's *Arctic Hysteria* (Denmark, 1996) and Kleemann's *Arkhticós Doloros* (*The Arctic in Pain,* Greenland, 2019; a collection of Kleemann's poetry with the same title was published in 2021) share affinities. Reading bodies into the Arctic landscape has also been a component of postcolonial representation in Black British video art, including as part of Isaac Julien's *True North* (UK, 2009), which remediates the story of Black explorer Matthew Henson, who helped lead Robert Peary to the North Pole, in collaboration with Greenlandic Inuit experts. What unites these video works is their interventionist ethos. For Arke and Julien, the challenge is aimed at a century of uncontested masculinist, heteronormative, colonialist, racist, and pseudoscientific ethnographic mythology about the Inuit North, propagated by Peary and many other explorers. Kleemann, and other women filmmakers working in Greenland, such as Jørgensen and Pipaluk Knudsen-Ostermann, the protagonist of *Silent Snow* (Jan van den Berg, The Netherlands, 2010), challenge twenty-first-century dominant discourses of similar pervasiveness. Rather than explicitly addressing the legacy of scientific exploration and heroic masculinism of the nineteenth century, these films engage with, mediate, and in some instances support the many recent inquiries into problematizing the masculinist "systems of scientific domination" that have characterized glaciology and climate science since their inception. The films further inquire into how these systems have marginalized not only white and Western women, but also Indigenous participants, especially Indigenous women (Carey, Jackson, and Rushing 2016, 773–80).

Pia Arke's "Ethno-Aesthetics" (1995) is the first Danish-Greenlandic post-colonial theoretical work. In it, Arke describes herself as a "mongrel," born to a Greenlandic mother and Danish father, growing up in both east and west Greenland, settling in Copenhagen as a teenager and training there as a visual artist and scholar. Arke is important in the Danish-Greenlandic context, since she was one of the first visual artists to interrogate dominant Danish paradigms about Greenlandic self-determination movements, before and after self-governance, through a lens of critical inquiry. In *Arctic Hysteria,* Arke imposes her naked body on her own photograph of a partly snow-covered Greenland landscape. She crawls on, rolls over, and caresses the photo enlargement, while making guttural noises and sniffing gestures. She concludes by tearing the picture to pieces and moves out of the frame on all fours. Remaining on the screen are shards of photographic paper. Arke cannot merge with the landscape in *Arctic Hysteria;* nor does she seem to seek her place, symbolically, within the topography. *Arctic Hysteria* can also be understood as a contribution to a long history of feminist film and video making that explores embodiment and representation of the female body (see, e.g., Juhasz 2001). Her work is also a direct riposte to more than a century of explorer narratives, photos, and film documentation, including those by Robert Peary, which feminized and demeaned Inuit cultures and linked them to prevalent European "hysteria" narratives. The term *Arctic hysteria* was coined and popularized by Peary during his multiple trips to northern Greenland in the early 1890s. The pseudoscientific term refers to Inuit—women in particular—who would, according to Western explorers, doctors, ethnographers, and scientists, scream and lose self-control, at times engage in "dangerous" acts, and then suffer from amnesia. The Inuit themselves do not recognize this as a "disease," and many have written about how the condition came to be described through colonialist anthropology (see, e.g., Dick 1995). Arke's work turns the lens around to expose the explorers' and scientists' fascination with the Arctic as a form of "hysteria," providing a commentary on masculinist pseudoscience defining the experience of the Arctic and the Inuit, particularly Inuit women (see also MacKenzie and Stenport 2019b).

The works discussed herein can be seen as an artistic component of creating alternative knowledges, and what has been proposed as a feminist glaciology stands in contrast to the dominance of Western masculinist scientific paradigms. Mark Carey and colleagues write: "Crucially for feminist glaciology, feminist political ecology argues for the integration of alternative ways of knowing, beyond diverse women's knowledges to include—more broadly—the unsettling of Eurocentric knowledges, the questioning of dominant assumptions, and the diversification of modes and methods of knowledge production through the incorporation of everyday lived experiences, storytelling, narrative, and visual methods" (Carey, Jackson, and Rushing 2016, 773). This line of reasoning marks a departure for the disciplines of history of science, glaciology, and climate change. The works of video artists and filmmakers in Greenland are becoming central to bringing these power dynamics to the forefront.

Kleemann's *Arkhticós Doloros,* made twenty-five years after Arke's work, provides an example of a feminist glaciology artist practice that relates climate science to postcolonial critique. Taking place on the Ilulissat Ice Sheet, *Arkhticós Doloros* shows Kleemann unfurling a long rectangular sheet of black plastic with a string of light bulbs attached to it. She wraps the plastic around her body, lets the wind take hold of it, and then binds herself with rope and envelops herself with the black fabric. If Arke's work reflects on the place of the Indigenous woman's body—and its agency—within the history of Arctic science and explorer representation, Kleemann's work comments not only on the effects of global warming on melting glaciers, but also, and more pertinently, on who gets to articulate, historicize, and construct the significance of Greenland's glaciers in a global context. In so doing, her efforts contrast starkly with the group of spectators surrounding her on the glacier—glaciologists, we are to infer—who have been dropped at the top of the ice by helicopter and who are standing passively by as she moves her body in and through the black plastic, ultimately wetting her face and hair in one of the meltwater pools. With her body enveloped in what can be seen as a shawl of petro- and electro-modernity, her performance, in part, laments and mourns the rapid changes taking place in the environment through anthropogenic climate change. In this interpretation, the work's title gains new resonance: it personifies the Arctic environment and specifically the glacier as a living being that suffers pain, along the lines of other Indigenous cultures' belief systems (Cruikshank 2005). In other ways, the performance must be seen as a challenge to the discourse of climate determinism and Anthropocene stereotypes of big data, assessment, and scientific analysis—including the many ice core drillings by scientists from around the world that document millennia of climate evidence—that are ubiquitous in many of the documents about climate change, including those that lay the foundation for the work of the Intergovernmental Panel on Climate Change (IPCC) and SWIPA ("Snow, Water, Ice and Permafrost in the Arctic," a part of the Arctic

Monitoring and Assessment Programme). In *Arkhticós Doloros,* Kleemann insists on performing an alternative epistemology of ice.

Kleemann's performance is part of an international interdisciplinary research project called "On the Moraine," which provides a counterpoint to the lack of women and Indigenous participation in glaciology. Indeed, as Carey and colleagues note, among "studies probing the discipline of glaciology, only a tiny subset analyze gender.... Fewer still recognize indigenous knowledges, local perspectives, or alternative narratives of glaciers" (Carey, Jackson, and Rushing 2015, 773). The "On the Moraine" project explores the effects of climate change on one particular region of Greenland—Disko Bay—and the retreat of the Ilulissat Glacier. The moraine is what is visible once the glacier has retreated. It is often configured as a visible scar in the landscape, posited as unequivocal evidence of global warming (this is certainly the case in Jeff Orlowski's 2013 documentary *Chasing Ice,* discussed in chapter 2). As such, *Arkhticós Doloros* provides a forceful commentary on the many fly-in-fly-out (mostly male) international glaciologists and scientists who have, for decades, promulgated the global image of "disappearing Greenland" after spending time on the ice sheet and recording its seasonal variations. This representation of Greenland, whether in scientific journals or through the many "science communication" efforts that scientists are expected to perform, also plays into a long history of assumptions derived from over a century of anthropological and ethnographic fieldwork in the Arctic with communities who live on and near ice and glaciers. Namely, they assume that these groups are "disappearing" in parallel to the melting ice, thus replicating in an ethnographic framework what Mike Hulme (2009) calls climate determinism.

Ivalo Frank's video *ECHOES* (Denmark/Greenland, 2010) provides a complementary vantage point to *Arctic Hysteria* and *Arkhticós Doloros.* Having spent part of her childhood in Greenland, Frank has become a significant spokesperson for the diversity of Greenlandic cinema through the traveling film festival mentioned at the beginning of this chapter, Greenland Eyes—whose title evokes a long history of documentaries about Greenland that foreground glaciers and coastal sea ice as well the "eyes" that gaze upon Greenland. *ECHOES* addresses complementary aspects of Greenlandic history and the Anthropocene, namely the debris left by Cold War military installations (figure 7.2). Set on Kulusuk Island in southeastern Greenland, the work presents inhabitants, especially children, of the settlement Ikateq, as well as expansive landscape shots of the seemingly pristine nature that

FIGURE 7.2.   Military detritus in Ivalo Frank's *ECHOES* (Denmark/Greenland, 2010).

surrounds the village—in counterpoint to the shots of rusting oil barrels, abandoned vehicles, and debris left behind at a former US military airfield and the US radar-site DYE-4, Greenland's easternmost station of the DEW Line.

Juxtaposed with the landscape representation, *ECHOES* includes an interview with the East Greenlandic drum dancer Anna Kûitse in which she reflects, in imperfect English as well as in Danish, on a migrant life. The film does not show Kûitse performing a drum dance to an audience, but it integrates her dance movement as a reflection on Greenland's position during the Cold War and after (see also Frank 2019, 337). *ECHOES* can therefore be productively situated as part of a strong emergent video art tradition by women artists in and about Greenland, a tradition that reflects specifically on the gendered constructions of the Greenlandic environment in the context of the climate crisis and the Anthropocene.

Jørgensen's fiction feature *Anori* also engages in some of the aesthetic forms found in video art, linking gendered representations and performativity with landscape. The film remediates two Greenlandic myths, one about the wind (Anori) and the other about greed and revenge (Kagssasuk). The film operates in four realms, two realistic and two mythopoetic. The realistic scenes are crosscut between the present, starting with a boat accident and the transfer of Anori's partner Inuk to New York City, and a series of flashbacks recounting Anori and Inuk's love story in Greenland. The first of the two mythopoetic registers depicts forms of psychological trauma and horror, spanning a series of shots with Anori, Inuk, and Malik in a Greenlandic mountain landscape of ice and snow, and dream-like dance scenes in New York City streetscapes. This crosscutting foregrounds Anori's traumatic experiences of Inuk dying and learning that Malik was not the good friend she and Inuk may have thought. The second series of mythopoetic scenes represent aspects of psychological interiority, where *Anori* returns repeatedly to the expressiveness of a woman dancer's body, which, as in the works of Arke and Kleemann, is connected to the Greenlandic environment. Unlike in their work, however, the presence of a performing woman's body does not provide a meta-commentary on colonization or environmental change. Instead, it roots Anori in Greenlandic cultural heritage and a cherished wintry mountainous environment. While she is still in New York, these mythopoetic scenes reoccur, just as her mental anguish is acted out through similar dance sequences in the city's streets and on the Brooklyn Bridge. Her experiences are thus both constructed and mediated through Greenlandic ways of knowing. Snow and ice scenes mediate between physical and psychological experiences. By having her experience mediated though Greenlandic ways of knowing, construed as both physical and metaphysical, *Anori* functions as a form of interventionist historiography, whereby the world of New York—coded in part as that of "modernity" and globalization—does not determine how she understands her experience.

THE GLOBAL ANTHROPOCENE AND THE CLIMATE CRISIS FUTURE

Spanning the public and private sectors, with shifting emphasis as governments have changed, a dominant—but by no means unchallenged—perspective on climate change, resource extraction, and independence has emerged during the past decade. Since 2010, subsoil and offshore extraction rights are Greenland's alone; resource control is thereby foundationally linked to both actual self-determination and its positively construed imaginaries, just as the potential financial gains from extraction and "hyper- and mega-industrialization" are posited as one of few means to achieve a viable economy independent from Denmark (Sejersen 2015; see also Lindroth 2019). In official discourse, global warming has been presented as an opportunity. Representatives of business and government interests in Greenland have downplayed risks and instead framed climate change as a positive opportunity (Nuttall 2017; Rastad Bjørst 2019). Strongly articulated counterarguments to this resource and climate change optimism include the societal risks of succumbing to a mono-economy and the environmental effects of industrialization and ensuing infrastructure (ports, drill rigs, hydroelectric plants, etc.) on fragile ecosystems. Counterarguments also address cultural and community impact, including detachment from Inuit and Indigenous perspectives and relationships to and with the ice, land, and ocean (see, e.g., Körber 2017). These debates have put Greenland into a "state of potentiality and elusiveness [and] anticipation," Lindroth argues, since a major extractive industrial project has yet to be realized (2019, 22).

Aka Hansen's documentary *Green Land* (*Nuna Qorsooqqittoq/Grøn Land,* Greenland, 2009) intervenes in the competing political and economic futurity discourses. Like many of the films produced in Greenland around the time of the Self-Government Act's passage, Hansen's focuses on the perspectives of youth, emphasizes the use of the Greenlandic language, and foregrounds urban contexts that downplay the visual coding of pristine landscapes. Interviews with five young Greenlanders bring about a multivalent impression. Lill-Ann Körber notes that "two different sources of knowledge about climate change thus become apparent: one from international discourses that the five learn about at school and in the media, . . . another from individual and collective memory" about local conditions (2017, 153). As one of the first Greenlandic works to link debates about resource extraction, independence, and climate change, *Green Land* was selected to screen as part of COP 15 (the fifteenth meeting of the UN Conference of the Parties to the Convention on Biological Diversity) in Copenhagen in 2009. The film brings together Greenlandic perspectives on globalization, climate change, and future-oriented agency. Rather than portraying Greenlanders as passive victims of the "crisis narrative" of climate change and melting ice, in which Inuit are "at risk"—a standard approach in early twenty-first-century media depictions by non-Greenlanders (Sejersen 2015, 21)—Hansen's work counteracts the climate crisis determinism in Greenpeace productions such as *Greenland Thin Ice* (Andreas Rydbacken, Denmark, 2006), which presents Greenlanders, Thisted argues, as "ghosts or zombies: living dead, robbed of the culture that made their lives meaningful" (Thisted 2013, 23).

Hansen's *Green Land* also counteracts other dominant modes found in climate change documentaries about Greenland made by non-Greenlanders, such as the presentation of pro-industry, pro-resource-extraction, and pro-climate-change perspectives in *Greenland Year Zero* (Anders Graver and

Niels Bjørn, Denmark/Greenland, 2011). *Winter's Yearning* (Sidse Torstholm Larsen and Sturla Pilskog, Denmark/Greenland/Norway, 2019) looks at the questionable outcomes of the pro-industry, pro-extraction politics by focusing on the small western Greenland town of Maniitsoq, where, in 2006, US aluminum giant Alcoa purchased the rights to establish a US$3.5 billion smelting plant. These plans have not come to fruition, and the documentary examines a sense of suspended life (see also Sejersen 2015, 129–63). Inhabitants continue to be invested in the futurity and globally connected promise of the plant, while the film promotes a view of Greenland, well known in documentary history, as characterized by a lack of access to high-quality education and work and by the presence of alcoholism and abuse. Hansen's *Green Land* played a key role in shifting the Greenlandic media debates of the time beyond narratives of marginality, adaptation, or opportunism, insisting on agentic youth perspectives to shape alternate futures of a globalized Greenland. This made her work a significant contribution to the debates regarding the Self-Government Act.

Like *Green Land,* the multimedia works by Inuk Silis Høegh present Greenland as part of an interconnected set of climate change and resource extraction debates. For over a decade, Silis Høegh has presented alternative visions of Greenland's futurity and global connectivity imaginaries. Key examples include the extended land-art and film-art video installation *The Green Land,* where the single-channel component first screened at the CPH:DOX film festival in Copenhagen in 2021. *The Green Land* features an extensive collaboration with sound artist Jacob Kirkegaard, whose audio recordings capture the sounds of wind, stone, fire, plants, and roots. Silis Høegh's mode of operation is expansive, engaging film and video art as well as massive installations. For example, in *The Tip of the Iceberg* (2009), an installation made for COP 15, the artist draped the North Atlantic House—the home of the Greenland government's representation in Denmark as well as the historical nodal point for all shipping traffic to and from Greenland—in a digital photomontage of iceberg images collected along the coast of Greenland. *Iluliaq* (2013) was a site-specific work at the National Gallery of Canada, part of the gallery's *Sakahàn: International Indigenous Art* exhibition. *Iluliaq* covered the building's Great Hall with a twenty-one-meter iceberg installation that appeared as if it were melting as the edifice's windows were slowly replaced throughout the installation. An audio component of the work replicated the sounds of melting ice.

Silis Høegh also contributed to Denmark's pavilion, "Possible Greenland," for the Venice Biennale in 2012. Cocurated by geologist Minik Rosing and

NORD Architects in Copenhagen, the exhibition explored the potentials of a globally connected Greenland of the future, in terms of urban planning, resource extraction, the built environment, and art. Silis Høegh's contribution, *Connecting Greenland: AIR+PORT,* created with photographer Julie Edel Hardenberg, imagined the future potentials of shipping and transportation in an Arctic transformed by climate change, while catering also to the presumed needs of migrant mining workers. Juxtaposing two extremes—a Greenlandic future impacted by global warming and the origin of life found in the country's prehistoric past—Ivalo Frank's *The Last Human (Siunissaq— Det sidste menneske,* Denmark, 2022) addresses the findings of Greenlandic geologist Minik Rosing and the hopes of Greenlandic youth. The feature-length documentary earned an award at CPH:DOX and includes captivating landscape photography of the distinctive Greenlandic environment, with numerous interviews with young Greenlanders and extensive on-camera commentary by Rosing as he discusses his finding of the first traces of life on Earth in the Isua fjord in southwest Greenland. These works all engage in articulating, developing, and designing new futures for Greenland. Significantly, these works are exhibited not only in Nuuk, Sisimiut, and Ilulissat, but in globally situated cities and events: capitals, international exhibitions, and UN conferences. As such, they demonstrate, again and again, that discourses of climate crisis determinism, colonialism and marginality, victimization and adaptability, and industry and resource extraction opportunities must be challenged, reconfigured, and circulated in a global context to posit alternate versions of the future.

Following the advent of self-government in 2009, a multitude of artistic forms and cinematic inquiries have charted diverse pathways for Greenlandic self-expression. The works address national and cultural identity, language and Indigeneity politics, the Anthropocene, the climate crisis, LGBTQ2S experiences, and decolonization. In short, this body of work situates Greenland in a global context. Moreover, the creators' use of popular genres such as horror, romance, and melodrama, and their work in documentary, video art, digital media, art cinema, and installation, attests to the variety of representational modes that have flourished in Greenland in the twenty-first century. The works also draw attention to the intersection between Indigenous experiences and modern art practices, reconfiguring both at the point of intersection. Greenland's discourses of futurity highlight not only the changing nature of the emergent country, but new ways forward for a political aesthetics that functions as interventionist historiography.

EIGHT

Greenlandic Reconciliation Cinema, Self-Determination, and Interventionist Historiography

SEEKING TO ADDRESS HISTORICAL and ongoing traumas in Greenlandic culture and society—as a foundational component of the newly independent nation—during 2013–17, the Government of Greenland tasked the Greenlandic Reconciliation Commission (GRC) to focus on the lasting social and sociological effects of the modernization period of 1953 to 1979 (Forsoningskommissionen 2017). The GRC shed light on the following areas: forced relocations and urbanization; unequal education and employment opportunities between Danes and Greenlanders; and the loss or degradation of Greenlandic language and Inuit culture. The GRC's primary aim was *internal* reconciliation within Greenland and among all the various factions of the Greenlandic population, rather than with the former colonizer. The Danish government declined to participate in the process.

The GRC project emphasized, Kirsten Thisted writes, the ongoing and multiple legacies of the modernization and colonial period, with "the suggestion that the responsibility should not rest exclusively with the colonizers but that the Greenlanders too need to reconcile their own involvement in and responsibility for the process," including "discussion about the right to define and interpret the past—and to take responsibility for the past" (Thisted 2017, 233). Prioritizing historiography and the establishment of an inclusive and culturally diverse contemporary society, the GRC's focus was on "renarrating and mastering the past, rather than on establishing truth and pursuing justice in terms of accountability [and] reconcil[ing] Greenlanders with one another internally" (Andersen 2019, 242, 221). These priorities set the GRC apart from most international truth and reconciliation processes.

The GRC's priorities—storytelling and testimony for internal reconciliation, the power of historiography, the integration of goals for the future—are expressed in the title of its final report: "We Understand the Past; We Take Responsibility for the Present; We Work toward a Better Future" (Forsoningskommissionen 2017). In terms of cinema and other media, the GRC process led to a number of significant media and cinema projects that speak to an emphasis on fostering and securing an enhanced media sovereignty for Greenlanders, with Greenlandic filmmakers taking on the responsibility to tell the history of Greenland themselves, in Greenlandic. This reflects the GRC's interest in interventionist historiography, in deliberate efforts to answer Prime Minister Aleqa Hammond's call to action for the GRC, namely, "to regain historical sovereignty over Greenland's past and to make Greenlanders the masters of the narration of Greenland's history" (as paraphrased and translated from the Greenlandic in Andersen 2019, 215). Film practitioners and production companies whose films had set the stage for an independent Greenlandic film and media context around the time of the Self-Government Act's passage in 2009 were instrumental in addressing many of the GRC's priorities, especially in their emphasis on relating the past to the present and reconsidering the modernization period.

Cinema, video art, and informational films have been exceptionally important for the GRC, building awareness and support both in Greenland and abroad. At least two short works were made explicitly to further GRC-related conversations in Greenland as well as in Denmark. The first is Aka Hansen's short film *Half & Half* (Greenland, 2015), a bicultural identity work made for a seminar about the GRC held at Copenhagen University in 2015. The film situates Danish-Greenlandic cultural relations in the body of the filmmaker herself, as an example of what Alisa Lebow has called the "cinema of me" (2012; see chapter 6). The short experimental work juxtaposes shots of Hansen in Danish and Greenlandic garb. Traveling through Copenhagen, she explains in voiceover that in Greenland she is called a Dane, and in Denmark a Greenlander. In Denmark, she is asked why she does not look like a Greenlander, to which she then raises the question of whether there is a way to look like a Dane. She asks if the language you speak, or think in, is what marks your identity as fixed, and if you need to speak Greenlandic to be a Greenlander—can she be half and half? The director speaks Danish in the film, while the title and soundtrack are in English; *Half & Half* thus speaks directly to the GRC's interest in language politics (for more on Hansen, see chapter 7).

Ánorâk Film, one of the most active production companies in contemporary Greenland, produced the GRC's official informational film, used to invite participation in the public meetings and interviews constitutive of the process. Called *Saammaatta* ("Reconciliation," 2015) and directed by Inuk Silis Høegh, the five-minute film illustrates the GRC's emphasis on historiographic sovereignty, internal reconciliation of the past, and the acknowledgment—perhaps even championing—of contemporary Greenland as linguistically, culturally, ethnically, and geographically diverse. The short film integrates found footage, notably to represent a brief history of Danish colonization, as well as multiple interviews with a range of Greenlanders used to illustrate different perspectives on present-day life, indicating that the GRC process should lead to a stronger Greenlandic society for the future. In addition to *Saammaatta,* the GRC produced several short documentaries addressing key issues examined as part of the process, including sexual, alcohol, and drug abuse; the relocation of Greenlandic children to Denmark; and the multivalent linguistic and cultural identities among Greenlanders today, especially youth (Sammaattaa.gl 2017).

Ánorâk Film's next planned intervention in the global circumpolar colonization, truth, and reconciliation debates is the film *Twice Colonized* (Lin Alluna, Canada/Greenland, expected 2023), featuring lawyer, activist, and artist Aaju Peter. Coproduced by Alethea Arnaquq-Baril and Stacey Aglok MacDonald (based in Iqaluik, Nunavut) with Emile Hertling Péronard, the documentary brings the Danish and Canadian colonization regimes into dialectical tension with one another through the activism and personal experiences of Peter (see chapter 4). The film won the prize for best "doc-in-progress" at Cannes in 2022 and demonstrates the continuously strengthened transnational pan-Arctic film production collaborations enabled through the Arctic International Film Fund.

SOUNDS OF A REVOLUTION: THE RISE OF GREENLANDIC DOCUMENTARY AND *SUMÉ* AS SOCIAL HISTORY ROCKUMENTARY

In the GRC and post-self-government context, Silis Høegh and Péronard's rockumentary *Sumé: The Sound of a Revolution* (*Sumé: Mumisitsinerup nipaa,* Greenland/Denmark/Norway, 2014) stands out. Produced by Ánorâk Film, *Sumé* is a film of firsts: the first feature film made by Greenlanders about the history of Greenland and the country's first feature-length documentary to receive international distribution (figure 8.1). It thus speaks to

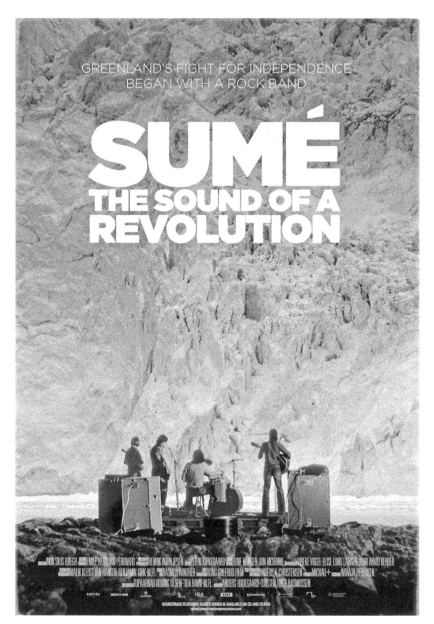

FIGURE 8.1. Poster for *Sumé: The Sound of a Revolution* (*Sumé: Mumisitsinerup nipaa,* Inuk Silis Høegh, Greenland/Denmark/Norway, 2014).

and furthers several of the key priorities of the GRC, especially in its connection to media sovereignty and emphasis on an interventionist historiography. *Sumé* is about the Greenlandic rock group Sumé (1972–76), whose success was built on a combination of political and anticolonial protest and popular music. Led by Per Berthelsen and Malik Høegh, with group members from different parts of Greenland, Sumé was formed while the band's members were studying in Denmark, the only place for Greenlanders to receive post-secondary education at the time (the University of Greenland opened in 1987). Based in Copenhagen, and writing all their lyrics in Greenlandic, the band released their first two albums, *Sumut* (1973) and *Inuit Nunaat* (1974), on the Danish socialist, anti-imperialist, and anti–Vietnam War label Demos. Their third album, *Ulo 1* (1976), was released on Ulo, a label they formed (after Sumé disbanded, band members and their producer Karsten Sommer continued their work with Ulo Records, releasing more than 150 Greenlandic albums). The sound of the revolution they promulgated seemed normal, expected, and anticipated—as it took the form of rock—yet they also undercut this normalcy by singing in Greenlandic, knowing that only some fifty thousand people could understand the language. The group sold ten thousand records in total, meaning that, on average, nearly every household in Greenland with a record player would have had at least one of their albums. *Sumé* tells the story of this influential group who mobilized Greenlanders through the popular idiom of rock music, while also, in the context of the GRC, uncovering additional layers of Greenland's journey to home rule in 1979, subsequently extended to self-government in 2009.

Sumé has had remarkable success for an Indigenous rockumentary emerging from a small national context. The only Arctic Indigenous rockumentary to precede it was Artcirq's thirty-six minute *Northern Haze: Living the Dream* (Derek Aqqiaruq, Canada, 2011; see chapter 4). *Sumé* has screened at over sixty film festivals, including the Berlin International Film Festival (the Berlinale, in 2015 and 2017) and at imagineNATIVE (winning two awards in 2014). International reception has been positive. For instance, reviewer Stephen Dalton praised the film, while remarking that the "skimpy length" of seventy-four minutes may prevent a deeper treatment of a poignant colonial history: "*The Sound of a Revolution* has all the sunny charm of a greatest hits collection, but it feels like a more gritty, challenging concept album may lurk within" (2015). Dalton's remarks are illustrative: the history of Greenland's colonial relationship to Denmark remains little known outside Northern Europe. *Sumé* walks a fine line in illustrating a contentious history without, it appears, wanting to take explicit sides in a political context where the ongoing GRC serves as an unstated backdrop and full independence remains a politically contentious concern. Indeed, the trajectory of the band members signals how conflicted this history is and how thorny the political mobilization was for Greenlanders, as well as band members, though this is addressed only slightly in the film proper. Lyricist Malik Høegh wrote unequivocally political statements against the history of Danish rule, while guitarist Berthelsen went on to become a parliamentarian in the new Greenlandic government, founding Demokraatit, the center-right party that is skeptical of Greenlandic independence and therefore favored by ethnic Danes living in Greenland. This tension in the band and their divergent political opinions represent broader trends in Greenland and Denmark, both in the 1970s and in the twenty-first century (see chapter 7). Attempts to reconcile these tensions are also key to the GRC process.

Sumé screened in Greenland's cinemas on general release, selling ten thousand tickets. This is remarkable, given that there are only three permanent movie theaters in Greenland, though public film screenings are common in community centers. The director and producer notably took the film on a tour of Greenland with a mobile projector. On its release in Denmark, it garnered overwhelmingly positive press reviews. Some Danish reviewers remarked on how little they knew about the history of 1970s Greenland, and one described how "exotic" the music of Sumé sounded when first broadcast on Danish public radio (Duus 2014). *Sumé* has also screened as part of a closed program for the Danish Parliament, with the purpose of educating politicians about shared Danish-Greenlandic history, while the GRC, without Denmark's participation, was taking place. The complex, and ongoing, colonial relationship depicted in *Sumé* also connects to its funding and production parameters. Greenland-based Ánorâk Film and producer Péronard could access some of the generous support schemes offered by the Danish Film Institute, because the film had a major Danish coproducer—in this case, Bullitt Film (the film's coproducers also included Nordic public service TV broadcasters DR, KNR, NRK, and SVT). This need for international coproduction funding is a reality for emergent Greenlandic filmmakers. The country has had limited public financing schemes, yet a strong and active filmmaking community with international aspirations has emerged over the past decade, thanks in part to the rise of relatively inexpensive digital technology for recording, editing, postproduction, and distribution. These changes in technology and circulation enabled *Sumé,* and many other works emerging in Greenland, to be acts of Greenlandic media sovereignty (for more on these aspects, see Thorsen and Péronard 2021; Grønlund 2021d).

The success and circulation of *Sumé* make it a milestone film that has helped propel the Greenlandic film industry forward. Indeed, in 2016, the Government of Greenland passed a 65 percent increase in the annual state production funding for the film support program that had been in place since the production of *Nuummioq* (Otto Rosing and Torben Bech, Greenland, 2009). In 2019, the Danish government granted 10 million kroner (around US$150,000) for documentary filmmaking in Greenland and the Faroe Islands. The funding was opened to Greenlandic, Faroese, and Danish filmmakers, but Danish filmmakers had to coproduce with Greenlandic or Faroese filmmakers to obtain the funds. A recent documentary about Danish-Greenlandic relations, *The Raven and the Seagull,* discussed below, is an example of the outcome of these new funding schemes. The effects of the success and reach of *Sumé* should be understood in this context. Its popularity has additionally provided both an influx of capital for further Greenlandic filmmaking and increased awareness of Greenlandic film production internationally.

As an Indigenous rockumentary that addresses a colonial past through depictions of a set of activists that are also rock stars, *Sumé* foregrounds the ways in which popular music can make seemingly local concerns and the language of protest legible to an international audience. *Sumé* was at the forefront of a growing trend, and subsequently, other Arctic Indigenous rockumentaries emerged, including *Arctic Superstar* (Simen Braathen, Norway, 2016), about Sámi hip-hop artist SlinCraze, and *WE UP: Indigenous Hip-Hop in the Circumpolar North* (Priscilla Naungaġiaq Hensley and David Holthouse, USA, 2018), which features global circumpolar hip-hop artists from Alaska, Greenland, and Sápmi (including Finland, Norway, Sweden, and Russia). Historically, the dominant form of rockumentary emerges from direct cinema and cinema verité, from *Lonely Boy* (Wolf Koenig and Roman Kroitor, Canada, 1962) and *Dont Look Back* (D. A. Pennebaker, USA, 1967) to the faux cinema verité of *Truth or Dare* (aka *In Bed with Madonna,* Alek Keshishian, USA, 1991) (see Iversen and MacKenzie 2021; Stenport 2021). These direct cinema and cinema verité rockumentaries are mostly about performativity. Though seemingly realistic, such rockumentaries challenge key assumptions of documentary, including that subjects in front of the camera are not staged and do not act. The dominant forms of rockumentary foreground the explicit performativity of the stars on stage. This is a testament to the "authenticity of performance" convention of the genre, far more so than its potential to engage with or remediate cultural or political history, which is core to *Sumé*.

In its style, approach, content, and purpose, *Sumé* is thus part of a different strand of rockumentary. It belongs to an important and unidentified strand of the rockumentary tradition: "the social history rockumentary" (MacKenzie and Stenport 2017). *Sumé* and the works mentioned above form an Arctic Indigenous subset and are of particular relevance to Indigenous media sovereignty. At odds with the dominant direct cinema and cinema verité strands of the form, the "social history" rockumentary explicitly engages in telling larger political and cultural narratives through the history of rock, pop, country, hip-hop, and reggae. In these rockumentaries, music and band performances are grounded within larger social, cultural, and political contexts, and many openly address colonialism, class conflicts, the rise of the social and religious right, gender and sexual inequities, or racial oppression and desegregation. This break from realist documentary traditions also has a specifically Greenlandic context. As Silis Høegh notes, he wanted *Sumé* to forgo the aesthetics and modes of production found in Greenlandic television documentaries made at KNR; rather, he seeks to work within the language of transnational documentary filmmaking, appropriating and retooling this global form to place Greenland at the center of the proverbial stage (MacKenzie and Stenport 2015; Turner 2022). Like many social history rockumentaries, *Sumé* draws on the fascination with the potential of youthful political activism. By mobilizing the performance and exposition of rock and pop music for political purposes, the films, and the musical genres from which they emerge, assert and celebrate individual cultural identities and experiences. They do this through—perhaps paradoxically—the feeling of being a part of an alternative community. In the case of *Sumé,* that community experience is also one of political mobilization toward increased self-determination.

Sumé *and the Sound of a Greenlandic "Soft Revolution"*

Although *Sumé* is set in the diegetic present, the time recounted in the film stretches back to when the band was together during the mid-1970s. A recurring red amplifier—also featured on the film's poster—signifies the connection between past and present. The band's focus, using the international idiom of rock, was very much on contemporary Greenlandic politics, with lyrics highlighting the effects of Danish colonialism, industrialization, and modernization policies, including alcohol abuse, unemployment, and forced relocation. Starting out as an underground cult band, playing clubs and impromptu venues in Copenhagen to fellow Greenlanders in the diaspora and the occasional Dane, Sumé quickly became well known, especially in Greenland. They were successful and talented enough to play for audiences in other parts of Europe, and eventually secured an invitation to join the British band Procol Harum as an opening act on their global tour (Sumé declined). The group also performed, in Greenlandic, at political rallies in the colonial capital, Copenhagen, deploying the popular idiom of rock to challenge received notions of Greenlanders and the continued Danish presence in Greenland. Aqqaluk Lynge, a well-known Greenlandic youth activist and subsequent politician for the socialist and separatist party Inuit Ataqatigiit, describes in interviews in the film how the burgeoning anticolonial protest movement would pair up with Sumé performances to mobilize audiences in both Greenland and Denmark for political awareness and activism.

As Silis Høegh notes in a web interview for the Berlinale, the inclusion of the word *revolution* in the film's title is significant. What happened in Greenland in the 1970s, he emphasizes, was a "soft revolution" in the "hearts and minds" of the Greenlandic people (*Sumé Electronic Press Kit* 2015). At the time, there was no violence or rioting in the street, either in Denmark or in Greenland. Sumé thereby became even more significant, since it formulated a distinct language of protest—in Greenlandic—against Danish colonization and the effects of the modernization period on Greenlanders and Greenlandic culture. As the press release for the film states, the group's political songs were the first to be recorded in the Greenlandic language—a language that prior to Sumé didn't have words for "revolution" or "oppression" (*Sumé Electronic Press Kit* 2015). Silis Høegh similarly describes, in another Berlinale web interview, how the lyrics of Sumé were perceived by many as dictating the next steps in the political agenda of Greenlanders at the time (Silis Høegh 2015). This placed the voices of Sumé, and therefore a politically awakening Greenland, within dominant modes of Western pop culture expression in a way that working through "traditional" Indigenous forms would not easily accommodate. Instead of promoting Inuit cultural revivalism—for example, by using the Greenlandic drum or wearing traditional costumes during performances—Sumé engaged with and articulated Greenlandic political protest through a contemporary and global form of communication. Their use of the melodic three-to-four-minute rock song is significant. The band sings exclusively in Greenlandic and draws on Greenlandic vocalization patterns, with only one song on their third album including a distinctive rhythm performed on a Greenlandic drum. Sumé also employed a standard rock lineup of stage performance with drummer, bassist, and keyboardist in the back, singer and lead guitarist in front. The sound, practice, and iconography of a 1970s hard rock band is thus a part of Sumé's appeal. It is not for nothing that the subtitle of *Sumé* is *The Sound of a Revolution,* and not, say, *The Voice of a Revolution.* It was the sound of the music, the synthesis of rock with the Greenlandic language, that gave the band its power.

Sumé's lyrics, and their popularity both with general audiences in Greenland and among Greenlandic student politicians in Copenhagen, also reflect the growth of pan-Indigenous, pan-Inuit, and pan-Arctic political mobilization in the 1970s. The initial First Circumpolar Arctic Peoples' Conference was held in 1973 in Copenhagen, with participation from Canadian and Greenlandic Inuit and Scandinavian Sámi, among other Arctic groups. The 1973 conference, and others that followed, formalized a set of shared priorities for Arctic Indigenous peoples seeking strengthened self-determination from settler colonialism. The Nordic countries became an important nodal point for the development of an Internationalist Indigenism (see Minde 2003). The contributions by the rock group Sumé, and their standing in the public sphere, especially in Greenland but also in Denmark, must be seen as an important contribution to an international history of Indigenous self-governance movements. The film *Sumé,* moreover, is one of few feature films in the Arctic context that revisit, historicize, and juxtapose the 1970s with present-day self-determination discourse and policy, thus articulating an interventionist historiography that is also of significance for understanding the history of pan-Indigenous and internationalist political mobilization.

Sumé's corpus—songs, records, performances, and media interviews—is historically significant in that it counteracts westerners' and Danish colonizers' privileged mode of historiography about Greenland: the printed word. Similarly, the documentary *Sumé* offers an interventionist historiography, not in essay or textual form, but through audiovisual moving images. This intervention speaks to key priorities in the GRC's emphasis on writing a new historiography for Greenland, as well as to the priority of achieving reconciliation among diverse groups of Greenlanders. A key difference with moving images and live music (and, in many cases, prerecorded music) is that watching and listening to these works is a communal experience, as compared to the solitary act of reading a novel or a book of history. This collective experience reinforces and heightens a sense of community in real time among the audience. This experience allows for the building of new histories and constitutes an important act of both interventionist historiography and, in the case of *Sumé,* media sovereignty. As such, the film and Ánorâk's oeuvre in general reflect what Indigenous filmmakers have repeatedly confirmed—namely, that oral storytelling, video, and performance are prioritized over the written word to express their own history and experience (Judell 2002; Arnaquq-Baril 2018).

FOUND FOOTAGE, COUNTER-ARCHIVES

Sumé's use of found footage differs greatly from the traditional function of found footage in direct cinema rockumentary. *Sumé* uses found footage not only to document the band's historical performances in concert and on television, but also to offer a visual archive of political, social, and cultural contextual images from the time. This practice correlates with the priorities of the GRC, which emphasizes the past's impact on the present. The film uses found footage to document interviews with the band on Danish television, to retrieve and redistribute previously lost images of recent Greenlandic history, and to remediate iconic colonialist images, such as scenes from the Leni Riefenstahl vehicle *S.O.S. Iceberg* (*S.O.S. Eisberg,* Arnold Fanck, Germany/ USA, 1933).

Among these different categories of found footage, the most important is the film's rediscovery and use of amateur films and home movies. In recent years, the importance of amateur archives, counter-archives, and collections that are not part of the gatekeeping system of state-run archives has become apparent. These archives offer oral and visual histories made by various publics, especially those that have been otherwise marginalized (Flinn 2011). The need for these amateur archives, especially for women, Indigenous peoples, racialized minorities, and LGBTQ2S communities, "reflect[s] the reality that cultural heritage has been unevenly cared for" (Chew, Lord, and Marchessault 2018, 5). *Sumé*'s deliberate and consistent collection, remediation, and use of amateur images from Copenhagen and Greenland in the 1970s provide images of lived experiences in stark contrast to those constructed at the time by the Danish media. Indeed, much of the retrieved amateur footage consists of Greenlanders filming Greenlanders in quotidian situations, including families enjoying celebrations and holidays, children playing at daycare centers, or teenagers cruising through settlement streets on mopeds. This found footage, therefore, is remarkably different from the official newsreel clips culled from public television (especially Denmark's state-run DR) or made by outsiders. Although some of the amateur footage features performances of the band itself, the "stars" of *Sumé* are only partially the band members (or producers and stage managers) and political activists; the "stars" are also the people who emerge in the found footage and fan interviews from the 1970s. This amateur footage reveals a history of media sovereignty of Greenlanders making images of their own in the 1970s. *Sumé,* by bringing these images onto public screens and into public consciousness, engages in an interventionist historiography parallel to the work undertaken by the GRC.

The amateur and found footage was obtained through solicitation over social media and through informal networks in Greenland and Denmark. *Sumé* thereby also functions as an expanded metahistory of Greenlandic film, responding implicitly to the GRC's call for Greenlanders to tell their own history while encouraging stronger awareness and acceptance of the parallel existence of multiple cultural experiences. Having received over seven hundred rolls of 8mm and 16mm film from their public outreach project (each between one and five minutes in length), Silis Høegh and Péronard subsequently made the footage available (much of it unused in the film itself) as a public archive for Greenlanders and beyond. It can be accessed, tagged, and shared on the Inuiaat Isaat ("Eyes of the people") website, which is connected to Ánorâk Film. The Inuiaat Isaat archival project also makes it clear that just because there were no Greenlandic-made feature documentaries prior to *Sumé* does not mean that there was no moving-image documentary history of Greenland filmed by Greenlanders. Just as *Sumé* tells a little-known history of Greenlandic politics and contextualizes that story as part of anticolonial and anti-oppression movements all over the world during the 1970s, the film also presents little-known material to a global public for the first time, thereby making a political statement about the significance of film production and film archives for nationalist purposes. The film and related amateur documentary footage thereby construct an interventionist historiography that speaks to the priorities of the GRC, just as these priorities illustrate tenets of the Self-Government Act, including prioritizing Greenlandic language.

Sumé not only tells the story of a band, as do many rockumentaries, but also offers accounts of the group's audiences in Greenland and Denmark at the time. In retrospective interviews, fans talk of how they feared that the group would perform badly—and thereby make Greenland look bad in the eyes of Danes. Interviewees also note their shock, at the time, that the political ideas forcefully expressed by the band were even being stated in public, in either country. Present-day fan interviews are equally important in relaying *Sumé*'s ongoing political significance. Many fans are interviewed, and they convey how the music has stayed with them during the long process toward self-governance. In addition, a young political activist (she was born after the 1970s) remarks that *Sumé*'s lyrics and political resonance remain as significant in the twenty-first century as they were then.

Sumé's perspective on the present Greenlandic political and cultural context is authentic in its recognition of the complexities of colonial history and the GRC's emphasis on internal reconciliation. At the same time, the documentary never takes a clear political stance either for or against independence. Rather, what *Sumé* seems to propose is the need for, if not constant revolution, then ongoing revisions to the historiography imposed by Denmark. Near the end of the film, Malik Høegh and Aqqaluk Lynge both state in different ways that change has not significantly occurred—that even though self-governance has been obtained, Greenlanders still live under the Danish nation-state. Much like the effects of the songs of Sumé on the Greenlandic audience in the 1970s, the goal of *Sumé* is to continue this struggle, through the social history rockumentary, to provide another set of voices in the public sphere about the necessity for change—and the ways in which the voice of change can be found within what, in the first instance, might seem the highly unlikely form of the pop song. In addition, *Sumé*'s international success as a social history rockumentary provides an opportunity for meta-reflection on the significance of documentary as a form of political protest and nation building in the Arctic and beyond. *Sumé,* in parallel with the GRC, combines two significant narratives: the story of popular protest against colonialism, and what the film's production history and dissemination signal about the ongoing struggle of Greenlanders to tell their own stories in a form that engages local and global publics. *Sumé* is a compelling example of the social history rockumentary, embedding such complexity at the levels of both form and content as an example of media sovereignty.

The documentary *The Raven and the Seagull* (*Lykkelænder,* Denmark/
Greenland, 2018), directed by Lasse Lau, offers additional perspectives to
those of *Sumé* on the long-standing, complex discourses about the relationship between Greenland and Denmark. It also foregrounds the role of film
and media, questioning agency, performativity, individual and collective
responsibility, and political mobilization in a self-reflexive mode. Composed
as a series of vignettes, the film comprises a few, mostly static-shot scenes
from various towns and regions of Greenland, as well as from Denmark,
including several iconic colonial and public locations in Copenhagen, like
the National Museum and the North Atlantic House. Neither the montage's
juxtaposition of the distinct locations represented nor the meaning of each
sequence on its own is explained through voiceover or titles. Therefore, audiences are left to make their own connections. While the priorities of the
GRC are not acknowledged in the film itself, they seemingly permeate its
content and form. Released as a documentary and winning the Nordic:Dox
Award in 2018, the work combines art video, performance, found footage,
satire, irony, and self-reflexivity to attain the status of an "authentic"
document(ary) about Danish perceptions of Greenland. Conceptualized and
directed by Lau, a Dane, *The Raven and the Seagull* was coproduced by the
Danish Film Institute and the Government of Greenland. Greenlanders
Pipaluk Jørgensen and Péronard were coproducer and executive producer,
respectively. The film was made with substantive input from Greenlandic
artists, intellectuals, and public figures. To this end, the film is a variant of,
or commentary on, Greenlandic media sovereignty. One of the GRC's priorities was for Greenlanders to "understand the processes in the historical development of and the cultural interaction within Greenland and in the relationship between Greenland and Denmark" (Forsoningskommissionen 2017,
part 5.3.) The film also reflects the GRC's emphasis on Greenlandic priorities
of internal reconciliation in multivalent ways. *The Raven and the Seagull* is
the outcome of what Lau calls an extended research project into postcoloni-
alism (Rasmussen 2015; see also Grønlund 2021d).

The filmmaker turns the GRC's investigative lens away from Greenland
and onto the Danish imaginaries of Greenlanders and Greenland to illustrate
the construction of Greenlanders as a homogeneous collective identity by
Danes and Danish media. The film's original title in Danish—*Lykkelænder*—
is significant in its illustration of precisely these competing discourses. On the

one hand, *Lykke* means *happy* and *-lænder* is a suffix meaning "inhabitant," as in *grønlænder* (Greenlander). Literally, the film's title means "an inhabitant of happyland." The working title in English was indeed "Happy Native" (Kran Film Collective n.d.), reflecting what Søren Rud affirms to be a pervasive trend in "Danish media, film and literature," namely, to depict Greenland and Greenlanders in "stereotypes: noble or primitive people; or lost and rudderless in the modern world" (Rud 2017, 145). The literal translation of *Lykkelænder* can also be read as a reverse commentary. After all, Denmark has landed one of the top spots in the annual "World Happiness Report," issued by the UN Sustainable Development Solutions Network, every year since the index's beginning in 2012. So the title might just as well refer to the Denmark described by Rud—happily oblivious of its colonialist practices and engaging in "naive celebration of exceptional Danish benevolence" (2017, 1). The English title refers to the myth of the Raven and the Seagull. In the introduction to ethnographer Lawrence Millman's travelogue *Lost Places: A Journey in the North,* Paul Theroux highlights the myth, as told to Millman in Qaqortoq in southern Greenland:

> Once a raven and a seagull got into a fight over a piece of meat. The raven was on the Inuit's side, the seagull on the side of White Man. They fought for days, for weeks, even for months. Whoever won, his side would be the stronger. They tore and bit furiously at each other. At last the seagull won: White Man would be stronger and more plentiful than the Inuit. But by the time he flew away with his piece of meat, it had become quite rancid. (Theroux 2000, vii; see also Millman 1990, 164–65)

In *The Raven and the Seagull,* the myth is told by an anonymous male voiceover as a scene plays out in which two dark-haired Greenlanders (assumed to be of Inuit rather than Danish descent) practice Taekwon-Do. While the myth is an old one, its presence in the film, juxtaposed with the martial arts battle, is construed both as an allegory for Greenland's present and a questioning of the value of the prize to be won at the end of the battle. What remains, exactly, have been spoiled? The Danish assumptions of Greenland as a tasty piece of food to be devoured? The chagrin that although presently self-governing, the country of Greenland has been despoiled by colonialism? The myth itself demonstrates that this symbolic battle has lasted much longer than the days, weeks, and months the story tells. This audio/visual juxtaposition functions in a similar manner to the image juxtapositions that run throughout the film.

This central scene is another example of how *The Raven and the Seagull* is political in its uses of experimental techniques. The film repeatedly turns the tables in terms of directorial identification: it is as if Lau is aiming to destabilize an authorial colonizer's presence, as if it were possible to construct the film from an insider's or a colonized perspective, while maintaining a keen sense of irony in the process. This is, of course, an ethically questionable stance—a fact the film embraces. Aspects of absurdity and alienation are a key part of Lau's approach, illustrating a precarious position of directorial auteurism, an effort that was challenged in some Danish reviews of the film. Some of the questions asked by the film are whether the Greenlanders portrayed, interviewed, and observed are "documentary subjects," imbued with independent agency, or, in some aspects, belittled and patronized. The film's self-reflexive and collaborative process of funders and contributors seems to ask how Greenlanders believe Danes see them. Yet, in so doing, *The Raven and the Seagull* draws on the visual sovereignty techniques described by Michelle Raheja, whereby Indigenous groups are actively reshaping the ethnographic and documentary legacies to, as she writes, "laugh at the camera" (Raheja 2007, 1160; see chapter 3).

The Found Footage Tradition and the GRC

The purpose of *The Raven and the Seagull* is to provide a Greenlandic response to Danish representations of Greenland during the past century. As such, it aligns with the priorities of the GRC. The film notably takes old colonial images and claims—in part, at least—media sovereignty over them. This stance is signaled at the very opening. Unlike most films by Danes or outsiders about Greenland, this one does not begin with images of arrival by ship or plane. Instead, the film starts out in an embedded manner, with a jeep in Arctic winter driving through a remote landscape in an extreme long shot that lasts more than two minutes. *The Raven and the Seagull* contains four pro-filmic realities: images shot in Greenland; images shot in Copenhagen and Denmark; remediated images of the first and only voyage of the Danish ocean liner MS *Hans Hedtoft;* and a myriad of Danish and European found footage, from ethnographic film shot in 1913 to *The Eskimo Baby* (*Das Eskimobaby,* Heinz Schall, Germany, 1918) to newsreels and postwar documentaries, representing Greenland in black-and-white footage. Like *Sumé, The Raven and the Seagull* is also about film and media history, specifically addressing Danish and outsider representations of Greenland through ethnographic, documentary, and fiction film. For instance, the use of films like *The Eskimo Baby* speaks to the ways in which images produced through colonial acts function as a way to create grand narratives about peoples for the colonizers themselves. The found-footage films are remediated through their placement in a new work that defamiliarizes the images while creating a new meaning for them. That includes the ways in which the found footage illustrates the priorities of constructing Greenlandic history from within and furthering understanding of, and reconciliation between, different Greenlandic demographics—the priorities of the GRC.

Interspersed throughout the film is remediated footage of the *Hans Hedtoft,* which was built to provide year-round service to Greenland from Denmark. The ship sank and suffered casualties on the return of its first voyage to Greenland on January 30, 1959 (and is the last known ship to have sunk after hitting an iceberg). Returning from Greenland, the *Hans Hedtoft* also contained 3.25 tons of Greenlandic archives, which were to be preserved in Denmark and then returned. The documents in the archive, many from missionaries and the Church, included birth records and other documents tracing the history of Greenland and its colonial heritage, dating as far back as the 1600s. Using remediated images of ships and ice, with recordings of distress calls presumably from the ship (though they may be reconstructions), these scenes foreground Greenlandic lost history through acts of colonialism, even ones such as this, which could be read as benevolent (the restoration of the archive, not the manner in which the archive was produced). A persistent undercurrent in the film is the challenge to the view of this "lost" colonial history.

In contrast to the found footage material, many of the contemporary shots in *The Raven and the Seagull* are static, as if the stationary camera is observing what is taking place—yet these "observations" are called into question through the many obviously self-conscious and self-reflexive stagings and character performances that take place within the scenes. These quasi-observational shots and staged performances are never explained or contextualized in terms of locations or agency; it is up to the viewer to discern which reality (or misrepresentation of reality) is on screen, as the film juxtaposes them in turn. While there are formal juxtapositions between one shot and another, there are also juxtapositions within the mise-en-scène itself. In one such example, a pink plastic pig on wheels continues to pop up in the Greenland scenes, much like the recurring prop of the red amp in *Sumé.* These internal juxtapositions place *The Raven and the Seagull* outside the context of Indigenous Greenlandic storytelling; instead, the film is functioning within the paradigms of high-modernist European art cinemas.

There are other kinds of internal location juxtapositions in *The Raven and the Seagull,* which focus on characters. The most significant are roles played by several well-known Greenlandic actors, who are listed as "themself" in the IMDB list credits and simply as "actors" on Lau's promotional website. These include Angunnguaq Larsen, known for his role as the fictitious Greenlandic prime minister in the first three seasons of the Danish television serial *Borgen* (Adam Price, Denmark, 2010–22), whose character here, in a shirt and tie, talks of going *"qivittoq"* in Copenhagen, and how Danish tourists are seen as *qivittoq* upon their arrival in Greenland. He describes this while sitting in a kayak in downtown Copenhagen by North Atlantic House—which, among other things, houses the permanent representation of Greenland and is situated beside the Greenlandic Trading Square, which was the center for trade with Greenland for two hundred years—and then paddles off in the Copenhagen waterfront. This kayak scene also gestures to the fact that many historical Danish films about Greenland begin at North Atlantic House.

Other actors playing some version of "themself" include the internationally recognized Nukâka Coster-Waldau, who portrays a stereotypical—in the eyes of Danes, we are to assume—drunken Greenlander protesting Danish colonialism outside the Parliament buildings in Copenhagen. We see her again later, in the Knud Rasmussen exhibition at the National Museum of Denmark, all the while calling out the tenets of the Jante Law (*Jantelagen*), which includes rule 6: "You're not to think *you* are more important than *we* are." Derived from Dano-Norwegian Aksel Sandemose's satirical novel *A Fugitive Crosses His Tracks* (*En flyktning krysser sitt spor,* 1933), the Jante Law is often understood as shorthand for a particular mode of Scandinavian sociological and psychological repression and suppression that underpins social structures and (anti-)hierarchies. Coster-Waldau's presence in the National Museum's Rasmussen room also resonates with the interventionist, activist, retrospective screening of Pia Arke's *Arctic Hysteria* (Greenland, 1996) in that space in 2010 (see MacKenzie and Stenport 2019b).

Also playing "themself" is Vivi Nielsen, a lead in *Heart of Light* (*Lysets hjerte,* Jacob Grønlykke, Denmark, 1998), the first fiction feature film shot entirely in Greenland. In *The Raven and the Seagull,* Nielsen plays a Greenlander, we are invited to assume, forcibly displaced to Denmark. In her sequence, she calls out "Rasmussen" as Angunnguaq Larsen lands on Denmark's sandy shore in his kayak, dressed like Knud Rasmussen—just as

FIGURE 8.2. Greenlanders challenging Danish perceptions of Greenland in *The Raven and the Seagull* (*Lykkelænder,* Lasse Lau, Denmark/Greenland, 2018).

she did to the male lead actor in *Heart of Light,* in a nod to the many stereotypes about Greenlanders propagated in and through that film. Other actors include Makka Kleist, who performs a drum dance in hospitals and other institutions, dressed in a costume like that worn by Asta Nielsen in *The Eskimo Baby.* In another sequence, Kleist performs a soliloquy about the thin surface lining of Christianity on the bodies of Greenlanders. Yet another scene shot in Greenland features a young artist wearing only white and remarking that a new history of Greenland can be written upon her clothing. These scenes blend and ironically comment upon many of the ethnographic documentary tropes known in Arctic cinemas, including the restaging and reenactment of "Indigenous" or other identities that are central to both *Nanook of the North* (Robert Flaherty, USA, 1922) and Knud Rasmussen's contributions to *The Wedding of Palo* (*Palos brudefærd,* Frederick Dalsheim, Denmark, 1934). The staged performances by Greenlandic actors thus directly relate, in many instances, not only to performativity in general, but explicitly to the history of Greenlandic representation in film history. What *The Raven and the Seagull* seeks to do, in the context of the GRC, is to offer a counter-history of Greenland that stands in opposition to the one written and imagined on screen by Denmark, thereby creating a space where Greenlanders can both tell and represent their history as their own (figure 8.2).

GREENLANDIC GLOBAL CULTURE AND DENMARK'S VIEW OF GREENLAND (AND VICE VERSA)

Kirsten Thisted has argued in several publications that Greenland's engagement with globalization and global popular culture is a distinguishing aspect not only of the self-determination movement, but also of the government's priorities, developed in close collaboration with the country's private sector (see Thisted 2013, 243–46). These efforts have often included a clever and ironic play on words, such as adopting the slogan "Pioneering Greenland" as an organizing principle for the "Branding Greenland" campaign, playing both on the Indigenous foundations of Greenlandic culture and on futurity- and economic-development-oriented priorities that are central to the national independence discourse. These discourses and imaginaries play a recurrent role in *The Raven and the Seagull*.

The film proposes that the center of global culture is not the Danish or European nation-state, or the United States, but rather Greenland. Transnational sports, such as Taekwon-Do and soccer, feature prominently. These sports point to a recurring theme about global citizenship and challenge an outsider's assumption that Greenland might not actively participate in global culture formation. Young Greenlanders ridicule comments made by Japanese tourists that they didn't know that Greenland had iPhone 6s access. Another group of Greenlandic youth discuss the lyrics one of them has written for an anti-Danish hip-hop song. The film then cuts to the finished version of the song booming out, as we see an image of a Greenlandic harbor, with a military boat patrolling, juxtaposing the Greenlandic landscape with Denmark's continued power over foreign affairs and international security. The film then cuts to the young boy recording the lyrics, foregrounding how his voice has been remediated and amplified through the final production of the song just heard. This act of portraying media sovereignty in the making is yet another example of the reclaiming of one's history and narrative in the GRC context.

While Greenlanders may be interconnected with global culture, Danes are, in *The Raven and the Seagull,* presented as in no way connected to Greenland. Danish tourists, for example, are positioned as foreigners abroad. They are shown walking around towns and villages, mostly shot from afar; this framing and the distance from the camera make them seem docile. Coming off cruise ships, they wear blue raincoats as a kind of uniform—provided by the cruise line to protect against inclement weather—and are portrayed visiting churches and cemeteries. Contemporary Greenlandic culture is of no interest to them. Indeed, at times they choose not to "see" Greenland at all. On a cruise ship, two women don't look at scenery passing by, but instead stare at devices. A man peers through a telescope; another man is on a stationary bike in the gym. These Danish tourists are not represented as if they see themselves as "at home"; they are very much "abroad," and Greenland is not a constituent part of their Danish imaginary. Again, this foregrounds the ways in which Danes cannot—or refuse to—see Greenlandic life and history except through their own "benevolent" colonial lens. In the context of the GRC, what is significant here is the need for Greenlanders to claim their own stories through media sovereignty, as the Danes themselves create histories and images of Greenland that leave Greenlanders outside the frame.

The GRC Reveal: Irony and Empathy

As interventionist historiography, *The Raven and the Seagull* is concerned, within the frame of the GRC context, with remediating, reenacting, and performing an alternative to the Danish grand narrative about Greenland and Greenlanders. The film does so in several different ways. One is by problematizing the status of Lau as director, whose role oscillates between that of a colonial master and a postcolonial subject; second, by having Greenlandic actors "play themselves" or play in relation to Greenlandic film roles that made them well known in Greenland and Denmark; third, by restaging and reenacting stereotypical "Indigenous" or colonizer and tourist identities; and fourth, by explicitly repurposing Danish and European film and media history about Greenland. A significant component of this aspect is the "big reveal" moment at the end of the documentary, where Angunnguaq Larsen and Vivi Nielsen meet in a Copenhagen movie theater; we see them both watching Asta Nielsen perform the "uncivilized" Greenlander on the stage of a university or a science academy in *The Eskimo Baby.* The two chuckle and joke about how the (unnamed) director of *The Raven and the Seagull* had them both revisit roles they had played before, and how ironic this gesture felt to them. These sly intertextual, metatextual, ironic, and self-reflexive approaches are part of the film's project of historical activism and interventionist historiography. They also complicate its gestures to an appropriation of media sovereignty, given that Lau, in fact, is Danish.

The Raven and the Seagull's use of interventionist historiography is different from that of *Sumé*, which strives to tell a story to Greenland, to Denmark, and to an imagined global audience, welcoming viewers into the work through idioms with which they are familiar: the realist documentary and the rockumentary. *The Raven and the Seagull* is in a different register, not offering "footnotes" or realist transparency, not explaining the past and the present through a profusion of talking heads. The use of found and archival footage is not deployed to transparently illustrate the past. The film, then, addresses the audience with an implied message: you must either already know some of this, or want to know, for the documentary to make sense. Unlike *Sumé*'s filmmakers, Lau posits it as entirely appropriate if some knowledge in *The Raven and the Seagull* is opaque to some and not to others, which is another salient feature of its high-modernist political aesthetic. This approach might also be seen as the privilege of the colonizer's perspective. The film's integration of multiple stories, characters, locations, languages, and media forms—as well as the Greenlandic governmental funding sources and active participation by Greenlandic artists and intellectuals in the filmmaking process—address many of the key priorities of the GRC. These include the emphasis on telling new histories of Greenland that are grounded in multivalent experiences of colonialism and self-determination combined with the integration of and reconciliation between different Greenlandic perspectives.

As GRC-oriented films, *Sumé* and *The Raven and the Seagull* reflect the notion that reconciliation is a process that happens both internally and on a global stage. These works also emphasize that global media and popular culture are important components of any process of truth and reconciliation. The different strategies used in the two films point to the various modes of interventionist historiography that come into play in these processes, a key aspect of which is recognition of both internal dynamics (e.g., the oblique references in *The Raven and the Seagull*) and global visibility (e.g., the band Sumé's history having relevance to global popular culture). The ongoing attention to truth and reconciliation in the global circumpolar North is thus becoming constitutive aspects of media sovereignty.

NINE

Russia's Contemporary Arctic Cinema as Geopolitics

THE ARCTIC HAS PLAYED a role in the national imaginaries of Russia and the USSR distinct from that of any other Arctic country. Unlike northern Scandinavia and North America, the USSR began its mass resource extraction and industrialization of the Arctic in the mid-1930s (for a contemporaneous account, see Smolka 1937). This increased industrialization was accompanied by a surge in Arctic exploration—part of building the national mythology of the Soviet Union. The Soviet Far North was a central component in Stalinist collectivist planning, from mining to concentration camps. Therefore, unlike the Arctic in Western Europe and North America, the Soviet Arctic became a patchwork of industrialized and urbanized space surrounded by vast expanses of wilderness. This process, like so many other aspects of the USSR's collectivization project, led to the expropriation of—in this case, Indigenous—*kulak* land and livestock (Slezkine 1994, 198–99). The Soviet policy, broadcast through the state propaganda apparatus, was to conquer the Far North (Hønneland 2020, 35), a goal made up of equal parts actual colonialism and fantasies regarding Soviet technological mastery.

After the fall of the USSR, many of the military sites, mines, and other industrial infrastructure in the Russian Arctic were abandoned and left to decay. Over the following decade, as resource extraction and geopolitical concerns again became prominent, Vladimir Putin's government made the Arctic a renewed priority (Laruelle 2014, 4; see also Lagutina 2019, 10). Russia's national narrative also changed, because of the loss of USSR territory from Ukraine, Belarus, and the Baltic States to Kazakhstan and Uzbekistan

All translations from Russian in this chapter are by Deirdre Ruscitti Harshman.

(the longer-term effects of this motivated Russia's 2022 illegal invasion of Ukraine, based in no small part on Putin's sense of loss of this quasi-mythological greater Russia).

The sanctioned national imaginary of a Russia that was incomparably vast turned toward re-expansion in the North in the post-Soviet era. As Lilya Kaganovsky notes, this prioritized view of the Arctic has been a component of Russian media for nearly a century. She argues that the

> Russian North has always represented both the heart and soul and the *other* of Russia. It has had to be perpetually colonized, and yet it remains, at once, the bulk of the country and an unassimilated, uncivilized, unconquered space. That may be one of the reasons why the Arctic/Russian North has also been making an appearance in contemporary cinema, both documentary and feature films. . . . These films illustrate how the Arctic region has once again been mobilized for political purposes. (Kaganovsky 2017, 181; see also Lajus 2013, 110–13)

Unlike European and North American cinematic representations of the Arctic from the 1920s to the 1940s, the key films made in and about the Soviet Arctic envision an already industrialized world interconnected with other regions.

Dziga Vertov's *A Sixth Part of the World* (*Shestaya Chast Mira,* USSR, 1926), shot by remote camera teams throughout the USSR, documents the country's rich agricultural and livestock resources, while also serving as a paean to incipient industrialization (for more, see Kaganovsky 2019, 46–67). Vertov uses this display of the new USSR's diversity—both cultural and economic—to argue for "complete socialism" in contrast to the decadent West, which he depicts at the beginning of the film with a Black minstrel show. In the Arctic, his team documented the Samoyedic people on Novaya Zemlya Island in the Barents Sea, linking them via montage with citizens from Bukhara to Leningrad and, in the process, creating an imaginary Soviet people. *Turksib* (Viktor Alexandrovitsh Turin, USSR, 1929) documents the construction of the Turkestan-Siberian Railway, delineating how the remote regions of the USSR were becoming latticed together through industrialization and nascent urbanization. What both these documentaries portray is how mechanization (including the mechanization of the camera) led to the interconnectedness of the Soviet Union's disparate corners, including the Arctic.

Vertov's and Turin's Russian Constructivist cinema gave way to Stalinist socialist realist films, many dealing with Arctic explorers, such as *Conquerors* of the Night (*Pobediteli nochi,* Adolf Minkin and Igor Sorokhtin, USSR, 1932) and *Chelyuskin: Heroes of the Arctic* (*Geroi Arktiki, Chelyuskin,* Yakov Poselsky, USSR, 1934). These works focused on "great" men accomplishing "great" things for the Soviet state. In the 1950s and '60s, melodramatic historical narratives came to the fore alongside the rise of an ethnographic film movement documenting Arctic Indigenous peoples. Films such as *Boarding School for the Peoples of the North* (*Internat dlia detei narodov Severa,* USSR, 1958), *The Tale of the Komi Land* (*Skaz o zemle Komi,* Yuri Mogilevtsev, USSR, 1967), and *Feast of the North* (*Prazdnik Severa,* Rafail Gol'din, USSR, 1970) (see Sarkisova 2019, 231–53) mobilized ethnographic practices to again champion an expansive Soviet state and its national myths. These Soviet-era works differ greatly from those that followed the fall of the USSR. While Soviet-era films—whether constructivist, socialist realist, melodramatic, or ethnographic—tell stories that draw on various iterations of Soviet (and most often masculinist) heroic myths, the Russian Arctic works made after the fall of the USSR often display various kinds of human and material detritus.

This chapter examines films, media works, and media events that address the Russian Arctic in the post-Soviet, Putin era—a body of works less stylistically unified than those of the Soviet period. However, as a subset within the overall project of building a new cultural and political imaginary, with the Arctic reincorporated into the national and international imaginary of contemporary Russia, these films remain haunted by the Arctic's role in the Soviet past, especially during the Cold War. The chapter addresses how state and quasi-state organizations use moving images to reclaim the Arctic as a significant aspect of contemporary Russia's political context.

CLAIMING THE NORTH POLE? THE TWENTY-FIRST-CENTURY RUSSIAN ARCTIC MEDIA IMAGINARY AND COLD WAR IMAGINARIES

We begin by addressing film and media production in the period following the largest Russian Arctic media event in recent history: the planting of a Russian flag on the seabed of the North Pole in August 2007 by explorer and scientist Artur Chilingarov, a public relations coup that elicited a storm of news coverage (figure 9.1). Chilingarov's stunt took place on an expedition dive with a submersible; it was connected to a research expedition funded by a Swedish philanthropist and organized by a private group. Neither an offi-

FIGURE 9.1. Artur Chilingarov planting the Russian flag on the seabed at the North Pole, August 2007.

cial government-sanctioned action nor connected to the Russian military and its fleet of nuclear submarines, the stunt made Chilingarov Russia's Arctic spokesperson, as a profile published in *Science* confirms: "Little known outside his homeland at the time, Chilingarov ... was transformed into Russia's public mouthpiece on the Arctic. Vladimir Putin, then president, awarded him the Hero of Russia medal, to add to the Hero of the Soviet Union gong he earned for organizing the rescue of a ship stuck in polar ice in the Arctic Ocean under the Pole" (Parfitt 2009, 1382, 1384). Chilingarov's "hero" status runs through the contemporary history of the Russian Arctic imaginary and many of the works discussed in this chapter.

The Chilingarov event catalyzed several domestic and foreign meta-narratives about Russia's relationship to its northern regions and the global circumpolar North, drawing on polar explorer visual imagery, Cold War militarization rhetoric, Russian national identity narratives, and discourses related to the contemporary climate crisis. This congeries of discourses prompted a surge of anxiety internationally concerning Russia's geopolitical intentions in the High Arctic. Furthermore, scientists reported record-breaking loss of sea-ice cover on the Arctic Ocean in the early fall of 2007, and this became a global media event (see Christensen 2013; Wormbs 2013), indicating another aspect of the pervasive anxiety about geopolitical and environmental outcomes in the Arctic, especially the increasing visibility of climate change.

At this time, terms such as *New Cold War, Arctic Meltdown,* and *Arctic Cold War* proliferated in the media (Antrim 2011, 107). Predictions of a largely ice-free Arctic Ocean affording passage over the North Pole and burgeoning resource extraction opportunities, alongside lingering Cold War geopolitical tension over rights and access to the northernmost parts of the world, soon came to a head. *Time* magazine called this confluence of events the "Fight for the Top of the World" (Graff 2007), while others have called it the "scramble for the poles" (Dodds and Nuttall 2015). A phrase in Russian referring to the coming of "a new 'Ice Cold War'" also began to circulate at the time (Laruelle 2014, 113).

The 2007 flag planting had different connotations internationally and domestically. The *Financial Times,* the *New York Times,* and several other international media outlets wrote extensively about Chilingarov's action in ways that further linked it to Cold War paradigms; most Arctic studies scholars also designate it a significant moment, with many noting its "new Cold War" ambiguity (Hønneland 2020, 54–58; Dodds and Nuttall 2015, 80). In fact, Russia adopted its first comprehensive Arctic strategy in 2008, just a few months after the flag-planting incident. However, the country's foreign policy, security, and geopolitical priorities emphasized transnational economic cooperation, the general modus operandi in the Arctic since the end of the Cold War. This policy does not, in fact, contradict Russia's adherence to the principles of UNCLOS (UN Convention on the Law of the Sea) or motivate military action in the Arctic beyond posturing (see Laruelle 2014; Lagutina 2019; Hønneland 2020).

Multiple international documentaries featured the 2007 event and "new Cold War" geopolitical associations it provoked. For instance, the CBC/Channel 4 documentary *The Battle for the Arctic* (Canada/UK, 2009), though ostensibly attempting a realist, "balanced" approach, represents the Russians, and Chilingarov in particular, as staging a belligerent spectacle, while the Canadians, in contrast, seek to prove their claims to the North Pole through science, augmented by the "objectivity" of peer review. *Arctic Circle: The Battle for the Pole* (Wally Longul, Atsushi Nishida, Takashi Shibasaki, and Yoichiro Yamamoto, Canada/Japan, 2009) is an international coproduction that takes a similarly oppositional stance, emphasizing climate change research in its first part and Russian and Norwegian oil-extraction politics in part two. Another film, the Danish documentary *Turf War in No Man's*

Land (Suvi Andrea Helminen, Denmark, 2009), deploys talking-head interviews intercut with climate change findings, resource extraction assessments, and geopolitical statements in voiceover, with Danes claiming a few weeks after Chilingarov's flag planting that the North Pole is actually Denmark's. This existentialist road movie—interspersed with Helminen's ruminations on her relationship with the sublime Arctic landscape—follows a Danish research group on a Swedish icebreaker trying to identify possible Greenland-extending underwater ridges ranging to the North Pole. Narrative tension hinges on the question of whether a high-efficiency, nuclear-powered Russian icebreaker will aid this attempt at geographical mapping. The Russians cooperate, to some extent, before withdrawing their assistance without explanation, with nationalist protectionism suggested as a reason.

If *Turf War in No Man's Land* is a form of road movie, Maarten van Rouveroy's *Black Ice* (The Netherlands, 2014) is a didactically activist work. The film chronicles a group of Greenpeace activists who tried, in the fall of 2013, to scale the *Prirazlomnaya* drilling platform, located in international waters in the Barents Sea (Vidal 2013). Greenpeace's actions result in their ship, the *Sunrise,* being seized by Russian authorities, making the Russians the unequivocal villains. The so-called Arctic 30 (an international Greenpeace crew hailing from eighteen countries) were subsequently arrested by the Russian coast guard (who drew weapons on the unarmed protesters), charged with piracy, and jailed in Murmansk. *Black Ice,* an activist documentary unbounded by nation-state interests, does not attempt to find a balance between environmentalism and resource extraction. Instead, the Arctic 30's story takes a critical position on nation-state investments in the hydrocarbon industry—in this case, those of the Russians in the Arctic. *Black Ice* delineates the fact that the stakes of resource extraction and climate change are not and should not be bounded solely by the politics of individual nation-states.

Geir Hønneland argues that Russian domestic reporting of the North Pole flag planting in 2007 correlates with ambivalent and ambiguous media constructions of the Arctic in post-Soviet Russia, casting it both as a region that has been "'forgotten' (economically neglected and politically marginalized)" and as Russia's "'future' (the country's most important reserve, economically and spiritually)" (Hønneland 2020, 58; see also Gritsenko 2016; Rowe 2013). In terms of Russian domestic politics, Cold War imaginaries may continue to inform the agenda, but the focus since 2007 has been on increasing fossil fuel resource extraction, supporting a more diversified economy in the North, and expanding industrial and transportation infrastructure. These aspects have been consistently emphasized in the 2008, 2013, and 2020 Russian Arctic strategic plans (Lagutina 2019).

The 2007 flag planting also resonates with a notion of the Arctic as a theater of military tension reaching back to the Cold War. Long-standing "geographical meta-narratives" have carried over from Soviet Cold War ideology into contemporary Russian identity formation, Laruelle argues, since "Russia's territory is *larger* than other countries in the world and forms a specific continent (Eurasianism); Russia is going *higher* in the universe (Cosmism); and Russia is going *farther* in the north (Arctic mythology)" (2014, 39). Yet it is important to note that these narratives are dynamic, reflecting geopolitical and domestic priorities of the moment. An extensive analysis of Russian media coverage in the period 2013–16 shows that "Russia's official narrative on the Arctic has changed several times over the past decade and a half. In the first half of the 2000s, its Arctic policy followed the foreign policy principles of Putin's first two terms as president, particularly the ideas of reviving Russia as a great power, restoring its military might and maintaining its status as an energy superpower" (Klimenko, Nilsson, and Christensen 2019, 3–4, 28). These preexisting, conflicting, complementary, and ambiguous narratives are evident in domestic and foreign media coverage about Chilingarov and the 2007 flag planting while also informing cinematic production and the circulation of media and films about the Arctic—documentary as well as fiction—made in Russia during the past two decades.

The close alignment of media censorship with state interests is a known factor in Russian civic life. During Putin's regime, the media has been "seemingly browbeaten into subservience to the president's authoritarian, patriotic agenda," such that "television is a 'totalitarian dream'" (Beumers, Hutchings, and Rulyova 2009, 1, 6). As scholars note, television and cinema are where "state efforts to control public opinion are most pronounced" to reaffirm "a continuous narrative of a 'Great Russia' . . . constantly under threat from domestic and foreign enemies" (Wijermars 2016, 1). Indeed, some argue that federal Russian television in the Putin age "is reminiscent of Soviet television of the 1970s," in the ways it "delivers official state messages with various degrees of professionalism" (Fossato 2006, 4). We note this aspect of contemporary Russian media, television, and cinema censorship and politics because they correlate with a long tradition of representing the Arctic in Soviet and Russian media.

These sets of constraints are evident in twenty-first-century media reporting about the Arctic: "Between the Kremlin's media hype concerning the 'Arctic Race' and the articles of experts published for limited circulation in specialized academic journals, the public does not have much to read" (Laruelle 2014, 12). The myth of the Russian Arctic is therefore large and pervasive, but with only a limited, and tightly controlled, body of discourses feeding this imaginary. This lack of material points to the fact that much of what is available is not only in line ideologically with the state, but—like any effective propaganda—repeats its points over and over again, as evident in images of Chilingarov, the North Pole flag-planting event, and the near-complete absence of Indigenous representation and coverage in contemporary Russian film and media.

INDIGENOUS CINEMA AND MEDIA IN CONTEMPORARY RUSSIA

As has often been the case in the Arctic, narratives of conquest, extraction, transnational economic cooperation, and international science and technology collaboration (however ambivalent or ambiguous they may be) tend to effectively obscure and marginalize Indigenous populations in the High Arctic. Twenty-first-century Russian media discourse and national identity narratives concerning the Arctic mostly follow that route, as does a significant amount of the scholarship about Russia's Arctic policies and engagements. In previous chapters, we examined Indigenous media in the North, films about Indigenous issues made by non-Indigenous practitioners, and the circulation of Indigenous media within the global circumpolar North through film festivals such as Skábmagovat and imagineNATIVE. We foregrounded the rise of Indigenous media sovereignty in relation to self-governance and self-determination, elaborating on Faye Ginsburg's influential concept (2016) and relating this to issues of sovereignty and self-determination, alongside the rise of creative hubs such as the Arnait and Isuma collectives, FILM.GL, and the ISFI. To date, there are neither similar movements nor cinematic production of equivalent cultural, social, or political impact in Russia. Funding for Indigenous production is exceedingly difficult to access, and Indigenous filmmaker practitioner networks are small and scattered. Similarly, Russian domestic media tends to disregard the perspectives of Russian Indigenous populations in the Arctic almost entirely (see Klimenko, Nilsson, and Christensen 2019).

The Indigenous history of Russia's Arctic is the story of twenty-six distinct population groups—"The Small Peoples of the North"—who account for about one-quarter of the population of two million in the Arctic Zone of the Russian Federation, according to the national census of 2010 (Lagutina 2019, 22). Implemented in the 1920s by government agents, "the imperial practice of placing the circumpolar hunters and gatherers in a separate category was never questioned" (Slezkine 1994, 1). At no time have Russia's Arctic Indigenous populations possessed unequivocal legal protection of their ancestral lands. During the Soviet era, reindeer-herding grounds were increasingly overtaken by resource extraction facilities and the industrial and transportation infrastructure necessary for refining and transporting Russia's wealth of rare-earth minerals, timber, oil, and gas. The manpower needed for Arctic resource extraction largely consisted of forced labor until Stalin's death in 1953; after that, workers from all over the USSR were incentivized to move north, although these enormous enterprises were perpetually short staffed. Seen as "irrelevant to the industrialization effort," Yuri Slezkine writes, "most aboriginal groups were an irritating distraction and a financial drain, all the more exasperating because their 'inefficiency' seemed to be the result of sheer obstinacy" (1994, 339, 340).

As in most other Arctic Indigenous regions during the second half of the twentieth century, traditional cultures and livelihoods based on nomadism, reindeer herding, or seal and walrus hunting were severely disrupted. Massive relocations into "urban" centers and prefab housing, abandonment of traditional settlements and occupations, and the suppression of native languages and customs ensued, as did the compulsory removal of children to state-run boarding schools. To give one example, by the 1980s, only one Sámi village remained officially classified as Indigenous on the Kola Peninsula, though its population was almost entirely non-Native (1994, 340).

Most Indigenous Arctic Russian films made over the past two decades—and only a small number exist (see Sarkisova 2019)—are ethnographic and observational, whether narrative or documentary. The award-winning *The Lord Eagle* (*Toyon Kyyl,* Russia, 2018) by Eduard Novikov of Yakutsk falls partly into this category. Set in the 1930s, the film follows a season in the lives of an elderly couple whose isolated home on the northern taiga is visited by an eagle, whom they do not wish to disturb because of its magical qualities. Novikov is part of an emerging group of Indigenous filmmakers connected to Sakhafilm in Yakutsk, one of the fastest-growing film production hubs in Russia, despite its isolated Siberian location in one of the coldest regions on earth (see Haynes and Roache 2020; McGinity-Peebles 2021). The film, while devoid of any social criticism or commentary on post-1930s environmental destruction, screened at the Moscow International Film Festival and won the audience award at imagineNATIVE in 2019.

Production in Sakha has grown at such a rate (at least twenty-two films over the past few years) that the portmanteau *Sakhawood* has emerged. These works are gaining global recognition: for example, the 2017 Busan International Film Festival held a retrospective entitled "Sakha Cinema: World of Magical Nature and Myth." This popularity arises in part from the intersectional nature of the work produced, as Adelaide McGinity-Peebles notes: "Sakha cinema is at the intersection of multiple identities and film cultures—Asian, Arctic, Russian, Indigenous—so it is able to appeal to diverse audiences and film networks" (McGinity-Peebles 2021). Sakhawood's partnerships with Arctic and European film networks have been impacted by the Russian invasion of Ukraine, eliminating or drastically diminishing the opportunity for collaboration and festival screenings.

Another trend is exemplified by the work of director Aleksei Vakhrushev, a Chukchi from Anadyr, Chukotka. Over the course of a substantial number of documentaries, Vakhrushev has filmed the practices, environments, and beliefs of the Chukchi and Inuit Yup'ik peoples, focusing in *The Book of the Sea* (Russia, 2018) on sea mammal hunting in the Bering Strait. Like Alethea Arnaquq-Baril in *Angry Inuk* (Canada, 2016), Vakrushev combines documentary footage and experimental animation techniques. In this case, clay animation is used to illustrate the pan-Inuit mythology that connects life on land with aquatic life, humans with animals, and spiritualism with twenty-first-century communication technology. Like *The Tundra Book: A Tale of Vukvukai the Little Rock* (*Kniga Tundry, Povest'o Vukvukae malen'kom kamne,* Russia, 2011), Vakhrushev's best-known film to date, *The Book of the Sea* has received international distribution and inclusion in many documentary and Indigenous film festivals. *The Tundra Book* is one of only a few films by Indigenous directors to examine the environmental, social, psychological, and linguistic effects of the Soviets on nomadic tribes, "addressing the traumas of forced resettlement, repressions, and collectivization that resulted in the destruction of Indigenous cultural practices" (Sarkisova 2019, 237), including the state policy of forced relocation of children to government schools.

Resource extraction has generated pockets of wealth among Indigenous populations in Russia's Arctic, but this is unevenly distributed and tends to be concentrated among those close to industry or the administration of the *okrug* (region). Opportunities for graft and corruption are plentiful. Although companies are required by law to compensate herders for expropriated land, "in-kind" or paternalistic arrangements abound, with herders reporting an increase in land appropriation (Tysiachniouk, Petrov, and Gassiy 2020, 13–16). The Yamalo-Nenets Autonomous Okrug, home of the Nenets people, is one of the few Indigenous Russian regions to maintain a strong connection to nomadic reindeer herding. In contrast to many other Arctic regions, the number of herding nomads in Yamal increased during the period 2003–19, from 13,300 to 16,300, though access to land remains an issue (Magomedov 2019b). A Yamal journalist reports: "We do not know the inner workings of how the fuel and energy complex acquires those tribal lands, via a tender or an auction. . . . There is a deficit of land to use, but where is the land? That is the question" (cited in Magomedov 2019b). Other activists allege that the reconciliation process has stalled: "Resolution of the land issues in favor of the small indigenous populations of the North has practically stopped. . . . We want . . . to become equal participants in all sociopolitical and ethnocultural processes in the country" (cited in Magomedov 2019b). These land-claim conflicts are not unexpected given Russia's interest in furthering resource extraction and industrial installation across the entire northern part of the country.

The Yamal Peninsula Indigenous social media activist group Voice of the Tundra (Golos tundry) formed in response to a lack of land for reindeer and the marginalization of several Indigenous leaders (Magomedov 2019b). Founded by Eiko Serotetto, the group is led by a young reindeer herder and not by the typical activists and professionals so often found in such positions of leadership in urban centers (Magomedov 2019b). Arbakhan Magomedov outlines how this group of "rural protestors" uses VKontakte (VK)—the Russian social media platform run out of St. Petersburg, whose five hundred million accounts make it the most popular website in Russia—as a ground-up protest platform. The Nenets, then, use the largest media network in Russia to organize among themselves and make their voices heard within the broader, mostly state-controlled, Russian public sphere. Moreover, the 2013 arrest of RAIPON (Russian Association of Indigenous Peoples of the North) leader Dmitry Berezhkov in Norway, at the behest of the Russian Federation, and RAIPON's subsequent restructuring have, in aggregate, led to the acknowledgment that the government places oil extraction above all local community needs (Magomedov 2019a). Voice of the Tundra serves an important role in mobilizing new Indigenous political voices.

Voice of the Tundra began as a social media petition to the Russian president, arguing for the stable maintenance of reindeer herd numbers, before subsequently growing as a voice publicizing the Nenets' concerns. As Magomedov argues, Voice of the Tundra "poses interesting questions about the intersectional nature of resistance and self-awareness" (2019a); we can infer from this a realization of resistance being possible beyond the level of the individual. At the same time, Magomedov's informants feared repercussions, including being branded as foreign agents by the state (2019a). Moreover, the magnification of a Nenets collective voice by appropriating a dominant media platform instantiates the making of a minority group—in this case the Nenets—in more visible ways than is typically possible in more monological or unidirectional forms of activist media practice.

New film festivals also play a part in an emergent Russian Indigenous counternarrative. For instance, the Arctic International Film Festival "Golden Raven," held annually in the Chukotka Autonomous Region, shows a wide array of work produced both in the Arctic and globally. However, per its mandate, this festival is focused specifically on topics relevant to the Arctic, including ecology, interethnic relations, generational continuity, preserving Indigenous tradition and culture, and preserving and popularizing the cultural heritage of Indigenous peoples in the North and Far East. Along with promoting tourism to the spectacular Chukotka Autonomous Region, the festival searches out and supports talented young filmmakers working in the Arctic region and helps promote the development of regional film-production infrastructure in general. The festival functions as a global conduit for exchange through the "creation of a communication channel and strengthening of cooperation between Russia and the world community through the territories and cultural space of the Far North of the Russian Federation" ("Golden Raven" 2019), coprogramming with imagineNATIVE and Skábmagovat to showcase work from around the global circumpolar Arctic to build community and strengthen practitioner networks.

Public and state-organized events include the recurring science, business, and cultural festival called Arctic Days, launched in 2010. In its earliest versions, the festival deployed film and media to advance the interests of the state, emphasizing resource extraction, exploration, and settlement, more than Indigenous land claims and cultural rights or the impact of fossil fuels on global warming. As a constellation, these different emphases portray a diverse cinematic context that is closely connected to the mediatized construction of environmental protection matters, while also revealing divergent narratives, especially in regard to the emerging (or at least partial) media sovereignty of Indigenous groups. In so doing, the case studies of this chapter illustrate a non-monolithic portrayal of Russia's Arctic Region and media-tized political interests, which also provides a cinematic set of connections with other Arctic regions.

Russia's evolving Arctic priorities can be discerned from its decade-long investment in the Arctic Days event, held annually in Moscow since 2010. The 2020 event was postponed because of COVID-19 and has since returned without any artistic or media content. Chilingarov was selected to chair the 2021 event (Arctic Days 2021). Here, we trace the emergence of the event and emphasize the significance of the program of the Third International Festival of Nonfiction Films during "The Arctic Days in Moscow" in 2014. Organized by the Ministry of Natural Resources and Environment of the Russian Federation, with support from a number of state-owned corporations and resource extraction companies such as Gazprom Neft (oil), Novatek (natural gas), and Noril'skii Nikel' (nickel), Arctic Days combines scientific conferences, business seminars, political speeches, and cultural events (Arctic Days n.d.). For example, in a public comment in 2018, Putin described the purpose of Arctic Days as "reinforcing our presence in the region, implementing scientific research initiatives and large programs for commercial and economic development, and turning our ambitious infrastructure projects into reality," while also noting an interest in preserving the environments and ecological diversity of the Arctic and recognizing Indigenous groups of the Far North (Kremlin 2018). The state's interest in the public-facing and PR-related aspects of Arctic Days for enhancing political power is thus an integral part of the event.

The film festival component of Arctic Days is an example of a larger political, media, and economic paradigm within which Russian representation of the Arctic must be situated. The films presented between 2010 and 2014 represent state interests, to the exclusion of other issues and stakeholders in the Russian Arctic. While Arctic Days is not generally well known outside of Russia, and certainly not in North America, it is a significant event, demonstrating the confluence of economic, political, scientific, and cultural discourses and interests in the Russian Arctic. In 2010, Putin lauded its public education and outreach campaign in the following manner:

Concerned with steady, balanced development in the Russian North, we seek to strengthen our connections with our neighbors in our common Arctic home, as we believe that preserving the Arctic as a zone of peace and cooperation is of the utmost importance. It is our conviction that the Arctic area should serve as a platform for uniting forces, for genuine partnership in economics, in the sphere of security, in science, education, and the preservation of the cultural heritage of the North. (Putin 2010)

Evoking an internationalist rhetoric that privileges collaboration in the name of science and security, Putin's proclamation simultaneously celebrates the Russian nation-state, while his oblique reference to "cultural heritage" should be understood as an acknowledgment of Russia's Indigenous populations in the North. In the past, Arctic Days paid particular attention to the Yamalo-Nenets region, collaborating to some degree with cultural offices and organizations in the capital Salekherd. This focus stems from the fact that the Yamal Peninsula, apart from being home to large Indigenous and partly nomadic reindeer-herding communities, holds the largest reserves of natural gas in Russia, with numerous pipeline, infrastructure, and resource-extraction megaprojects in place or under development (see Harding 2009). Moreover, the Yamalo-Nenets region also provided an "image" of Indigenous populations in Russia, especially the representation of reindeer herders, stretching back to cinematic representations of Russian Indigenous peoples from the early Soviet era (Sarkisova 2019, 234–36).

In his public discussion of Arctic Days in 2010, Putin signaled the centrality of film and media to the public education campaign and noted the collaboration of the Yamalo-Nenets Autonomous District of the Russian Federation (Putin 2010). These statements are part of a propaganda move, giving cover to the political priorities of Arctic Days under the guise of promoting culture. The larger state goal is clearly the normalization of resource extraction. The Arctic Days program of 2012 also emphasized Russia's strategic geopolitical and resource-extraction interests in the North, camouflaged beneath the state's carefully articulated emphasis on cultural heritage, the Yamal region and Nenets group, and the mobilization of cinema and media. The PR for the 2012 festival emphasized how the "unique" event gathered nonfiction and animated films dedicated to the Arctic "in order to reinforce cooperation and development of the new forms of interaction between the Arctic countries, to facilitate cultural exchanges in the Arctic region," and so the festival "is called to become an indicator of a reasonable and careful attitude to this unique area" (Arctic Days 2012). Arctic Days' 2012 edition was also notable for including an event entitled "Yamal Days" that foregrounded Nenets culture and reflected the PR material's emphasis on collaboration and cooperation. Yet, that same year, the Russian government's hostile stance toward Indigenous groups reached the point that it closed down RAIPON (see Tarasov 2012). As mentioned above, former RAIPON vice president Dmitry Berezhkov was arrested in Tromsø in June 2013, on a Russian extradition warrant, before being released after Norway found that no crime had been committed. Russia's hostility toward RAIPON is indicative of the country's undercutting of Indigenous peoples' organizing in the North. *Nunatsiaq News* reports: "Russia's move to suspend RAIPON's activities on the grounds that its activities are 'illegal' [is] receiving criticism from Canada, Norway, Greenland, and international non-governmental organizations, including the Arctic Council, which also asked for RAIPON's reinstatement" (*Nunatsiaq News* 2013). The ban on RAIPON was removed, but only after a new leadership election—with pressure from Moscow. The reality of Russia's disposition toward Indigenous community organizing is clearly at odds with Putin's public celebration of the Yamalo-Nenets in his 2012 pronouncements about the festival.

In 2014, the public forum about the Arctic was conjoined with a newly created polar holiday the previous May, after Putin issued an executive order turning an existing holiday in some parts of northern Russia into an annual Polar Explorers' Day, or *den' poliarnika:* "The executive order establishes the Polar Explorers' Day which will be celebrated on May 21. The first drifting station North Pole-1 was set up on this day, May 21, seventy-six years ago," in 1937 (Putin 2013). Tying into this announcement, the 2014 public forum in Moscow included photo and art exhibits about polar explorers, several film screenings, and a workshop of documentary and polar expedition filmmakers called Dialogue—Arctic Documentary: Development, Problems, Priorities. The establishment of a new holiday in recognition of Russia's polar explorers, and a program in arts and cinema organized with its commemoration in mind, signals the close ties that have evolved in government-funded initiatives seeking to link nationalist rhetoric about Russia's interests in the Arctic region with the deployment of moving-image cultures, discourses of cinephilia, and the direct inclusion of filmmakers.

The 2014 iteration of Arctic Days deserves attention due to the significance placed on film and media, especially given the decreased emphasis on cinematic representation in the programs released for 2016, 2018, 2020, and 2021. The largest and most ambitious of the Arctic film festivals in Russia to date, it featured over a hundred new films, with fifty-four Russian films listed in the official program (Arctic Days 2014). In contrast to earlier Arctic Days programs, all films listed in the 2014 edition were of Russian origin. Many of the works screened were medium- to long-form documentaries made for broadcast on news channels, often the state-owned station Russia-1, although shorts were also included, some only two or three minutes long, such as several animated films about the Arctic for children. Because television documentaries and children's films around the globe do not get wide-ranging distribution deals, only a few of the films screened in 2014 have been distributed outside of Russia (though a few can be found on platforms like YouTube). Many of the films screened at the festival—like almost all film and media works made in Russia—received state support, being produced either by state-supported news channels like Rossiia-Kul'tura (Russia-Culture) or Pervyi Kanal (Channel One) or through subsidies provided by the Ministry of Culture. A substantial percentage of these films have women directors or screenwriters; in addition, several of the films were made by Indigenous filmmakers or depict Indigenous cultures. Most of the 2014 documentaries focused on the Far North, near the Arctic Sea; a few, however, were filmed farther south. This variation in geographic focus lent an amorphous image to how the Arctic has been defined, geographically and spatially, in the state-sanctioned Russian imagination.

Over thirty-five hundred people attended the Arctic Days events in 2014—among them, two thousand scientists from around Europe, who presented research on resource extraction, geology, and other related topics (Arctic Days 2014). The festival was reportedly covered by some five hundred journalists, including members of Russian and foreign media companies such as Pervyi Kanal, Russia Today, Rossiia-24, TK Zvezda, TK Moskva 24, Izvestiia, Kommersant, Vedmosti, Ekspert, Interfaks, MIA Rossiia Segodnia, Der Spiegel, Radio France, RAI-Radiotelevisione Italiana, Xinhua, Al-manar, and Le Courrier de Rossie (Arctic Days 2014). State politicians also made up a strong presence, with Putin, Dmitry Medvedev, Moscow mayor Sergei Sobianin, Chilingarov, and a European representative of the United Nations Environment Programme attending. In addition to the film festival, Arctic Days in 2014 also featured, alongside the previously mentioned exhibition on polar explorers, a photography exhibition entitled *The Arctic,* documenting contemporary Arctic life, nature, science, and the everyday life of Indigenous peoples. Minister of natural resources and ecology Sergei Donskoi stated at the exhibit's opening: "We want to present the region to guests and participants of Arctic Days in all its richness, beauty, and diversity. The Arctic region is strategically important for our government—it is not only the 'resource crown' of Russia, it is a special 'place of power,' inspiring man [*sic*] with its amazing energy" (Fotovystavka "Arktika" 2014).

Donskoi's statement, echoing Putin's address at the 2010 Arctic Days launch, signaled Russia's various interests in the Arctic—from resource extraction to geopolitical dominance—and thus demonstrated the close alignment of cultural activities such as Arctic Days and its related film festival with Russian state interests in an Arctic open for investment and extraction. This interwoven relationship reflects the contemporary Russian media practices described above, in which the media is a quasi-extension of governmental discourse. Two films that screened at the 2014 Arctic Days film festival—Oleg Soldatenkov's *The Polar Prize* (*Poliarnyi priz,* Russia, 2014) and Vladimir Tumaev's *White Moss* (*Belyy yagel,* Russia, 2014)—illustrate the complexity of these priorities.

Arctic Days 2014: Polar Exploration and Indigenous Ethnography

One of the most significant films to address the reemergence of Russia's multifaceted interest in the Arctic is the Russian documentary *The Polar Prize,* made for the state-owned news channel Russia-24. Directed by Russian television documentarian Oleg Soldatenkov and shown at the Arctic Days film festival in 2014, *The Polar Prize* focuses on the contemporary international rivalry for control of land and resources, a power struggle intensified by a ways in which climate change is presented as opening up more of Arctic circumpolar coastlines for extraction, shipping, and settler habitation. The press packet from Russia-24 states that the documentary covers the "history of the international rivalry for control of the Arctic, a struggle involving about 50 countries." Beyond this history, *The Polar Prize* also covers the effects of climate change on Indigenous peoples in Russia, Greenland, and Alaska, the need for international cooperation, and the accumulation of garbage and plastics on Russia's Arctic coastline. Alongside this global focus, the documentary also addresses perceived threats felt by Russia in the Arctic, echoing Putin's speech cited above concerning the centrality of the Arctic to Russian pride and the country's national imaginary.

The Polar Prize foregrounds the fact that no country has taken military action in the Arctic, along with Russia's desire for it to be a region of peace and global cooperation—with the caveat that militarized rhetoric to defend Russian interests is needed. This points to a duality that Geir Hønneland outlines in reference to the work of Nancy Ries: "While Russians feel ashamed of their country abroad, as one of [Ries's] respondents puts it, 'within themselves they are all very proud that they are Russians, that they come from such a country, which has such a strange history'" (Hønneland 2020, 40).

Indeed, alongside the Chilingarov flag-planting media event of 2007, *The Polar Prize* is part of that very rhetoric, representing the polarity of Russia's internal (normally self-denigrating) and external (proud and strong) discourse. Indigenous accounts and interviews with Russian, Danish, Canadian, and American politicians, policymakers, academics, and explorers (including Chilingarov) comprise the bulk of *The Polar Prize.* Yet the documentary has some self-reflexive components, combining talking heads (with B-roll shots of "behind the scenes" cameras and filters) and archival and found footage, animation, images from video games, and time-lapse photography. Mobilizing these various strategies renders the film more formally complex than many earnest, talking-head environmental documentaries; these aesthetic strategies, along with the humorous use of found footage, also serve to downplay geopolitical tensions, working to undercut the sensation of watching an instance of straightforward propaganda.

Featuring detailed attention to competing Arctic claims of different countries (specifically Denmark and Greenland, the United Kingdom, Iceland, the United States, and Canada), *The Polar Prize* appears to argue that national interests are the strongest factor and that there is no unified front among the Western powers. There is some ambiguity in the way geopolitical changes in the Arctic are framed in the film. For Russia, the region is viewed as a barrier to the outside influences and powers that surround it—as one interviewee says in the first two minutes of the documentary, "the Arctic to Russians is both a shield and a dream." Thus, an intertwined narrative about national pride and external threats pervades the documentary's first half. This intertwining relates to Hønneland's typology of Russian self-narrating as satirical, tragic, or comic, as discussed above (2020, 13). All three modes are used in *The Polar Prize,* with satire and comedy present in the first half of the documentary, specifically when it deals with geopolitics; tragedy comes to the fore when *The Polar Prize* turns to global Indigeneity and climate change.

While avoiding overt discourses of militarization, *The Polar Prize* contains one tongue-in-cheek exception: an analysis of the video game *Naval*

War: Arctic Circle (2012). Its plot is based on the notion that melting ice caps have opened up the region for resource extraction. While the film states that it is a Canadian game, reinforcing Hønneland's account of Russian narratives that pit Canada as a key antagonist in the Arctic (2020, 50–51), the game was designed by the Norwegian development company Turbo Tape Games and distributed by the Swedish video game publisher Paradox Interactive. A single player can choose between adopting the perspective of either Russia or NATO. That the game can be played from the Russian perspective is not noted in the film; rather, the game is taken as evidence, through the scrim of pop culture, of the Arctic as a locus of climate change and resource extraction within the Western geopolitical imaginary.

The Polar Prize's take on this game is part of a larger narrative about the array of international interests aligned against Russia in the Arctic, which Marlene Laruelle presents "as the scene of a new race among great powers [that] makes it possible to portray Russia once again as a besieged fortress, caught in a vice-like grip, by the advance of NATO" (2014, 10). Laruelle goes on to note how Putin and Medvedev's stakes in Arctic adversarial discourses are part of a Cold War memory that pervades the contemporary Russian imaginary of the Arctic (2014, 10). To this end, the documentary also addresses Western claims about Russian activity in the Arctic, which the film portrays both as misrepresentations of its actions and as a source of pride, especially by presenting Chilingarov's North Pole flag planting as a proud achievement, akin to America's lunar flag planting. Yet *The Polar Prize* also makes rhetorical claims about how international cooperation ought to function in the Arctic, outlining the Russo-Norwegian agreement reached on Arctic fishing rights as mutually beneficial (although many Slavic Russian politicians saw this agreement as an embarrassing defeat; see Hønneland 2020, 105–17). Implicitly, *The Polar Prize* portrays Russia as the adult in the Arctic geopolitical conversation.

While claiming that Russia desires international cooperation, *The Polar Prize* argues that each of the Western powers puts national interests before international cooperation. This claim is humorously illustrated through an animated segment depicting the Canadian and Danish navies and politicians continually returning to the contested but uninhabited Hans Island (Tartupaluk) between Nunavut and Greenland. There, sailors leave bottles of Canadian whisky or Aquavit for the others to find, demonstrating the intrinsic absurdity of trying to claim this small, inconsequential, uninhabited island. In 2018, Canada and Denmark announced a joint task force to resolve this territorial dispute, with the possibility of shared governance. Hans Island took on a new significance in June 2022, when Canada and Denmark resolved their dispute by placing a border down the middle of the island, as an act to demonstrate international geopolitical diplomacy in a riposte to Russia's invasion of Ukraine (because of this, North America and Europe now share a land border; see Breum 2022). This "battle," while humorously illustrated in the film, with the supposed national tensions diffused, is not portrayed as an act of friendly Arctic partnership/partisanship between Canada and Denmark.

Against this backdrop, *The Polar Prize* provides insight into Russia's state-sponsored priorities. While Russian politicians and scientists discuss how their disputes around Arctic claims could be resolved through UNCLOS, this international endeavor is portrayed as having been thwarted by American intransigence. Other claims are made about the symbolic remilitarization of the Arctic by including footage of military or militarized coast guard ships—specifically, a battleship flying the Danish flag. If the first part of the film is concerned with geopolitics and the role of nation-states, the second part of *The Polar Prize* acknowledges the environmental risks of climate change, addressing the dangers imperiling Russia and the Arctic region through the increased regularity of natural disasters and extreme weather. Yet there are also "benefits" from changes in the Arctic climate. As the film explains, natural resources are now more easily accessible and warmer weather has facilitated transportation, both internationally and along the Russian coastline. This dual messaging—climate crisis *and* potential benefit—in *The Polar Prize* illustrates the inclusivity of Arctic Days as a conference addressing both the perils and the benefits of climate change.

Against this discursive backdrop, *The Polar Prize* then shifts to Alaska, a place the film alleges is one of those most affected by climate change, with temperatures reaching as high as 30 °C (86 °F). In the film, members of a Yup'ik community address how a changing climate has affected fishing and altered animal behavior and migration patterns in their vicinity. Moreover, rising tides and erosion are also changing habitat and living spaces in Yup'ik coastal communities. It is striking, but perhaps not surprising, that while Indigenous voices from the United States have a presence in *The Polar Prize,* none from Russia are included in the documentary. Although Alena Efimenko, advisor on Indigenous affairs to the Russian Federation's delegation of the Arctic Council, addresses the changing patterns of reindeer and how this affects Indigenous peoples in the Russian North, it is notable that this brief account is offered by a member of the Russian government rather than by an Indigenous spokesperson.

The Polar Prize's global Arctic perspective next turns to the changes taking place on the glacial shelves of Greenland. Their disappearance would be a catastrophic global event, but substantial changes have already taken place within Greenlandic Indigenous communities. This may be the last generation of children who grow up in this sort of environment before it changes dramatically, the narrator intones, echoing the argument made by Inuit politician and activist Sheila Watt-Cloutier in *The Right to Be Cold* (2015; see chapter 3). The film then turns to larger questions about the environment, addressing the rise of plastics and other garbage that now floats up to the Arctic from other regions. The film pointedly notes the irony that those most reliant on nature end up dealing with other civilizations' garbage. This focus is related to a main argument presented in the documentary *Silent Snow* (Jan van den Berg, The Netherlands, 2013), which traces the accumulation of toxic substances in Arctic marine mammals from pesticides and other pollutants used in the global South. In *The Polar Prize,* Jim Hamel, director of the International Association of Aleutians, notes that if the problems of pollution affect Indigenous communities' ability to gather food, it is a matter not just of nutrition, but of losing one's culture.

The Polar Prize concludes by returning to Russia, offering an account of the ship *Mikhail Somov,* which began an expedition to the Arctic in 2013 from the city of Arkhangelsk, with the goal of conducting a series of experiments during its three-month voyage. One of the *Somov* expedition's goals was to clean up human-made waste in the Arctic, especially near remote weather and scientific stations (where researchers have no way to properly dispose of their waste). In this concluding part of the film, we move away from the experimental and at times humorous techniques deployed in the first half to a more traditional documentary form, implying that geopolitics can be laughed at—its battles in equal measure important and ridiculous— whereas global climate change does not have a frivolous side to it. Rhetorically, this shift positions Russia as the responsible party when addressing the serious consequences of human activity. However, it's worth noting that resource extraction itself (Russia's key priority in the Arctic) is not foregrounded as the cause of these two modes of pollution. *The Polar Prize* ends on a sentimental note, stressing how those who have become involved in issues of the Arctic carry a piece of it with them afterward (though the Arctic means different things to different people) and struggle to save it. This is brought home by the inclusion of comments by Iakov, a Russian Orthodox priest.

The Polar Prize, then, is a retort to the West and its perceived demonization of Russian Arctic interests, making the claim that the Arctic ought to be a new testing ground for peaceful cooperation and calling on the Arctic Council to play a key role. Perhaps invidiously, the model proposed by the film is that of Antarctica: the Arctic ought to be open to all for research and transport (for instance, China, South Korea, and Japan are all interested in exploration, even though they possess no Arctic borders), but the film states that there must be rules for exploration in place for this to happen, presently thwarted by the West's competing national interests. One can therefore see how this proposed model of peaceful globalization made *The Polar Prize* a documentary that played well in the context of the 2014 Arctic Days, where it was screened as part of the International Festival of Nonfiction Films. The documentary fits into the larger narrative of how the Russian state views the Arctic, which Laruelle describes as two-sided:

> The state-produced narrative on the role of Russia in this region of the world is ... like Janus, double faced: on the one hand, the rhetoric designed for domestic public consumption relies on older ideological sources, inspired by the Soviet legacy and the Cold War decades; on the other, that aimed abroad seeks to capitalize on the Arctic as a brand. This brand enables the Kremlin to position [itself] as an actor in touch with the international community, to renegotiate bilateral relations with the other Arctic players, and to advance mechanisms of legitimization based on soft power. (2014, 3)

This leads to two geopolitical strategies: "security-first" (military and nationalist) and "cooperation first" (economic and open to collaboration with other Arctic nations and interests) (2014, 7). These are mutually exclusive positions that coexist simultaneously, a Janus face invoked in the name of both environmentalism and national interest. And contrary to Russia's stated interests, *The Polar Prize* takes an ethical stance on drilling, warning of the possibility that, were an event like the Deepwater Horizon explosion in the Gulf of Mexico to happen in the Arctic, the effect would be far more catastrophic. The film ends on an ominous note: "The Arctic has become closer and better understood, but only time will tell if it has become less dangerous."

ETHNOGRAPHY AND/OF THE STATE: *WHITE MOSS, SEVEN SONGS OF THE TUNDRA,* AND *ANGELS OF REVOLUTION*

Only one fiction film was screened as part of Arctic Days in 2014: Vladimir Tumaev's *White Moss.* Shown at the end of the first day of the festival as a keynote, *White Moss* was marked specifically as an "artistic film" (Arctic Days 2014), using the tropes of international art cinema, ethnographic practice, and melodrama to tell its story. Based on two short stories by Nenets author Anna Nerkagi, its romanticized ethnographic depiction of a nomadic reindeer-herding tribe of Nenets on the Yamal Peninsula foregrounds stereotypical clashes between "modernity" and "tradition," young and old, man and woman, and city and country. Tumaev, himself an ethnic Russian, transposes Nerkagi's stories set in the 1980s to a contemporary twenty-first-century context of snowmobiles and smartphones. This juxtaposition of remoteness and sublimity (standard Arctic visual tropes) with technological change—as when two of the film's main characters climb a rickety wooden tower to access a faint cellular signal while remarking on the sky's beauty—is depicted using the aesthetics of international art cinema (in which the camera lingers on "beauty" as opposed to simply driving the plot forward), which speaks to why the film became popular outside the Russian Federation.

White Moss has circulated through several different contexts. In addition to the special screening at Arctic Days in 2014, it was selected as the audience favorite at the 36th Moscow International Film Festival the same year, going on to win a slot in the 2015 Rome Independent Film Festival and an LA Film Fest World Fiction special mention. Until 2021, it was available in the United States through Amazon Prime. The film's broad circulation is one example of how often antiquated depictions of Arctic Indigenous populations continue to shape public perceptions, both in Russia and internationally—aided in this case by visually stunning cinematography, including many panoramic shots of the tundra landscape bathed in the aurora borealis.

In the context of the 2014 Arctic Days festival, the inclusion of *White Moss* also signals the Russian Federation's close connections to the Yamalo-Nenets okrug leadership in Salekherd, known for its relatively strong—within the Russian context—Indigenous cultural programming and its substantial gas reserves and extraction infrastructure projects. Clearly, Arctic Days' privileged representations of the Yamal Peninsula and Nenets nomadic reindeer herders are political choices. Most often, Indigenous participants in Russian state-sanctioned events are portrayed as passive and primitive, devoid of agency or the motivation to achieve self-determination. Indeed, *White Moss* eschews any mention of resource extraction's environmental impact, telling instead a traditional, melodramatic love story involving two young Nenets. Alesha (Evgeniy Sangadzhiev) pines for his former girlfriend Aliko (Seseg Khapsasova), who stayed in the Russian city where she was sent for forced education, while Alesha returned to the land and, against his will, entered an arranged marriage with Savane (Erzhena Buyantueva). When Aliko returns after her mother's killing by a wolf, Alesha hopes for a reconnection and is rebuffed. Aliko, who is coded as urban with her high heels and new technology, has lost any connection to her past and has no interest in her Indigeneity. She states that what happened between them was long ago, prompting Alesha to go on a drunken binge. During this time, his community moves to find a new source of white moss for their reindeer. In a moment of sobriety, Alesha sees a baby and realizes his paternal desires, perhaps signifying a reconnection with his wife and community. He heads to the now departed encampment, intent on reuniting with Savane. When he finds them gone, he sets out on his snowmobile, coming across Savane just as she is being attacked by a wolf. He comes to her defense, and both Alesha and Savane are injured (the wolf is killed) before consummating their marriage and proclaiming their love on the frozen, bloody tundra.

In contrast to its melodramatic art-cinema plot and aesthetics, *White Moss* is punctuated by a retrospective voiceover in which Alesha offers an "insider" account of Nenets culture that traces his "journey" to understanding and living in the contemporaneous Nenets culture, and the competing pulls of urbanity versus living on the land (although Alesha himself never betrays a desire to relocate to the city). At odds with his actions in the film's diegetic present, the voiceover functions both as an ethnographic entry point into discussions of Nenets tradition and as a way of communicating his coming to terms with his own life. The voiceover allows the viewer greater empathy, as one can hear through Alesha's self-narration how he has grown since his self-centered youth.

Shot mostly on location, *White Moss* has been described by its director Tumaev, who notes that the difficulties of filming in Arctic winter temperatures were significant, as an aesthetic ethnography (K. Walsh 2015). The film thereby echoes the accounts of nearly one hundred years of Arctic filmmaking, particularly the difficulty of filming in the harsh tundra environment with equipment ill-designed to function at low temperatures. Yet Tumaev also states that he would have abandoned the project if it had to be shot anywhere else (K. Walsh 2015). Authenticity is thus, for Tumaev, about location—a conventional way to see the Arctic, foregrounding the sublime beauty of a natural environment that appears to be mostly uninhabited. Although authentic locations are used (besides some interior shooting done at Mosfilm studios in Moscow), other questions of authenticity arise in the dialogue. Only one of the main actors is a Nenets speaker from the Yamal region; others had to learn the local Nenets dialect phonetically. This production decision speaks both to the common disregard for Indigenous languages and to the complexity of language politics writ large in Russia.

White Moss, then, has a tense relationship with questions of ethnographic or Indigenous authenticity. It is the land, not the people, that Tumaev sees as a source of authenticity. In a disconcerting manner, according to Yamal regional television, Tumaev was interested in the shape of people's eyes (Yamal Region TV 2013), and he cast the film largely on the basis of physical appearance, not the ability to speak a Nenets language, let alone the one appropriate for the film. Because of this decision, many cast members are members of Russian minority groups (Evgeniy Sangadzhiev is from Kalmykiia; Alina Aiunova is from the capital city of Kazan in Tatarstan, central Russia). The focus on landscape and on its ocular beauty decontextualizes Arctic spaces, which become interchangeable from an outsider's perspective. *White Moss,* nevertheless, does deal with aspects of Indigenous experience under colonialism. In an interview with producer Svetlana Dal'skaia, the topic of young people leaving their communities for "civilization" (the interviewer's word) is addressed and is a key component of the film. It is mostly young women leading this exodus, Dal'skaia avows, as their encounters with technology make them less interested in a life full of labor-intensive household chores like sewing and cooking. Consequently, young men leave because of the scarcity of marriageable women. For both genders, Dal'skaia says, quoting a Nenets saying used in the film, leaving their communities opens romantic opportunities: "There is no love in the tundra, only a continuation of the family" (Dal'skaia 2014).

White Moss also addresses, albeit obliquely, other issues of central importance to Arctic Indigenous peoples living in a transnational context, raising debates about politics and sovereignty, along with matters broached in many truth and reconciliation commissions discussed in previous chapters of this book. These include forced removal (the youth must be relocated—via helicopter—to Salekherd to study); broken intergenerational ties (a consequence of that relocation); and alcohol abuse (vodka is sold to both Alesha and his father by Pavel, an ethnic Russian shopkeeper). The film, however, downplays these traumatic experiences and the oppressive agents that maintain Russian colonialism. The power relationships between ethnic Russians and the Nenets are naturalized—presented as "just the way things are"—and are not a matter for reflection either by the film's characters or through any formal strategy employed by the film. Neither are the diegetic actions of the characters placed within a broader contextual frame. This lack of political self-reflexivity explains *White Moss*'s enthusiastic reception within Russian official circles.

White Moss traces the many conflicts that emerge between Nenets urbanized youth (with their fixation on consumer goods) and elders (who bear witness to the loss of their language and culture because of globalization). Predictably, the role of urban space is ambiguous in the film. While the young adults who live in the cities and towns seem unconcerned about life on the tundra (as with the young man who has his father's reindeer killed so that he can get his "inheritance," calling reindeer "money"), the figure of the daughter is portrayed sympathetically as someone trying to find her place in the world. Yet it is not just the youth who settle in urban areas that are pulled away from tradition. Alesha is very attached to his snowmobile, not to the reindeer virtually every other member of his community uses for transportation. Cell phones are present, with a purpose-built tower to extend the signal. Although Pavel the shopkeeper is portrayed as an ethnic Russian racist who dispenses alcohol, his characterization may be interpreted as stemming from a lack of cultural understanding. That said, given that Pavel (and his racism) functions as an individual character within the narrative, he is not necessarily a stand-in for the Russian nation-state.

The oblique critique of Russian colonialism that viewers may read in the film gives us some insight into why *White Moss* was screened at Arctic Days. A review in *KinoKultura* states that "the film is not social or political critique, it's not about the annihilation of a people or a culture but deals with the larger themes of being true to one's love and family, of retaining one's humanity, way of life, and respect for tradition as technology and the siren-song of modern, urban life reach into the most distant of places" (V. Johnson 2014). By using the aesthetic and narrative tropes of art cinema, *White Moss* avoids directly confronting the colonial role of the Russian state, diffusing the social justice themes so often found in other "ally" films produced in the circumpolar Arctic. Thus, the film's aesthetic choices tend, in and of themselves, to weaken any critical stance vis-à-vis the Russian Federation's role within Arctic Indigenous communities, rendering it polysemous and equally open to critical or uncritical readings of the nation-state as a colonial power. This is evidenced by various mainstream reviews of *White Moss* in Russian media from across the political spectrum. The response to the film in Moscow and St. Petersburg was mostly laudatory, while in *Amur Pravda,* a local publication from the Russian Far East, a critic praised the film generously: "It will make audiences consider what is most important, and will possibly change their lives. It should be shown throughout Russia and beyond"— although, as the author notes, it was being screened only in Moscow at the time (*Amur Pravda* 2014). The reviewer in *Rossiyskaya Gazeta* (Russian Gazette), published by the Russian government, dubbed the film a serious contender for the Golden St. George, the Moscow International Film Festival's highest award, calling it an "artistic feat" (Kichin 2014). *Zavtra* (Tomorrow), an extreme-right newspaper combining ultranationalist and anticapitalist views, commended the film for showing the social problems of Russia's remote regions, especially alcohol abuse and the rapaciousness of urban capitalism, while praising it for being "completely devoid of postmodernism" and for its incorporation of "Soviet" conceptions in a contemporary exploration of ethnicity (Belokurova 2014). The turn against postmodernism and concurrent celebration of Soviet conceptions points to the aesthetic and political continuity between the USSR and Russia in the Putin era.

White Moss also offers a portrait of tradition versus settler culture that is different from, yet complementary to, other contemporary Arctic cinemas. For example, Arnait's *Restless River* (Marie-Hélène Cousineau, Canada, 2019; see chapter 4) is an analogous example of a work made by a nation-state settler director rendering a sympathetic account of Arctic Indigenous peoples. The plots and themes of the two films parallel each other in terms of intergenerational trauma, relocation, and forced schooling—although in *Restless River* these are cast as central to contemporary Indigenous experience and not glossed over as in *White Moss*. Some parallels can also be drawn with *Sami Blood* (*Sameblod,* Amanda Kernell, Sweden, 2016; see chapter 5), in which the story is told from the point of view of a young woman who strives to integrate into the dominant settler culture, only to return to her Sámi homeland in old age—wholly estranged from the community—to attend the funeral of her only sister, who stayed behind. Both *Restless River* and *Sami Blood* are told from a woman's perspective, opening up an interesting counterfactual: How would the story of *White Moss* unfold if told from the point of view of Aliko? The gender politics of forced marriage and limited educational opportunity are not addressed in *White Moss,* which is not surprising given its general adherence to state political ideology, whereby patriarchy and heteronormative gender roles are usually affirmed.

The Arctic Days event and its related film festival have generally eschewed broad and inclusive representation by Indigenous peoples, reflecting the Russian state's interest in limiting self-governance and Indigenous rights movements. Nomadic Nenets on the Yamal Peninsula have also come to represent Russian Arctic Indigenous peoples to international and domestic film audiences during the past decade, as seen in Russian television documentaries such as *Children of the Tundra (Deti tundry,* RTD, Russia, 2013) or Edgar Bartenev's *Yaptik-hasse* (Russia, 2006), an ethnographic observational documentary that has achieved international circulation. Like *White Moss,* these kinds of depictions disregard the traumatic effects of Soviet rule—including Cold War militarization and environmental degradation of the Yamal Peninsula—as well as more recent mega-industrial projects and the loss of herding-territory activists, such as Eric Serretto, who participated in the Voice of the Tundra social media protest movement. These films are also remarkably different from the kind of media discourse and production that became possible among northern Indigenous populations during the post-perestroika years. Galina Diatchkova describes that media presence as having "flourished and supported Aboriginal mobilization," including Indigenous information communication campaigns and mobilization of media as a political vehicle—signature initiatives of RAIPON at the time (Diatchkova 2008, 215, 217).

White Moss and most other films about the Yamal Peninsula Nenets also differ markedly from the internationally distributed feature films of Anastasia Lapsui, one of the most well-known Nenets filmmakers to date. Born into a nomadic tribe and educated through the boarding school system—and later at Ural State University—Lapsui worked as a journalist in Salekherd before emigrating to Finland in 1993, where, with funding from Finnish sources, she has been codirecting, with Finnish director Markku Lehmuskallio, Indigenous-oriented films ever since. These films are explicitly critical of Soviet colonization, Stalinist pogroms, and the ongoing attempts to disenfranchise Nenets from their culture, traditions, and livelihoods. Lapsui's *Seven Songs of the Tundra (Seitsemän laulua tundralta,* Finland, 2000), for instance, was the first fiction feature film made in the Nenets language, recovering histories lost in the "official" narratives of the Soviet era. Similarly, *Pudana: Last of the Line (Sukunsa viimeinen,* Finland, 2010) addresses the traumas of forced removal to boarding schools and the plight of nomadic children sent far from home.

Not simply tracing the history of colonialism, Lapsui's films also show the ways in which Nenets narratives have been elided within the Soviet/Russian imaginary, and indeed in that of the global Arctic itself. However, these elisions may still be a necessary evil—a precondition for securing the funding necessary for filmmakers to make films about Nenets in Russian territory. While Lehmuskallio and Lapsui's films unearth untold narratives of the past, certain topics, such as post-Soviet narratives about the present-day condition and persecution of Nenets, remain impossible to tell without resorting to the kinds of allegory found in much of Soviet-era Eastern Bloc cinema. Unable to speak directly about current political conditions, filmmakers behind the Iron Curtain instead used historical settings, metaphors, and forms of silence in addressing the contemporary moment.

In addition to the context provided by Lapsui's Finland-produced films about Nenets on the Yamal Peninsula, we address a Russian experimental art film that engages with the Soviet conquest of the Indigenous North. Aleksey Fedorchenko's *Angels of Revolution* (*Angely revolyutsii,* Russia, 2014), made the same year as *White Moss,* screened at the Moscow International Film Festival and internationally, winning several awards. It was not selected to be part of Arctic Days. Loosely based on the Kazym Rebellion of 1933, *Angels of Revolution* tells the story of Russian revolutionary war hero, Polina Schneider (Darya Ekamasova), commissioned to civilize and modernize the Khanty and Nenets tribes through avant-garde Soviet artistic strategies (figure 9.2). Ironic and satirical, the film is set within the historical context of the establishment of dozens of Soviet culture centers (*kultbazas*) in remote locations in the early 1930s, a program initiated to accelerate collectivization processes by building the boarding schools, hospitals, veterinary clinics, community centers, and trading posts that would transform nomadic Indigenous populations into "good" Soviet communists (the USSR's need for natural resources was often an underlying reason). In the case of the Kazym *kultbaza,* the acculturation project originated with securing fishing access to the sacred lake Num-To. Although the Kazym *kultbaza* was later described as a "hearth of socialism" (Toulouze, Vallikivi, and Leete 2017, 209; see also Rogatchevski 2020a, 134–36), boarding schools were considered especially colonial by Indigenous peoples and became an early locus for opposition by the Khanty. Numerous scholars have discussed the ways in which *Angels of Revolution* appropriates and subverts Soviet ideology and art (see Lipovetsky 2019; Mikhailova 2019; Prokhorov 2019; Roberts 2019). Fedorchenko represents Schneider and her compatriots as incompetent state missionaries blinded by

FIGURE 9.2. The parodic ethnographic mode in *Angels of Revolution* (*Angely revolyutsii,* Aleksey Fedorchenko, Russia, 2014).

ideology. They are all killed at the end, with only Schneider appearing to have preserved her faith in the project.

Fedorchenko's film furthermore exemplifies a trend in twenty-first-century Russian cinema given international release—namely, the erasure of Indigenous agency, especially when the films are ostensibly about Arctic Indigenous groups. In *White Moss,* the perspective is melodramatic, ethnographic, antiquated, and patronizing. In regard to *Angels of Revolution,* on the other hand, Fedorchenko has expressed his distaste for essentializing or further colonizing these populations through the cinema. Indeed, he has referred to *Angels of Revolution* as a "documentary" (Mikhailova 2019, 256, 258), raising the question: A documentary about what? To scholar Andrei Rogatchevski, it appears as if "Fedorchenko has consciously chosen to imitate, in an overblown manner and from an ironic distance, both the iconic imagery of the Russian revolutionary avant-garde and the stock imagery representing indigenous peoples as savages, because the film's aim is not to demythologize the Russian Revolution in its relation to the indigenous cultures of the Soviet North but to remythologize it" (Rogatchevski 2020a, 145–46). Tatiana Mikhailova, meanwhile, argues that *Angels of Revolution* "inconsistently slips into a visual rhetoric indicative of the 'imperial gaze,'" which avoids depicting "any characters, or even stories, of individual Khanty and Nentsy; they appear as an undistinguishable 'group,' which suggests the perspective of the colonizer rather than a mediator or interpreter" (Mikhailova 2019, 258). For both Rogatchevski and Mikhailova, then, *Angels of Revolution* both recasts and questions aspects of the past that are nevertheless reified into new, equally problematic, mythologies of the present. However, the aesthetic register of the film and its self-reflexivity throw into question the process of mythologizing itself and spectacularizing the past.

Given that Fedorchenko spent a decade researching avant-garde Soviet artists, examining photos, historical documents, and archives, *Angels of Revolution* should not be seen as an "ethnography" about the Khanty and Nenets. Rather, Fedorchenko's work turns the table on the customary ethnographic subjects of Soviet cinema, making the artist-revolutionaries the ethnographic object rather than the Indigenous population. One could ask whether the subject of this quasi-documentary is the failure of artistic production in support of revolution, or the failure of revolution itself. At the same time, Fedorchenko slips into an explicitly ethnographic mode in the film's final scene, in which an elderly woman sings a Soviet fighting song in her modern apartment before a stationary camera. The woman, it turns out, was the first baby born in the Kazym *kultbaza*'s hospital. It is worth noting how *Angels of Revolution* (particularly in this reference to documentary practice) relates to some of the priorities of Arctic Indigenous cinema we have addressed so far—namely, the function of interventionist historiography about Arctic settlements and colonization—and how such films have both aided and countered this process. While *Angels of Revolution* addresses many of the topics that run through contemporary Arctic Indigenous cinema, including forced acculturation, colonial and imperial quests for natural resources, and the erasure of Indigenous voices and agency, it remains tied to the perspective of the colonizer.

The Arctic continues to loom large in the Russian media and political imaginary. Geopolitical tensions have increased under Russia's chairmanship of the Arctic Council in 2021–23, with Sweden, the United States, Canada, Denmark, Finland, Iceland, and Norway all boycotting the council in March 2022 in response to Russia's war against Ukraine (Friedman 2022). The emphasis on increased resource extraction and disempowering of Indigenous populations is central to Russia's politics and to furthering the country's long-standing myths about its northern frontier. These aspects are evident in Chilingarov's continued hero status and his appointment to chair the 2021 Arctic Days event. Film and media continue to engage with these questions, from the representation of Chilingarov's flag planting in *Polar Prize* to the experimental fiction of *Angels of Revolution.* Contemporary Russian film thus speaks to the multiplicity of political and state priorities related to the Arctic, circulating across media and through various internal and external publics. As such, Russia's contributions to an Arctic global media context deserve further attention.

———

TEN

From the Cold War to the Climate Crisis

THE RUSSIAN NORTH AND UTOPIAN SVALBARD

DURING THE COLD WAR, the remote, ice-covered geography of the Arctic served as a symbol of geopolitical stalemates. As Sanjay Chaturvedi argues, the militarized Arctic space came to be "perceived and treated throughout the Cold War as an inanimate, passive chessboard on which geostrategic moves and countermoves were made with very little reference to ecological considerations" (2000, 454). Twenty-first-century environmental degradation and the accelerating climate crisis stem, in part, from this militarized geopolitical imaginary. A perception of the Arctic as blank slate, pristine environment, or vast open space has frequently been foregrounded in the imaginaries of polar explorers, settlers, and scientists who have traveled north to chart, conquer, and investigate. In the Russian Arctic, we find that additional environmental (or what Chaturvedi calls ecological) concerns such as mass industrialization, mega-infrastructure projects, and ongoing resource extraction have shaped the region from Stalinist times through the Cold War and into the Putin era. As Veli-Pekka Tynkkynen argues, this long-standing set of fossil energy priorities not only drive the national economy (in 2014, more than 50 percent of budget revenue and 70 percent of exports), but also influence "the spheres of culture and identity" (2018, 49). These aspects of Russia's economy, society, and political priorities also reflect the fact that Soviet climate scientists were influential in shaping world knowledge about anthropogenic warming and its effects in the Arctic throughout the twentieth century. This includes contributing to the first Intergovernmental Panel on Climate Change (IPCC) in 1990, though Western historians of science

All translations from Russian in this chapter are by Deirdre Ruscitti Harshman.

and technology have overlooked contributions by Soviet and Russian climate scientists (see Ashe 2018; Doose and Oldfield 2018). This disregard, intentional or not, speaks to the quite distinct narratives about climate change engaged in not only by nation-states, but by scientists themselves.

Climate change in the Arctic is not new in the Russian imagination. A Soviet textbook from 1960, for instance, asserts that an "end goal of meteorology" is to "artificially change the weather and the climate in a direction preferable for humans [and to] reduce their dependency on the weather and on climatological conditions" (Doose and Oldfield 2018, 19). Soviet scientists during the Cold War had identified a warming Arctic and were interested in the potential of warming it further. Influential oil engineer Petr Mikhailovich Borisov advanced this notion both to Soviet interests and internationally during a fifteen-year period. In scientific papers and the comprehensive book *Can Man Change the Climate? (Mozhet li chelovek izmenit' klimat?,* 1973), Borisov charts the warming in the Arctic during the late nineteenth century and the first part of the twentieth and presents plans to warm it even further, by building a fifty-five-mile dam across the Bering Strait, which would divert warm Atlantic waters to heat up the polar region and thaw the ice, thus making the region more habitable and available for economic development and resource extraction.

One of Borisov's main concerns is that, because the Arctic is so cold, it impedes the modernization and development of not only the Soviet Union, but also the rest of the world. He argues that the frozen Arctic disproportionately affects the USSR. The reasons for Borisov's proposal to heat up the Arctic are meticulously outlined in *Can Man Change the Climate?,* demonstrating the economic, social, and political benefits, as well as the climatological, oceanographic, and atmospheric sciences rationale, behind the plan. Macro-engineering technology can make it happen, he writes, given political will: "The project is technologically feasible, technically clear, and simple in concept. It can regulate the sea advection of heat practically to any extent desired. The modern level of engineering—the building industry, mechanical engineering, power engineering, automation and so on—is equal to the task. It is all up to the international cooperation between scientists and governments, particularly in the USSR, USA and Canada" (Borisov 1973, 170). Borisov thus considers whether the Arctic region has potential for improving geopolitical relations between the superpowers. As with much of Soviet planning, Borisov looked at the proposed Arctic warming as something that could be adapted to the use of the Soviets, but—given the international impact of the project—also as something that could lead to a political thaw. What Borisov points to is that climatological manipulation in the Arctic, especially its potential warming, should lead to positive results—a starkly different tone compared to today's climate change discourse.

Borisov is, however, prescient about climate change being a global phenomenon: "Any large-scale climatic amelioration cannot be confined within the borders of any one state. Hence, the problem of the major climatic changes inevitably overrides national boundaries and occupies a place among the urgent international social and political problems" (Borisov 1973, 172). Again, for Borisov, this is positive: "The amelioration of the Arctic will undoubtedly open a new chapter in international relations. It concerns not only the USSR, Canada, Greenland, Denmark, USA (Alaska), whose territories are situated in the planet's coldest regions, but also many countries in tropical and desert zones" (Borisov 1973, 172). Borisov's plan, formulated at the height of the Cold War, was noticed in both political and popular scientific cultures in the United States. Following Nikita Khrushchev's visit to the USA in September 1959, the US Joint Publications Research Service translated a document about the plan.

The utopian aspects are front and center in the contemporary reporting about the Borisov plan. The premise of uniting the USSR, the USA, and Canada in a macro-engineering project for the benefit of these countries and the rest of the world comes through clearly:

> "And here," concluded Peter Mikhailovich [Borisov], "is what I would like to dream about: the heads of the two states met in Washington; somewhat later, they will meet in Moscow, in order to continue their discussion, begun in Washington, of many problems interesting to all mankind. The main problem which is on the minds of N. S. Chrushehev and D. Eisenhower is that of warming up the political climate. When this warming up occurs, and the ice of the cold war melts, broad vistas for teamwork and warming up the eternal ice of the Arctic Ocean will open too. How close together the common struggle for such a great humanitarian cause as discovering for mankind new powerful sources of warmth and life will bring our peoples." (Lyubimov 1960, 8)

Borisov's ideas circulated internationally during the Cold War, including to popular audiences in the United States through an article in *Popular Mechanics* (Windsor 1956), which dismisses the proposal of a Bering Strait dam as simplistic and egregious Soviet propaganda. An article in the *Bulletin of the Atomic Scientists* (March 1969, 43–48), however, is based on a translation from the Russian by the Canadian Defense Research Board. It presents the project in neutral terms and as reflecting broad interest: "Soviet climatologists are vitally concerned with the problem of ameliorating the climate of Siberia and other northern lands as a means of developing these regions for an expanding population" (Borisov 1969, 43). This attempted engineering feat was, more importantly, an attempt at the feat of constructing, during the Cold War, an alternative geopolitical world order. Borisov's plan is a preposterously large-scale project, also foreshadowing the kinds of interconnections between questions of climate change and ideology that permeate Cold War environmentalism through the differing lenses of East and West.

INTERNATIONAL COLD WAR ENVIRONMENTALISM AS SCIENTIFIC SOLIDARITY AND DISSENT: AN ARCTIC THAW?

The discussion about the Borisov plan to "warm the Arctic" points to the fact that discourses about climate change in the Arctic and potential environmental destruction as part of Cold War aggression were mobilized in different registers both in the West and in the East. Environmental protests against a possible nuclear winter on the horizon cemented the frosty relations between the Cold War superpowers. The Arctic, perceived as an ice-covered theater of war (and one of the locations of increased nuclear militarization, such as submarines under the polar ice cap), became a convenient symbol for these tensions.

Environmentalism—an interest in preserving a livable planet for all—became one sanctioned channel for the articulation of dissent in the USSR. As Douglas Weiner argues, environmental scientists were more often perceived as pacifist naturalists than as real dissenters, just as calls to "nature protection" were mobilized by Soviet citizens "to forge or affirm various independent, unofficial, but defining social identities for themselves" (Weiner 1999, 20; see also Ashe 2018, 9–11). Borisov's engineering project can be understood partly in this light. According to J. R. McNeill and Corinna M. Unger, "Modern environmentalism . . . is, among other things, a child of the Cold War. . . . Fears of radiation poisoning and nuclear winter scenarios helped tilt popular culture in the direction of ecological thinking," while other "segments of the population, more committed to the vigorous prosecution of the Cold War, often viewed environmentalism with equal suspicion"

(McNeill and Unger 2010, 11). Environmentalism itself, then, was seen as traversing the East-West Cold War barrier because of its necessarily globally interconnected nature.

Meteorologists and geophysicists in both the East and the West notably shared internationalist and environmentalist counter-politics: "Internationalism was a very old element of many sciences, and the sensitivity to international relations promoted by the Cold War spurred a trend toward even greater collaboration across borders. Sometimes, as in the case of the physicist-led 'One World' movement against nuclear weapons, political goals were overt. More often, those goals remained in the background; as they saw it, scientists simply wanted to continue their traditional internationalism and openness" (Edwards 2010, 224). In this way, environmentalism became a vehicle of "soft" power to articulate geopolitical tension. The frame for understanding climate change was somewhat different in the Eastern Bloc; for "the Soviet Union and several of its Eastern European satellites, environmentalism eventually served as one of the few—sometimes the only—permissible form of critique of the state and the Communist Party" (McNeill and Unger 2010, 12). By the end of the 1970s, environmentalism and a growing recognition of climate change thus coexisted with a threat of nuclear winter (see also Mowat 1976). These competing conceptual frames—warming of the planet versus freezing of the planet—are thus constituent parts of the Cold War media, political, and rhetorical register.

As an outgrowth of 1970s environmentalism in both the East and the West, in 1982, the United Nations undertook the first global inquiry into the relationship between societal and economic development and the ecological crises that were becoming apparent in the wake of two hundred years of industrialization and colonization. Named after the former Norwegian prime minister who chaired it, the UN's Brundtland Commission consisted of an international and interdisciplinary panel of experts, including members from around the world: Algeria, East Germany, the Soviet Union, the United States, China, Canada, Indonesia, Guyana, Colombia, Zimbabwe, and Norway among them. A major outcome is *Our Common Future* (1987), a report published by the UN and Oxford University Press. This document spurred one of the most influential definitions of sustainability to this day. The Brundtland Commission report argues that humans and the environment should not be seen as separate, autonomous spheres, given that "the 'environment' is where we live; and 'development' is what we all do in attempting to improve our lot within that abode. The two are inseparable"

(Brundtland Commission 1987, xi). The commission, therefore, saw the environment as something beyond the simply physical, material realm, and as enmeshed within and inseparable from sociological and political frames. Moreover, like many current arguments about the impact of climate change, the commission's report moves away from development models that maintain a first- and third-world dichotomy, instead arguing for environmental models that can ameliorate the shared and global effects of climate change.

By the mid-1980s, as is clear from the Brundtland Commission report, climate change, thinning Arctic Sea ice and melting glacial ice, and the impact on the poles were being amply discussed, and in ways that signaled that these concerns crossed boundaries of East and West (see also Ashe 2018). In the late Cold War era, this was significant. Notably, the report enumerates arguments about the rising level of CO_2 emissions, 1.5–4.5°C of warming, and the effects of greenhouse gases on the poles and coastal areas—arguments that were still being made at COP 27 in 2022 thirty-five years later, without much progress having been made and now with less time to reduce emissions to a livable level. With respect to the Arctic, the Brundtland Commission model is a product of the Cold War: both dystopian and utopian. It is dystopian in its realization that the impact of climate change and ice melt may have deleterious effects on environments and societies in the Arctic in the face of a geopolitical Cold War stalemate. It is utopian in its emergent globalization discourse, advocating the perspective that the environment does not recognize ideological and economic divides and that cooperative, international strategies must be implemented around the world.

NEW RUSSIAN CINEMA AND CLIMATE CHANGE
ALLEGORIES: *HOW I ENDED THIS SUMMER, THE
WEATHER STATION,* AND *LEVIATHAN*

Environmentalism as a form of dissidence provided an outlet during the Cold War across the Eastern Bloc. Media representation of "ecological" matters was seen as less threatening. A similar context is discernible in the present. As television, the press, and other broadcasting remain tightly controlled in the Putin era, dissident voices are more often found in Russian fiction films. A special case concerns films made for the international arthouse market, which can deploy allegory to a far greater degree than the realist, state-sanctioned documentary can or than state-friendly fiction films such as *White Moss* (*Belyy yagel,* Vladimir Tumaev, Russia, 2014; discussed in chapter 9) do. Russian art cinema, by contrast, offers a different set of approaches and complexities in regard to the representation of the state, allowing somewhat critical voices to emerge, though not without potential consequences, especially when these works are recognized internationally:

> Russian art cinema is unique because it premieres outside of Russia proper at international film festivals such as Cannes, the Venice Film Festival, and Berlin Festival, to name a few. The first audiences are a mix of nationalities and probably comprise more Russian émigrés than citizens. As these films are premiered within international platforms and removed from the Russian national microscope, these films can pursue a specific idea, rather than strictly adhere to traditional topics in mainstream Russian film. Usually, current concerns and questions of identity are not neatly presented and answered for the audience within art films. (Dillon 2013, 3; see also Strukov 2015)

Because of its lower budgets and smaller audiences, contemporary Russian art cinema has been able to take liberties that popular cinema cannot: "The freedom of expression and choice in subject matter allows art house cinema to pursue issues and concerns that are not politically compliant. Russian art house cinema is a qualitative source due to its under-the-radar nature and low domestic popularity, coupled with the rarity of major, national theatrical releases" (Dillon 2013, 3–4). For these reasons, we now turn to three recent examples of Russian art cinema, with narratives and settings in the Arctic, that challenge dominant Russian narratives of resource extraction, nationalism, and the paradigms of the Anthropocene.

How I Ended This Summer (*Kak ya provyol etim letom,* Alexei Popogrebsky, Russia, 2010), set on a climate observation station in remote Chukotka (far northern East Russia), features two male protagonists—the older and experienced weather monitor Sergei (Sergei Puskepalis) and the younger, recently arrived Pasha (Grigoriy Dobrygin) (figure 10.1). Critics have read the film's intergenerational relationship both as part of a long tradition in Russian cinema foregrounding father-son relationships (though here the family relationship is symbolic and allegorical) and as an allegory for the strained patrilineage of Soviet and post-Soviet Russia (see Strukov 2015). A similar set of plot devices and character arcs characterize Lyubov Borisova's comedy-drama *The Sun above Me Never Sets* (*Nado mnoyu solntse ne saditsya,* Russia, 2019), filmed in the Yakutsk language and emerging from the growing film-production hub in Yakutsk, Sakha Republic. In this film, a young man arrives at a

FIGURE 10.1. Post–Cold War environmental and social decay in *How I Ended This Summer* (*Kak ya provyol etim letom,* Alexei Popogrebsky, Russia, 2010).

near-deserted Arctic island in northern Yakutia to join the elderly man who cares for the outpost there, and clashes in perspective and priorities unfold.

Both *The Sun above Me Never Sets* and *How I Ended This Summer* mobilize what can be called a quintessential Arctic environment: desolate, stark, forbidding, and captivatingly beautiful. In *How I Ended This Summer*'s long shots, indebted to both international art cinema and Soviet cinematic language, evidence of the Arctic region's militarization is ever present. The debris of defense and Cold War installations becomes vestiges of a past uncannily present, counterpointed by plot developments that pit Sergei and

Pasha against one another in a tense chamber drama of opposing interests, histories, and, ultimately, ethical choices about what information to divulge to the other. Wide-angle shots dwarf characters in the landscape and echo the expansive emptiness of many Arctic depictions, just as the desolate isolation seems to trigger mental instability. The film mobilizes a complex temporal juxtaposition, as long shots create one version of the present, which is juxtaposed with the mise en scène, replete with the still (radioactively) lingering debris of the past.

Cold War Debris, Digestion, and Dogma in How I Ended This Summer

Cold War debris is everywhere in *How I Ended This Summer:* defunct antennas, rusting oil barrels, and, most prominently, a portable nuclear power plant. Environmental degradation is wrought both externally and internally, with Pasha warming himself near the still-active generator during a cold night and frying a fish in it, which he serves to Sergei to poison him. This detritus speaks to the legacy of Chernobyl radioactive poisoning and to a Cold War fear of an impending nuclear winter; indeed, Pavel plays the video game *S.T.A.L.K.E.R.: Shadow of Chernobyl* (GSC Game World, Ukraine, 2007), a first-person shooter game about the disaster. The unforeseen legacy of the Arctic's use as a "safe remote space" for nuclear materials in the Cold War, like the legacy of Soviet Communism, lives on in the film, in which nuclear poisoning of the body echoes, allegorically, the poisoning of the environment through climate change. The threat of nuclear radiation is that it spreads through wind and rain, remaining invisible, its cancerous effects emerging only later. Similarly, the effects of global warming—ice melt, flooding, droughts, and so on—were hardly imaginable as concrete outcomes at the dawn of the industrial revolutions and in the Stalinist installations of mega-industrial projects in the Far North. The Cold War is made manifest in the twenty-first century in embodied ways (body temperature, nourishment), and the legacy of the Cold War directly kills one protagonist. Similarly, climate change is not directly noticeable. Unlike the pollution of industrial smokestacks that tower over landscapes, spewing smoke and soot that is visible miles away, it is pervasive, subtle, and not easily embodied.

At the plot level, failure to keep adequate meteorological records becomes one of the most contested issues in *How I Ended This Summer,* leading to intense animosity between the two main characters. Pasha is altering state meteorological records, presumably to be used for climate science purposes, to the horror of Serge. The film's emphasis on securing accurate climatological data provides an implicit correlation to twenty-first-century climate change anxiety, through the assumption that we must monitor, assess, mitigate, and correct the convulsive transformations in the world's environmental systems that are to come. But it also references a long history of the centrality of meteorology and climate science to geopolitics. Climate monitoring, including through various state agencies and university research centers— many of which collaborated internationally and competed for results with Western scientists—were a key science priority during the Cold War, with the State Committee for Hydrometeorology and Environmental Monitoring (Goskomgidromet) playing a powerful role in Soviet government (Doose and Oldfield 2018, 18–23). The desire for geopolitical climate politics was also present at the highest levels of government during the Cold War. Science and technology historian Paul N. Edwards notes:

> The best-known political event in the history of meteorology happened in the midst of this crisis [Bay of Pigs invasion in Cuba], and in response to it. On September 25, 1961, speaking before the UN's General Assembly, [President] Kennedy presented a series of arms-control proposals. He then proclaimed . . . "[W]e shall urge proposals extending the UN Charter to the limits of man's exploration of the universe, reserving outer space for peaceful use, prohibiting weapons of mass destruction in space or on celestial bodies, and opening the mysteries and benefits of space to every nation. *We shall propose further cooperative efforts between all nations in weather prediction and eventually in weather control.* We shall propose, finally, a global system of communication satellites linking the whole world in telegraph, telephone, radio and television. The day need not be far distant when such a system will televise the proceedings of this body in every corner of the world for the benefit of peace." (2010, 223)

Kennedy's statements to the UN also foreground how the centrality of media and communications to climate science was anticipated, providing a geopolitical connection to McLuhan's contemporaneous theories of a global village and the possibilities of recombinant communication technologies (1964, 7–16), and to his contention that the Cold War was a "hot war of information transferred to the domestic sphere" (1970, 69) through advances in media technology. Technology, and media technology alongside technologies of visualization more specifically, became the dominant tools of the Cold War.

To that end, meteorology continued, during the Cold War, to be one of the most international and internationalist-oriented scientific fields, which eventually led to the creation of World Weather Watch, a foundational network of climate and meteorology observation, recording, transmission, and communication systems that have shaped climate change research since the 1960s as a mechanism of East-West collaboration. By then, "Kennedy and Khrushchev had agreed on the value of global satellite meteorology and the principle of cooperation" (Edwards 2010, 225). In late March 1962, Khrushchev wrote—his rhetoric reminiscent of what Borisov proposes in his macro-engineering plan—in reply to Kennedy's suggestion: "Precise and timely weather forecasts will be another important step along the way to man's conquering of nature, will help him still more successfully cope with natural calamities and open up new prospects for improving the well-being of mankind. Let us cooperate in this field, too" (cited in Edwards 2010, 225). All these discourses, at the height of the Cold War, are utopian in their imaginary projection of the global possibilities offered by new technologies. After the fall of communism, a dystopian discourse emerged that centered on the detritus left by these technologies, both in terms of their material disintegration in the landscape and for the failure of the utopian optimism they inspired. These dystopias are mapped onto *How I Ended This Summer*. The film demonstrates and documents the detritus of these plans to monitor, control, and govern climate that were circulating at the height of the Cold War. In the background of its narrative are espionage and environmentalist cooperation in the name of science and for the benefit of the world, but the film also acknowledges that these attempts to monitor, control, and govern climate have failed.

The Weather Station (*Pryachsya*, Johnny O'Reilly, Russia, 2010) was released the same year as *How I Ended This Summer*. With only limited distribution outside of Russia, *The Weather Station* garnered few favorable reviews either domestically or internationally. The film begins in a scene reminiscent of *The Thing* (Howard Hawks and Christian Nyby, USA, 1952), with two men outside, one quite fearful, in the Arctic cold—then cuts to an intertitle stating "Two days later," which implies that the men there are all dead. Chief Detective Andrei and his deputy Slava arrive to investigate. The film flashes forward and back, telling two parallel stories. The weather station of the title houses two older men, Drozdov (Sergey Garmash) and Ivanov (Vladimir Gusev), who have worked there for many years; one spends his time building matchstick replicas of the Eiffel Tower and the Taj Mahal, the other trying to prove the existence of the Yeti. They are aided by their new assistant, Romash (Pyotr Logachev), whose job it is to do the cooking, but who often hides in the building. Another character, Vadim (Sergei Yushkevich), is a "man on the run," and Irina, his wife, functions as a film noir femme fatale. The character constellation can be read as a commentary on contemporary Russian stereotypes of masculinity and femininity.

The Weather Station places its two main characters in isolation, and the father-son and Soviet/post-Soviet dynamics of *How I Ended This Summer* also structure some of its plot. The cast is larger in *The Weather Station,* however, and the film's crosscutting between characters and temporalities both links them together and juxtaposes their differences to tell a broader story of twenty-first-century Russia. *The Weather Station* positions the Arctic as a site where the political and personal past of individuals and the country converge. As these different temporalities are deployed, they end up merging into one, while keeping the central trauma of the past intact until the second-to-last scene, which incorporates a flashback to the young orphaned boy (hidden in a washing machine) looking out over the destruction of the present. The two narrative trajectories are thus merged by the end in a sleight of hand, concluding with the orphan, as an adult, walking out into the white wilderness, alone, with no apparent future.

While the state wants to dismantle the station and turn it into a tourist site, catering to adventure seekers, Drozdov and Ivanov believe that it belongs to them and that they must protect it from such a threat. *The Weather Station* thereby makes an implicit argument relevant to the scientific monitoring of anthropogenic climate change. The film proposes that dismantling the weather station means jeopardizing the continuous recording of meteorological and climate data in the Arctic. A second component of that claim is that the dismantling of the station implicitly abides by a global market-economy framework of resource extraction in the Arctic that seeks to disavow any warming potential of those processes. To the elderly weather monitors, the state is forgoing its public responsibility to ensure adequate meteorological documentation. The nouveaux riches (a banker and his wife, the femme fatale) who come to explore caves as tourists, with lots of money, are representative not only of the "New Russia," whose citizens have benefited from oligarchies, corruption, nepotism, and lawlessness, but also of the willful destruction of the environment. Drozdov and Ivanov are upset that the weather station (an outpost of a science academy) can even be bought. This is another stab at the notion of impartial, objective, empirically based meteorological observations in the Arctic, and a jab at the corruption of Putin's

Russia. Yet the need to have rules in place is undercut by the scam of the Yeti, a red herring, whose story is faked for years by Drozdov as he tries to secure money from the gullible nouveaux riches to undertake an expedition, much like the polar explorations of old. Ivanov builds the great structures of the world on balsa wood, having never seen them. As a flight of fancy and escapism, Ivanov's tourism is imaginary, just as the story of the Yeti is imaginary. These registers are important to the film, since they posit Russia as a universe unto itself, which is a constituent part of contemporary social mythology.

Isolation leads to its own rules when set on the edge of the empire's control. Yet what happens at the edges is still linked to the urban bureaucracies, and indeed relies on them. As in *How I Ended This Summer,* long-distance radio communication is central in the Arctic Russian imaginary in *The Weather Station,* and indeed, "in all depictions of the Soviet appropriation of the Arctic the figure of the radio operator is of utmost symbolical significance, symbolizing the negation of distances and barriers and the interconnection of distant places" (Frank 2010, 117). In both films, the breaking of this link leads to doom and dismay.

Allegorically, the communication breakdown points to the fact that what happens in the Arctic—environmentally speaking—does not stay in the Arctic. History and the environment are intrinsically tied. In this and other Soviet Arctic weather station films, the meteorological station—like the radio operator of past Soviet films—is a privileged trope. It functions as an abject location, as the dark recess of the catastrophe that's impending but cannot be directly addressed. In *How I Ended This Summer,* climate change is implicitly but clearly acknowledged as being denied in many ways by the Russian state—or, at the very least, deemed not important in terms of its effects on the Arctic. Nuclear generators can be left to rust in the Arctic, and meteorological stations shut down and turned into tourist sites. The lingering effects of climate change are denied—implicitly a decomposing trace of the Soviet past—and are not a central state concern of the "New Russia." Both *How I Ended This Summer* and *The Weather Station* thereby foreground the fact that denial of the past—in family narratives, by the Soviet state, or in regard to the lingering climatic effects of resource extraction—lives on in present-day Russia and cannot be eradicated as easily as one might first assume.

How I Ended This Summer and *The Weather Station* are both overt environmental and climate change allegories, with a tension familiar to contemporary Russian cinema at its heart: the masculinist struggle of power between father and son, between the Cold War USSR and Putin's Russia. Some of these thematic tensions also shape the Oscar-winning and best-known recent Russian fiction film set in the Arctic: *Leviathan* (*Leviafan,* Andrey Zvyagintsev, Russia, 2014). Set in a fictional coastal town on the Kola Peninsula and filmed in Teriberka, a city located about a hundred miles east of Murmansk, *Leviathan* provides a commentary on environmental degradation, the failure of civic and democratic social structures, and the individual's fight against a soulless state and religious machine, both of which are presented as alienated from the natural ecosystem. On the Kola Peninsula, this kind of alienation has a long history. As Andy Bruno notes, a "1938 article in *Polar Pravda* exclaimed, 'By the will of the Bolsheviks under the leadership of the great Stalin, the land of fearless birds is now transformed into an industrial outpost of socialism, into an indestructible fortress of the USSR in a northern periphery'" (2016, 17). Intensive mining, clear-cutting of trees, the installation of multiple large hydroelectric power plants, forests denuded from acid rain, soil and air polluted by chemicals, and industrial-scale fishing of cod all characterize the Soviet and ongoing Russian exploitation of the Kola Peninsula. These developments have been part of the Russian Arctic imaginary for decades: the "Soviet rule had turned the Kola Peninsula into the most populated, industrialized, and militarized section of the global Arctic" (Bruno 2016, 4). The impact on Indigenous populations has been severe; few coastal Nenets remain and the Sámi population has been nearly decimated.

The connections between the Kola Peninsula and circumpolar Indigenous political activism are worth noting. As the transformative Sámi protests against the Áltá (Alta) dam were reshaping Indigenous self-determination activism and political mobilization in Scandinavian Sápmi during the late 1970s, Russian "authorities relocated reindeer-herding villages on the Voron'ia and Ponoi rivers to make way for hydroelectric stations." This led to "the flooding of ancestral homelands of the Sámi community" and a forced "move to the growing agricultural town of Lovozero with its new, if shoddy, high-rise apartments and socialist urban design" (Bruno 2016, 254–56). The continuum of Sámi experiences across the Kola Peninsula and westward throughout Scandinavia are worth noting. This continuity is implicit in *Leviathan* and never directly engaged. Yet the scars of environmental degradation and evidence of active resource extraction depicted in the film provide an opportunity to reconsider the history of circumpolar Indigenous political activism.

Leviathan serves as a useful example of how the Russian Arctic is represented in films that have readily—and to great acclaim—made it to the "outside" and influenced what Western audiences currently associate with Russian depictions of the Arctic. A similar set of representations of the Kola Peninsula, the natural environment, and mining are featured in the recent Finnish film *Compartment No. 6* (*Hytti nro 6*, Juho Kuosmanen, Finland, 2021). A particular kind of environmental symbolism is rich in *Leviathan*: the little man fights against the government's environmental intrusion in the Arctic. Kolya (Aleksei Serebryakov) lives with his family in a small house that is well integrated into the landscape, situated on top of a hill with a panoramic view of the coast. Throughout the film, this edifice symbolizes a scale and scope consistent with the ecological premise of the land. As such, it provides a formidable counterpoint to the mega-industrial installations that characterize most other parts of the Kola Peninsula. This structure does not, within the paradigm of the film, contribute noticeably to climate change, and the major resource-extraction infrastructure that the Kola Peninsula is (in)famous for is out of sight. At the same time, Kolya's family is torn apart by internal strife, including abuse, murder, abandonment, infidelity, and betrayal. The small house seems not quite able to contain all that is happening within it.

Notably, Kolya's house is posited in stark contrast to the ample evidence of abandonment, ruins, and destructive resource extraction surrounding it. As characters move around town, pipelines for gas and oil are repeatedly in the frame, but not remarked upon. They are there as a background to signal that the town is not a small, remote outpost divorced from the global connection points provided by fossil fuels in transit, but interconnected with them. Similarly, the film repeatedly includes shots that emphasize the litter and detritus present in Arctic landscapes; the characters themselves contribute to the waste by shooting bottles during an excursion, leaving the shattered glass in the landscape (figure 10.2). Abandoned properties, boats, and the carcass of a stranded whale all intermingle references to the contemporary Arctic. The environmental allegory of *Leviathan,* as a commentary on contemporary Russian Arctic policy (outlined in chapter 9), is thus interwoven with the character development and depictions of family and marital drama. In an interview, *Leviathan*'s cinematographer, Mikhail Krichman, says that Zvyagintsev "wanted a very realistic effect, in every sense of the word" (Graffy 2019, 308). To this end, camera movement is minimal, and shots are often long takes. This aesthetic resonates with an international-art-cinema set of priorities while also evoking Stalinist cinema, a combination

FIGURE 10.2. The "non-pristine" Arctic in Murmansk, from *Leviathan* (*Leviafan,* Andrey Zvyagintsev, Russia, 2014).

noted above with respect to *How I Ended This Summer* and also found in *Angels of Revolution* (*Angely revolyutsii,* Aleksey Fedorchenko, Russia, 2014), as discussed in chapter 9.

Leviathan's themes also resonate with dominant trends in Russian cinema, which has emphasized a center-periphery dynamic, with Moscow and St. Petersburg repeatedly offset against "the provinces." *Leviathan,* Julian Graffy notes, "makes constant and explicit reference to the pressures and tensions of life in the contemporary Russian hinterland" (2019, 310). At the same time, Nancy Condee argues—in an examination of the film's international circulation, the critical praise it received, and the debates it stirred in Russia—that it is a film of its moment, both anti-Putin and not (2015, 607).

Leviathan is also very much a Russian Arctic film of the twenty-first century. Like most contemporary Russian films, it provides no explicit criticism of Russia's Arctic resource extraction, environmental degradation, and contributions to the climate crisis. Similarly, no reference is made to Russia's Indigenous populations and their history of relationship with the land. But the takeover of the protagonist's land, the destruction and removal of his dwelling, and the rise of a large and foreboding church in its stead offer an implicit critical commentary on how environments are insensitively altered, how democratic processes are left by the wayside, and how hegemonic state power becomes manifested in the landscape in ways that connect tenets of Cold War imaginaries with the present day. Recent films about the Russian presence in Svalbard, the archipelago in the Arctic Ocean, raise similar issues.

An emblematic location in the history of Old and New Cold War geopolitics is the archipelago of Svalbard, governed by Norway through an international treaty and containing several important Soviet and Russian mining installations and communities. As an Eastern Bloc outpost in NATO territory, Svalbard also represents a Cold War imaginary. Mostly covered by glaciers and surrounded, for a good portion of the year, by polar sea ice, the location came to connote "a porous, dynamic, and partially impermanent border area between East and West" (Stenport 2015, 169). While Cold War militarization infrastructure projects dot many regions in the circumpolar Arctic, such is not the case in Svalbard. Yet coal mining in the archipelago ensured the USSR its westernmost Cold War outpost. Costly and inefficient mining operations have been maintained through the present day, which speaks to the geopolitical centrality of the archipelago for Russian interests.

During the Cold War, visiting the remote Soviet settlements Barentsburg, Grumant, or Pyramiden in any other way than through officially brokered public diplomacy stints (cultural exchanges and sports events, notably) were well-nigh impossible. At the same time, period documents affirm that both governing partners—Norway's governor, based in Longyearbyen, and the USSR's consul, based in Barentsburg—maintained a cordial relationship (see Haugdal 2020, 106–07). Yet Svalbard is also an area that continues, "more than elsewhere in the European Arctic, [to be] shaped, or at least influenced, by perceptions and policies formed in the Cold War period" (Åtland and Pedersen 2014, 20; see also Ash 2020). Russia's ambassador to Norway, Yuri Fokin, describes a rhetoric of opposition lingering in Svalbard politics between the two countries, a "legacy passed on from the Cold War: fear, mistrust, and suspicion" (cited in Åtland and Pedersen 2008, 227). Similarly, a "fear of a Western conspiracy often continues to prevail in the Russian readings of the Svalbard issue" (Laruelle 2014, 125). Thus, Cold War geopolitics continue in the post–Cold War world in Svalbard.

Two fiction film examples from the Cold War era foreshadow this ongoing tension and "the threat of the thaw" (see Stenport 2015). The Soviet-Italian coproduction *The Red Tent* (*Krasnaya Palatka,* Mikhail Kalatozov, USSR/Italy, 1969) frames the USSR as a friendly neighbor in the Arctic, as the Soviet icebreaker *Krasin* saves the crew from Umberto Nobile's catastrophic attempt to reach the North Pole from Svalbard via airship. By contrast, a Norway-USA coproduction, the action flick *Orion's Belt* (*Orions Belte,* Ola Solum and Tristan de Vere Cole, Norway/USA, 1985), presents Norway as covering up Soviet military activity to preserve its geopolitical status quo. In a history of cinematic representations of Svalbard, Andrei Rogatchevski notes that "many . . . Soviet films about Svalbard, on the one hand, and at least some Norwegian films about Svalbard, on the other, describe a kind of mutually observed parallel reality (each in their own way)" (2020, 169). As with any ideological positioning, they mirror each other, while offering distinct political narratives.

A remote area of the Arctic rich in evocative grandeur, Svalbard is Europe's "most marginal space; without an Indigenous or permanent population" (Ryall 2017, 232), thereby also representing some of the most cherished tropes of the blank-slate, ice-covered, remote white surfaces of the High Arctic, onto which cinematic images are projected (Ihle 2015, 259). Expanding tourism, international scientific collaboration in response to the climate crisis, and the continuing legacy of the Cold War dominate contemporary cinematic depictions of Svalbard. In both *Fortitude* (BBC/Sky Atlantic Television series, UK, 2015–18), about a fictionalized Longyearbyen, and *Operation Arctic* (*Operasjon Arktis,* Grethe Bøe-Waal, Norway, 2014), about a natural environment in need of protection, continued militarized threats and the impact of climate change provide thematic throughlines. Shot in Iceland, *Fortitude* is a climate crisis anxiety narrative, in which horror lurks in the glaciers, with disturbing effects on humans and fauna once the ice melts. The detective mystery centers on Svalbard as a destination for international tourists and scientists, all of whom seek to experience the pristine environment for different reasons (though, on a narrative level, they are all escaping from something), but sharing the common concern that because of climate change, whatever they are searching for may be rapidly disappearing, with traumatic environmental effects of destruction to come. The family film *Operation Arctic,* Norway's highest-grossing film in 2014, depicts the country as a fully militarized nation on standby, ready to respond to threats from its militarized neighbor to the East. Set in winter on Svalbard, *Operation Arctic* retells a contemporary story of Norway's imperiled small-nation status during the Cold War as a children's film, in which the resourceful big sister must save her younger siblings from invading polar bears and forbidding cold. The allegorical components are obvious: Russia may be on the prowl, a theme also rehearsed by the resource extraction series *Occupied* (*Okkupert,* Karianne Lund, Jo Nesbø, and Erik Skjoldbjærg, Norway, 2015–2019). These fictional stories recast the legacy of Cold War Arctic tensions in an age of climate change as part of contemporary genre narratives: the Nordic noir and the children's adventure film.

In terms of documentary, the Cape Farewell artist collective has made a number of films that bridge perspectives of artists and scientists convening in Svalbard, such as their *Art from the Arctic* (David Buckland and David Hinton, UK, 2006), documenting an expedition to the Svalbard archipelago. BBC Earth also launched a docusoap television series in 2016 called *Ice Town: Life on the Edge* (UK), which chronicles a group of international scientists and residents in Longyearbyen and their travails against darkness, climate, and malfunctioning research instruments, as well as interpersonal melodrama. Innumerable wildlife documentaries have also been shot in and around Svalbard over the past decade, demonstrating the relative accessibility of the explorer-support industry in this High Arctic location, which facilitates the practice for filmmakers and their crews. As an outcome of increased travel to the archipelago, a search on YouTube and Vimeo in 2021 indicates many dozens of films—some quite accomplished—being made by tourists who seek to capture both the environment and ecology, while also documenting the Soviet mining settlements.

In this context, we note a subset of recent Svalbard documentaries that address Soviet Cold War and present-day Russian resource extraction as a geopolitical strategy, its legacy in Soviet-established mining towns, and the ways in which the Cold War imagined past carries into the present to intersect with Anthropocene imaginaries. These films are *Grumant: Island of Communism* (*Grumant: ostrov kommunizma,* Ivan Tverdovskii, Russia, 2014) and *Dream Town* (Adrian Briscoe, USA/Norway, 2014), which examine Barentsburg, and *Efterklang: The Ghost of Piramida* (Andreas Koefoed, Denmark, 2012) and *Welcome to Pyramiden* (*Piramida,* Ivan Tverdovskii, Russia, 2016), which focus on Pyramiden. In these works, there are some similarities to Russian art-cinema depictions of the Arctic in *How I Ended Last Summer, Leviathan,* and *Angels of Revolution* (see chapter 9). Abandoned, defunct, and decrepit buildings and coal-mining infrastructure are extensively featured, thus provoking a stark contrast to Russian state-promoted discourse and policy of the Arctic during the past decade, which has foregrounded the reality that Russia's geopolitical priorities, and its ambition, are communicated through massive new industrial and fossil-fuel-extraction megaprojects. At the same time, as multiple scholars note—and as the films we discuss in this section reflect—the ambition is also ambivalent.

Russia's posturing, including Chilingarov's flag planting in the North Pole seabed in 2007, has not been matched by adversarial action toward Norway or the other Arctic 5 superpowers, either on Svalbard or beyond.

The supposedly pristine High Arctic wilderness of Svalbard is promoted by the government of Norway as "one of the best-managed wilderness areas in the world" (Norwegian Ministry of Justice and the Police 2009, 9). At the same time, the landscape is characterized by markers of the Anthropocene, including not only fossil fuel extraction and pollution, but also the Svalbard Global Seed Vault, the international community's response to anthropogenic plant extinctions, and the Arctic World Archive, a digital data preservation center. The most conspicuous of these Anthropocene connections is coal extraction. Mining began in Svalbard in the first decades of the twentieth century. Throughout the Cold War and into the present, Norwegian and Soviet subsidies ensured its continuation.

The ongoing operations at Svea Mine and Barentsburg are indisputable vestiges of the Cold War; neither site is profitable, and both are in continued operation largely to legitimize nation-state presence in the archipelago. In 2015, the UN's executive secretary on climate change, Christiana Figueres, requested that Norway shut down the coal mines on Svalbard since the operations counteract Norway's environmental priorities and are "incongruous with what the island actually stands for, namely climate research" (cited in Norum 2016, 53). Anthropocene evidence is thus an unequivocal component of the natural and built environments of Svalbard. It operates on a continuum that includes climate science research centers and installations, the globalization of mass tourism and itinerant labor populations, Cold War geopolitics of the coal economy, and stark symbolic optics. All Soviet and Russian buildings in Barentsburg, Grumant, and Pyramiden originated as utopian, or at least socially engineered, company towns of the coal era. They have been preserved not only by a cold climate, but also because removal of abandoned infrastructure was difficult and expensive during the Cold War and not considered a priority by Russia during the 1990s. Since 2001, the Svalbard Environmental Protection Act enacts a stringent historical preservation code to ensure that the "cultural relics" of the coal-mining infrastructure are protected (Catford 2002, 36). This legislation effectively prohibits Russia from diversifying its economic interests in Svalbard, preserves Anthropocene evidence of coal extraction, and also generates a seductive, nostalgic framework for the Soviet legacy encapsulated in the period buildings.

Grumant: Island of Communism is largely set in darkness, whether above or
below ground. The film's subtitle, "Island of Communism"—a name given to
it by tourists—also plays with the notion that, like other far-flung outposts
of the past, Svalbard is metaphorically the last to hear about regime change.
The film is infused with Cold War nostalgia and harkens back to the glorious
past, when coal was king and everything—from housing to food to entertainment (the community had five thousand films in an archive, now presumed lost)—was taken care of by the Soviet state. The film integrates several
strategies and can best be described as a poetic documentary. It includes
interviews with current residents; archival footage from the Soviet era; inte-
riors of mines, apartments, and community buildings; and landscape shots—
some harkening back to the sublime and exoticizing traditions of Arctic
representation. The film is organized around juxtapositions between the past
glory and current failure of coal, the past center of Anthropogenic industrialization and one that now lies in ruins, and the past glories of communism
and the community's present state.

Contrasts between the Soviet past and ambiguous ideologies of the
present are foregrounded by both the film's title and its framing. The town's
statue of Lenin both opens and closes the film (figure 10.3); similarly, there
are recurring shots of Lenin statues on the mainland, as miners leave their
homes and families for the journey north. In the opening moments, shots
loop back to Lenin and the hammer and sickle. In the film's concluding
scenes, there is an ironic and kitsch element to the Lenin statue, which is
juxtaposed with the (perhaps empty) rhetoric about peace and love from
Putin's New Year's Eve speech on the television screen. Inhabitants then sing
both the Russian Federation's national anthem and the Soviet Union's
national anthem, followed by people singing songs about international
understanding. These songs also harken back to the glory days of the mining
past, when Soviet and Norwegian miners would meet in good will.
Interviewees, many from Russian-speaking areas of Ukraine, spend a great
deal of time reflecting on the fact that during the Soviet era, coal miners
would have everything they needed. Now, in the post-Communist era, basic
survival is the mode of operation. Unlike in the Soviet past, when a whole
community infrastructure was in place, families are now displaced and connecting over Skype.

FIGURE 10.3. A remnant of the Arctic Cold War: Lenin statue in *Grumant: Island of Communism* (*Grumant: ostrov kommunizma,* Ivan Tverdovskii, Russia, 2014).

Despite all the threads of Soviet nostalgia, we also see how the Cold War past still plays a role in current aspects of the Anthropocene: Russia keeps resource extraction open in Barentsburg to ensure political presence in the Arctic during its transition from resource extraction to tourism. This is not the only Russian anthropogenic, resource-extractive site in the Arctic; however, it is the one that has the Soviet and Cold War past most consistently present, both as ruin and as the last vestige of the former state's operations. The film mobilizes an aesthetic of the Anthropocene in which the darkness of coal, of mining, and of the long Arctic winter nights occludes the devastation that fossil fuels generate. This aesthetic is juxtaposed with the film's diegesis, which foregrounds everyday life and occurrences. At the same time, the film emphasizes inhumane working conditions, isolation, and costs to human well-being and laments the deaths of so many coal miners, past and present. Indeed, throughout *Grumant: Island of Communism,* those who died in the mines, toiling at resource extraction, are memorialized. Although it includes no explicit display of resource extraction in general or of coal extraction in particular, Tverdovskii's film provides an implicit criticism of Russian state-sponsored hydrocarbon industries and their effects on humans and the environment, through the miners' accounts and through cinematographic strategies—in particular, filming mostly in darkness, with landscape and interior shots of the mines contrasted.

Unlike the bleak images that pervade *Grumant: Island of Communism,* Adrian Briscoe's *Dream Town* is an experimental documentary that, in the diegetic present, restages fantasies and memories of past events in the lives of current inhabitants, including miners, scientists, service personnel, and students. Briscoe's method of participatory documentary includes workshops on story development, videography, and editing. Participants then stage and film vignettes, which are interwoven with reflections on everyday life in Barentsburg, including Russian and Ukrainian miners commenting on the harsh work, glaciologists documenting retreating ice and thinning sea-ice cover, and a few longtime residents reflecting on the sociological experiment of the legacies of utopian mining towns in Svalbard, as an outpost of the USSR.

The juxtaposition of documentary images with the participatory recreations allows for the possibility of viewing the interior fantasy lives of inhabitants come to life against a landscape the film posits as bleak, forbidding, and seemingly at the end of the world. Outside their fantasies, most of the inhabitants are consumed with work in the service of resource extraction. This focus is contrasted with the work of the other main group of Barentsburg inhabitants: the international cadre of scientists who travel to Svalbard for empirical observation of the effects of climate change. What is revealed is that this socially engineered town, like so many Arctic cities constructed explicitly to support fossil fuel extraction and shipping to faraway places, continues to exist in a Cold War geopolitical spatial imaginary that *Dream Town* depicts as intrinsically connected to the Anthropocene.

While *Dream Town* gives voice to the individuals who underpin the exploitative global system of extraction, transportation, consumption, and pollution as unacknowledged cogs in this machine, the film also provides perspectives of the inhabitants' interiority. Briscoe constructs a participatory cinematic exercise to make visible the interior lives of Barentsburg inhabitants, especially when these do not conform to the ideals and ideas propagated by the Russian nation-state. In this way, participant interiority is made to contrast with the unequivocal causal relationship between coal extraction, as the ur-form of the Anthropocene, and effects of the climate crisis on Arctic environments such as the island of Grumant and beyond. Briscoe's approach to depicting contemporary life in Barentsburg is strikingly different from Tverdovskii's. Briscoe is concerned with interior lives and, by extension, with the ways in which the Anthropocene can be embodied. Tverdovskii is concerned with material ones, which underplay the human effects of the Anthropocene. Taken together, the films, by remediating Soviet-era history on Svalbard, confirm the legacy of Cold War Arctic geopolitics as a constitutive component of the present climate crisis.

One of the northernmost settlements in the world, Pyramiden is described by bloggers, travel writers, and tourist operators as both a ghost town and a "Soviet town frozen in time." Like Barentsburg, Pyramiden is an ideal Soviet city, replete with a Lenin statue, a culture palace, and a lawn whose grass grows on fertile soil shipped from Ukraine. Andreas Koefoed's experimental documentary *Efterklang: The Ghost of Piramida* depicts some of this detritus, while also repurposing it in a number of ways. Although the film documents some of the leftover remnants, the emphasis is on the sounds these objects and sites produce, and how those sounds are incorporated into new electronica tracks by the Danish ensemble Efterklang for their fourth album, *Piramida* (4AD, 2014). *Efterklang: The Ghost of Piramida* also operates on disjuncture and juxtaposition in terms of its visual montage, which brings found footage of the heyday of coal extraction at Pyramiden into tension with the musicians' experiences.

The film documents 8mm black-and-white and color home movies projected in the Moscow apartment of Pyramiden's former official photographer, Alexander Ivanovic Naomkin. The viewer is invited into his home, where Naomkin's nostalgic present-day voiceover about the beauty and wonder of life in Soviet Pyramiden accompanies the screening of footage from his personal archive. These sequences are intercut with the process of Efterklang's sound collection from debris strewn across the site and the landscape, with band members banging on cisterns, jumping on coal cars, tapping on empty bottles, and flipping through documents meant to produce eerie sounds that document Pyramiden's deserted state. As José Duarte notes, "the documentary seems to explore two different directions: first, the nostalgic view presented by Alexander [Naomkin]'s memories; second, the journey undertaken by Efterklang" (Duarte 2018, 132). Naomkin's wistful voiceover about the life he led at Pyramiden—emphasizing happy memories of community, family, and the Soviet state—is presented as divorced from the band's auditory documentation of industrial debris. Neither Naomkin nor the band members directly situate their engagement with the metaphorical or material spaces of Pyramiden as part of an Anthropocene or Cold War context, yet the geopolitical history of coal extraction is omnipresent in both forms of documentation, with one idealizing and the other suggesting that Pyramiden has been abandoned by Russia and turned into a "ghost town" (see Kinossian 2020).

A similar unresolved tension about the Anthropocene, the Cold War, and the Soviet Union underlies one of the most significant visual twenty-first-century artifacts related to Pyramiden: the photo book *Persistent Memories*

(Andreassen, Bjerck, and Olsen 2010). This is an interdisciplinary work in heritage studies and post-processual archaeology, which nevertheless presents a predominantly nostalgic view of the site and of Cold War culture. Assumptions are made that the Pyramiden available to visitors in the early twenty-first century is "authentic" and that the built environment has been preserved as it was when the mine closed in 1998. The very messiness and the scattered artifacts seem to support the notion of a hasty departure by a totalitarian state that disavowed the residents' right to gather their personal objects. Objects featured in public locations are allowed to speak for themselves as artifacts, representing a different social system, with their "thingness" predominant (2010, 23). The authors largely assume that Pyramiden serves as a counterexample to the notion of a "Heritage Ruin" as "a staged, neat and picturesque site that provides visitors with a disciplined and pure space" (2010, 138). They ask, "What are we to do with this rusting and crumbling place in the midst of pristine Arctic nature, Europe's last authentic wilderness?" (2010, 138). An alternative understanding of Pyramiden and its status as a site for tourism is that it is, in fact, a Heritage Ruin: one that is staged, designed, and disciplined in ways that erase its Anthropogenic connections.

There is, in fact, little "pristine wilderness" around Pyramiden. The valley and surrounding mountains, too, are monuments to the Anthropocene—the evidence is all over the landscape that this has been an industrial and resource extraction site for over a century. Clearly, the notion of Svalbard as the last true wilderness of the world is at least partly a result of a carefully orchestrated policy of the Norwegian government to maintain and increase control over Svalbard. *Efterklang: The Ghost of Pyramida,* beautifully shot and detached from historical context and the climate crisis impacting the Arctic in the twenty-first century, aligns more closely with this view. Its depiction of Pyramiden therefore prompts a discussion not only about what a Heritage Ruin may be, but about the very notions of a ruin and its antithesis, unexplored wilderness, in an Arctic characterized by geopolitical ambivalence that has, for a very long time, pitched alternating versions of nature and wilderness against exploitation, resource attraction, and military and political dominance. *Efterklang: The Ghost of Piramida* allows the Heritage Ruin to speak not only of a past, but of a present, through its use of sounds, which blend these temporalities together: some of the sounds come from decay, while others reverberate in the same way they would have thirty years ago. The projections of the amateur movies in Moscow also function, therefore, as a Heritage

Ruin: a document of the idealized past of Soviet utopia, never obtained, yet cultivated as a desirable goal offset against Anthropocene evidence.

Tverdovskii's second Svalbard film, *Welcome to Pyramiden,* takes a different stance. On the one hand, it is a far more pessimistic view; on the other, it effectively mobilizes irony and even satire to comment on Russia's present-day interests in launching the abandoned mining settlement, and what we can only read as an Anthropocene monument, as a tourist destination. The film has three protagonists, all of whom are working on developing aspects of Russian-based tourism. The protagonists must go to extreme lengths in the film to even obtain potable water, and despite the addition of a sign advertising "beer + food," there does not seem to be an overwhelming draw to the area. While the three protagonists are energetic in their endeavors, reflecting the growth in tourism to Svalbard and the Russian settlements over the past decade (see Rogatchevski 2020, 167), there nevertheless seems to be a subtle acknowledgment of their futility. The film is infused with a jaunty musical score, reminiscent of a travelogue, and a sardonic voiceover. Part of the irony and humor of the work is that Russian tourism is based on revealing a decaying monument to resource extraction and the Anthropocene, in order to justify a continuing Russian presence on the Norwegian island as resource extraction itself dries up. If Tverdovskii's previous film postulated a lost, nostalgic past in the Soviet era, *Welcome to Pyramiden* postulates that there is no future on the horizon. Yet the film's voiceover and its shot composition attempt to be humorous and whimsical about the apparent bleakness.

In contemporary cinema and media representation of Svalbard, Soviet presence is presented as kitsch nostalgia and therefore seemingly harmless: massive Cold War military installations and pollution of the environment are nonthreatening, relegated to an era when utopian imaginaries based on the notion of altering Arctic environments for the benefit of world peace were still possible. This nostalgic sentimentalism is arguably reassuring for Western tourists, as if Svalbard provides a ground zero of possibilities for a non-anthropogenic future. Similarly, in the Svalbard films, the naturalization of Lenin into a kitsch postmodern figure elides the crimes against humanity of the Leninist and Stalinist states. The image of Lenin persists in these films, and on the mainland, in ways that reflect an absence of a new national imaginary for Russia in the post-Soviet era (Gerlach and Kinossian 2016, 14–16). Therefore, returning to select vestiges of the Cold War era promotes a nostalgic history, very much different from the kind of interventionist historiography that we examine in other Arctic cinemas (see chapters 3–7).

An important component of contemporary Russian Arctic cinema is the ways in which the Cold War is remembered, memorialized, and repurposed. Environmental depictions are central to this approach, whether evoking such large-scale geoengineering projects as those envisioned by Borisov's 1950s proposal to dam the Bering Sea or illustrating the centrality of climate science to geopolitical relations between East and West. Many of the films discussed in this chapter function as cinematic sites of a post–Cold War space where Soviet infrastructure, practices, and buildings remain as visible evidence of that era. As such, they also function as reminders of Putin's totalitarian state and its consistent repurposing of Russia's Arctic explorer legacy. In addition, the films in this chapter are contemporaneous with Russian state narratives that promote an acceleration of fossil fuel and rare-earth mineral extraction all over the Arctic, from Murmansk to Arkhangelsk, and the reincorporation of the Eurasian Arctic into the national Russian imaginary after a decade of absence. The cinematic evidence discussed herein, however, provides no indication of such nationalist sentiment or ambitious industrial acceleration; instead, it presents detritus, ruins, and polluted and destroyed environments and communities. These works, then, document the ongoing and elided traces of the past and their resonance in the present.

ELEVEN

Looking Ahead

GLOBAL ARCTIC CINEMAS IN
THE TWENTY-FIRST CENTURY

DURING THE TWO-DECADE SPAN covered in this book, cinematic representations of an interconnected and globally connected Arctic, the climate crisis, resource extraction, and Indigenous agency and self-determination have undergone multiple iterations. Most of the works we have addressed are documentary, experimental, or art cinema, which rarely break through into global popular culture, though they have garnered significant attention in the Arctic and beyond through film festivals, Indigenous communities, art museums, streaming, and video on demand. Concurrently, several cli-fi and global catastrophe films have sought to bring explicit attention to the climate crisis and Arctic environments, from *The Day after Tomorrow* (Roland Emmerich, USA, 2004), *Snowpiercer* (Bong Joon Ho, South Korea/Czech Republic, 2013), and *Interstellar* (Christopher Nolan, USA, 2014) to more recent works such as *Greenland* (Ric Roman Waugh, USA, 2020) and *Don't Look Up* (Adam McKay, USA, 2021). One thematic strand ties these otherwise different films together: all are "big" planetary science films, connecting glaciology, astronomy, geoengineering, and the military with patriarchal valor and a male hero's journey to save the heteronormative nuclear family and, therefore, all of humanity. Another cli-fi work in global popular culture is Pearl Jam's video *Retrograde* (Josh Wakely, USA, 2020), a remediated animated work that combines images of the global flooding of major cities with images of calving icebergs and lonely polar bears as a not-so-subtle climate allegory, which includes Greta Thunberg appearing as a clairvoyant. The multiple discourses around ice mobilized by these works also resonate with Cold War tensions between East and West—and with the ways in which glacial landscapes of the Arctic were tied into those tensions—to signal a long-standing and variable anxiety of ice: it is both geopolitically connected and culturally constructed.

All these works—whether global Hollywood features, made for streaming platforms such as Netflix, or international art cinema—are intended for global audiences. Two documentaries with global reach discussed in chapter 2, *Chasing Ice* and *To the Arctic,* are versions of the cli-fi disaster film—one metrical and one affective. The first includes glorious representations of melting glaciers and calving icebergs that are to be saved by photographer James Balog—artist/scientist/explorer—through his photographic documentation of the disappearance of ice as visual evidence of the climate crisis. In *To the Arctic,* the affective narrative about disappearing ice and polar bear habitat also points to catastrophe while attempting to mobilize both anthropomorphism and empathy as catalysts for change. As such, films like these connect to both global change science and global media studies, as addressed in chapter 1. Specifically, they echo a range of anxieties around ice, which parallel historical depictions of the Arctic in which ice was an obstacle to the polar hero's journey: there was too much. A similar affective imagery is mobilized in global popular culture through movies such as *The Day after Tomorrow, Interstellar,* and *Snowpiercer,* whose doomsday scenarios feature worlds freezing over. Not surprisingly, in documentaries from the past twenty years, the anxiety is about disappearing and diminishing ice: there is too little. In Indigenous works, such as *Inuit Knowledge and Climate Change,* ice is an epistemology ignored by outsiders. It is knowledge revealed to and embodied by Inuit elders, who, sidelined by Western scientists, are filled with frustration that what is apparent to the Inuit remains invisible to those outside who think they know better.

In twenty-first-century documentary and news media, moreover, the lone, emaciated, shivering polar bear on an ice floe became the most iconic representation of the climate crisis. As Finis Dunaway argues, "polar bear imagery transcended the ephemeral conditions of weather to reveal alarming evidence of climate change, to make this long-term, gradually escalating problem appear immediate, located in the coeval present rather than in some vague, imaginary future" (2015, 260; see also Stenport and Vachula 2017). The image of the polar bear is quite often maternal, with cubs following behind. It is always a passive victim of climate change. This echoes the long-standing, but only recently examined, gendered knowledge assumptions of glaciologists (Carey, Jackson, and Rushing 2016) whereby glacial and sea-ice melting connotes passivity and victimization. The media works addressed in this book counteract that kind of one-sided representation, or "the passivity" in which ice "just melts" (Pollack 2012, A26). Our goal is to bring the diversity of

Arctic media works into a rich conversation with one another, allowing this dialogue to expand and reach broader contexts. Facts inform; stories transform—as the adage goes.

The polar bear on an ice floe thus becomes a key symbol as it connects some of the main arguments of cli-fi disaster films or popular documentaries: agency is limited, is likely futile, and, if exercised, is only individualized (see Ghosh 2016). The likely failure is twofold: the failure of the individual (whether the feminized polar bear or the male heroic explorer) to have any impact, and the failure to think beyond the individual. Opportunities for systemic change are thus squandered, and patriarchal narratives of heroism do nothing to stop this. So, there is a nihilist turn in global popular climate crisis cinema, according to which the Arctic and the ice have already disappeared and are not coming back, and thus the trauma of vanished ice can no longer be indexically represented as such on screen—the truth must be sublimated or suppressed. Although this nihilist turn, as John Marmysz notes, is "relentlessly criticized for overemphasizing the dark side of human experience, it might be equally true that this overemphasis represents a needed counterbalance to shallow optimism and arrogant confidence in human power . . . and that despite all of the accomplishments and wonders of civilization, humans cannot alter the fact that they possess only a finite amount of mastery and control over their own destinies" (2003, 159). He goes on to note the importance of humor to nihilism. "In thinking that is overly serious, one becomes inappropriately invested in a single way of deliberating about a subject" (2003, 161). The desire for things to change profoundly is combined with the recognition that, despite our efforts, they won't—and so, laughing (in a sense) in the face of doom becomes a way to process the climate catastrophe. These failures have given rise to a new approach to climate crisis representation, born out of exasperation and resignation: satire. *Don't Look Up* was panned by most film critics but valued by many climate scientists for the way in which it reveals and publicly magnifies their own frustrations with the ineffectiveness of media awareness campaigns and global political inaction and apathy (see Kalmus 2021; Oppenheimer 2021). The earnestness of previous films has been relinquished: instead, this film is a searing parody of the past fifteen years of inaction since the release of *An Inconvenient Truth* (Davis Guggenheim, USA, 2006). *Don't Look Up*'s use of satire reflects the disillusioned experiences of scientists while the film implicitly laughs in the face of imminent catastrophe, as no action seems to change anything. *Don't Look Up* is a critique of inaction while nihilistically laughing at the recognition of action's impossibility.

Over the past decade, perhaps one of the most significant global mainstream phenomena concerning the changing conception of "ice" in climate change and Arctic representations has come from Disney studios: *Frozen* (Chris Buck and Jennifer Lee, USA, 2013) and *Frozen II* (Chris Buck and Jennifer Lee, USA, 2019). In both *Frozen* films, ice is plastic and dynamic, while the film's "climate message" is opaque. On the one hand, Disney's environmental conception is not unlike that in *The Day after Tomorrow,* with the actual environmental threat facing the planet being turned on its head. Elsa's power causes the climate crisis—she makes the world freeze over—and only she can fix it. As in many previous climate catastrophe works, individual agency takes center stage, without any gesture toward collective action, beyond the team-work of a comedic menagerie of companions (a classic Disney device). At the same time, *Frozen* is a melodrama, one in which women are the protagonists and deep sisterly bonds, rather than heteronormative romantic love, resolves the narrative crisis. Yet Elsa's power also follows patriarchal norms: it cannot be controlled, and when unleashed it causes widespread destruction.

While ice is central to *Frozen,* the film does not have a coherent story to tell about climate change. Read through one lens, *Frozen* marks the death of salvage ethnography: Elsa's powers freeze over Arendelle; the kingdom becomes quite literally preserved for all time, like a fly caught in amber. In a way, reversal of the salvage ethnography paradigm is the goal of Anna, her sister, and her companions: the unfreezing of the kingdom enables it to leave stasis behind and move forward, implicitly echoing the critique of salvage ethnography launched by anthropology's "critical turn" in the 1980s and '90s. Melting becomes positive—just as the film promotes the idea that human agency over climate is paramount. This is demonstrated by the fact that *Frozen* was seen by US officials as a vehicle to address the climate crisis (Alter 2015). The film's mixed messages about climate, and the contradictions built into its allegory, signal the paucity of global popular culture addressing the issue. Seemingly a tale about women's empowerment—a counterpoint to the plot of most conventional fairy tales—the film portrays effective climate action as utopian and possible only when embedded patriarchal power structures of the Anthropocene are miraculously absent.

The inequitable power dynamics of global warming and resource extraction, including the Arctic paradox, are addressed in *Frozen II,* and this time specifically from an Indigenous perspective. The Sámi cultural context was present in *Frozen* for those who knew how to identify it; indeed, Sámi representatives felt that the "original *Frozen* failed to acknowledge that it had drawn some of its inspiration from Sámi culture" (McGwin 2020). Because of this, the International Sámi Film Institute and the Sámi parliaments of Norway, Sweden, and Finland became involved in an effort to place Sámi culture as an integrated aspect of the plot, characterization, production, and circulation of *Frozen II*. Disney and the Sámi parliaments entered an agreement on consultation, cultural sensitivity, and resource building, which Ojibwe writer Jesse Wente, chair of the Canadian Council for the Arts, stated was unlike anything else in terms of Indigenous collaboration, affirming that "I think it's a great precedent for how Indigenous nations might deal with a corporation the size of Walt Disney, as well as governments and other agencies, around the use of their cultural and intellectual property in popular entertainment" (Simonpillai 2019).

The underlying complexities of the *Frozen II* collaboration, notably the spectrum between celebration and commercialization, between recognition and exploitation, and between cultural diversity and whitewashing—has led to ongoing, and at times contentious, debates within Sámi communities (Kvidal-Røvik and Cordes 2022). After its North American release, *Frozen II* was translated and dubbed into North Sámi, screening across Sápmi. Part of the agreement between Disney and the Sámi parliaments was formalized with a "benefits sharing" agreement that would return a portion of the film's earnings to Sámi institutions, while other "cross learning opportunities" would be pursued ("Agreement" 2019). The ISFI's experience working with Disney on *Frozen II* also led to the development of the manual "Ofelaš—The Pathfinder: Guidelines for Responsible Filmmaking with Sámi Culture and People" (ISFI 2022).To this end, while *Frozen II* itself may or may not be an example of interventionist historiography on its own, the agreement between Disney and the Sámi parliaments, and the dubbing of the film and return of resources, can be seen in that light, as an (at least partial) example of media sovereignty.

Historical events significant to Sámi culture are integrated into *Frozen II,* with a recurrent emphasis on the negative environmental effects of settler colonialism. The dam is perhaps the most obvious example, standing in for the Álttá (Alta) dam political struggle of the late 1970s. Thus, Anne Lajla Utsi, director of the International Sámi Film Institute, described her favorite moment of the film: "The Norwegian government did build the dam, so it was nice to see it blown away in *Frozen 2*" (Roesch 2020). Here, Elsa's powers are ascribed to her

Indigenous power inherited from her mother, who she didn't know was Northuldra (the fictional Indigenous Sámi stand-in group). In a thinly veiled allegory, Elsa's father is the patriarchal, colonial, oppressive nation-state, whose detrimental impact on the Sámi and on the environment has been covered up by silence and complicity for too long. Disruption of the natural environment, the fragile Arctic ecosystem, and local communities is caused by these sublimated violent acts, which no one, even Elsa and Anna, know stem from within their own family. This is perhaps one of the darkest aspects of *Frozen II,* and not one that the film seeks to resolve. By calling attention to the unequal but intertwined relationship of settled and settler, colonizer and colonized, *Frozen II* can be read as interventionist historiography through global popular culture. But, as noted above, the strongest evidence for the film's status as interventionist historiography is found in the facts of its production, circulation, and language dubbing, all of which was influenced by the agreement between Disney and the Sámi parliaments, and the collaborative development of the Indigenous guide to filming that ensued. This is the first time that Disney explicitly collaborated with representatives of an Indigenous group when making a film about that group and its multilayered, historically complex experience. As Christina Henriksen, vice president of the Saami Council, stated, "*Frozen 2* isn't our story, but our story is a part of *Frozen 2*" (McGwin 2020).

The multilayered depiction of Indigenous women's strength and familial bonds in *Frozen* and *Frozen II* should not be underestimated; this is in contrast to earlier, exploitative Disney films such as *Pocahontas* (Eric Goldberg and Mike Gabriel, USA, 1995). This trait is shared with Amanda Kernell's *Sami Blood* (*Sameblod,* Sweden, 2016; see chapter 5), in which the bonds between the two sisters, and between the protagonist Christina/Elle-Marja and her mother, are ruptured by colonial Swedish culture to the extent that Elle-Marja disowns her Sámi heritage and native language. By contrast, Elsa and Anna are affirmed as Northuldra by the end of *Frozen II* and emerge strengthened from their perilous journey to set things right in Arendelle, arguably achieving self-determination. This resolution engages, on the one hand, what Rauna Kuokkanen describes as the pervasive myth in Sámi culture of the Strong Sámi Woman (2019, 172–74), which purports that Sámi culture is neither patriarchal nor misogynistic and is thus distinct from the Scandinavian settler state. But, on the other hand, perhaps Elsa's and Anna's actions and experiences in *Frozen II* reflect what "Indigenous feminist analysis has established," namely, "that Indigenous sovereignty struggles are always gendered" (Kuokkanen 2021, 312). As Kuokkanen argues, the "ethos of Indigenous feminist theory and practice of sovereignty struggles" differ from "conventional Indigenous politics and the rights-based, electoral politics driven approach" (2021, 311). Though defined within a Disney frame, the plot of *Frozen II* gestures toward self-determination and self-governance over Indigenous lands. It is the first film to reach global audiences that engages the Sámi and oppression by the Scandinavian nation-states, including through heteropatriarchy. To this extent, the two *Frozen* films can indeed be understood as interventionist historiography, as works that have placed and recognized the Sámi's colonial position in Scandinavia in front of a global audience for the first time.

It is also important to note the self-conscious shift by Disney studios with the Frozen series, away from their historical reliance on fairy tales in their animated feature films and toward the use of myths. Fairy tales function as didactic allegories, with black-and-white antagonists and protagonists and recurrent structural elements (Propp [1928] 1971). Myths, on the other hand, are expansive, shifting, generative, and often internally contradictory. Moreover, fairy tales are always understood to be fictional, whereas myths can be seen as fiction or as truth (Greimas [1985] 1992). Disney's shifting from fairy tales to myths leads to the potential for more complex characters and for proper care and respect toward the incorporation of Indigenous stories. The use of these myths allows Disney studios to address the climate crisis affecting the Arctic by, in part, acknowledging and incorporating Indigenous perspectives.

Disney studios brings this shift to the foreground with *Myth: A Frozen Tale* (Jeff Gipson, USA, 2019), originally made in virtual reality (VR) for Oculus Quest, and later released in 2D on the streaming service Disney+. *Myth: A Frozen Tale* can be seen as a concluding film of a trilogy, where the message about anthropogenic climate change is brought full circle. Unlike the features, this short has an abstract aesthetic much in line with Disney's *Fantasia* (James Algar and others, USA, 1940), with symbolic landscape renditions accompanied by an alternately soaring and bombastic musical score. There is little dialogue, but what there is subscribes to a mythopoetic mode and confirms that the fifth spirit—humanity—has thrown the earth's ecosystem out of balance, causing severe environmental and cultural destruction, and that only if it comes back into synchronicity with the other four spirits can earth achieve its original balance and harmony. The narrator (Evan Rachel Wood) concludes lamentingly: "But even a spirit out of sync can cause chaos. The fifth spirit lost its rhythm, and the harmony was broken. The elements raged out of control. Our world was fractured. When the chaos fell silent, the spirits had vanished, never to be seen again. But it is said that if the fifth spirit can find its rhythm, the others will return, bringing with them that beautiful harmony and the world can dance once again." *Myth: A Frozen Tale* deploys an aesthetic that veers from abstract animation to mythopoetics through VR to mobilize a sensory experience that foregrounds affect, moving away from narrativity as the means to engage audiences in the climate crisis, while moving toward immersive sensorial affect as a means of engagement.

<h2 style="text-align:center">THE ARCTIC WAS ALWAYS CONNECTED—
REMEDIATION AND DAWSON CITY: FROZEN TIME</h2>

While Disney's collaboration agreement with Sámi political and cultural institutions for the making of *Frozen II* is new, the connection between Hollywood, global popular culture, and the circulation of moving images to and in the Arctic is not. During the silent era, moving images quickly circulated through the Arctic. Giron's (Kiruna's) first cinema opened in 1912. Early films often addressed resource extraction and included depictions of settler colonialism. Thomas Edison's company shot *The Klondike Gold Rush* (Edison, USA, 1898) on the Canadian-Alaskan border in 1898 and went on to produce other *actualités* in the region, such as *Packers on the Trail* (Edison, USA, 1901). Multiple *actualités* featuring mining, infrastructure, and Sámi reindeer herding were made in Sápmi from the early 1900s onward, such as *En resa med världens nordligaste järnväg* ("A journey on the northernmost railroad in the world," Sweden, 1910) and *Malmbrytning å Kirunavaara* ("Iron ore extraction at Kirunavaara," Sweden, 1911). Colonial fantasies of the Arctic circulated as well, with Georges Méliès's *The Conquest of the Pole* (*À la conquête du pôle*, France, 1912) and Charlie Chaplin's *The Gold Rush* (USA, 1925) becoming canonical. Greenlandic-Danish explorer Knud Rasmussen wrote about his moviegoing experiences in Nome, Alaska, in 1927, when local Indigenous populations and settlers alike "went to the 'Dream Theatre,'" a three-hundred-seat movie theater that opened in 1913, "in the afternoons" (Ostermann and Holtved 1952, 65–66). Thus, the Nome population was well accustomed to the cinema and its conventions (see MacKenzie and Stenport 2021).

Addressing, in part, the silent era, experimental filmmaker Bill Morrison's compilation documentary *Dawson City: Frozen Time* (USA, 2016) expands the context of the Arctic's relationship to global media history. Morrison's practice is based on using found footage, often the most physically degraded examples of cinema's detritus, to build new works that reveal the centrality of cinema to twentieth-century visual cultures. *Decasia: The State of Decay* (USA, 2002), made entirely of decayed silent films, is a reflection both on the impermanent nature of cinema and on mortality. It was the first film of the twenty-first century to be chosen for the Library of Congress's National Film Registry in 2013. His *Spark of Being* (USA, 2010) draws on the entire history of cinema's detritus to revivify Frankenstein without using any images from Frankenstein films. *Dawson City: Frozen Time* takes his practice of remediation and turns it to a more properly documentary form. In the first instance, *Dawson City: Frozen Time* tells the story of the "Dawson find," when hundreds of nitrate films were found buried in Dawson City in 1979 (Gates 2022, 234–247). Morrison's work uses the rediscovery of buried films to tell a series of stories about resource extraction, environmental change, and the removal of Indigenous self-determination over the land. In so doing, he offers a heterogeneous series of contrasting narratives as an interventionist historiography of Dawson City and the Arctic in the early twentieth century.

Dawson, located just west of the Klondike mining site on the Indigenous territory of the Tr'ondëk Hwëch'in, was settled in the late 1800s as part of the gold rush and the mining that followed. Starting in 1903, the Dawson Amateur Athletic Association screened films in the city (Kula 1979b, 15), and over time, tons of them were shipped to Dawson for news and entertainment in the burgeoning city. While some of the films were disposed of in the Dawson River, as they were too costly to ship back, many were buried in the permafrost for nearly a half century, becoming landfill in a swimming pool then converted to an ice rink. This unknown archive of 533 nitrate reels led to the restoration of 372 Hollywood and Canadian films from 1903 to 1929 (Kula 1979b, 16). The works span everything from fiction features and shorts to newsreels, *actualités,* and early animation, including lost Canadian film production history, such as Montréal film pioneer Léo-Ernest Ouimet's *British Canadian Pathé News* (1917–22) and newsreels such as the *Montréal Herald Screen Magazine* (1919) (see MacKenzie 2004, 83–91). These newsreels contained material shot in Canada, the United States, France, and the Arctic (the latter from what was then the Northwest Territories), connecting Dawson City to the global media world of the time. Other film histories were found, too. As Sam Kula notes, while "many of the names are well established, as producers and directors, many more would have undoubtedly been better known if their films had been available when the [influential film] histories were written" (Kula 1979a, 146). About 75 percent of silent films ever made are lost. The Dawson find points to the contingent and partial nature of historiography; film history and its canon would have looked different if this find had been known. Yet, while these lost histories are explicitly addressed in the film, other implicit lost histories are present, too. Frozen and subject to environmental decay, the found works became something radically different than what they were at their moment of production: a remediation of the past. Morrison notes: "I think this . . . film has resonance beyond being a story about lost films found. It's a story of the twentieth century and the forces that were at play in shaping the world we live in today: Western expansionism and the rise and fall of capitalism as seen through the movies—as distilled a story of the hubris of capitalism as the Titanic is" (MacDonald 2016, 43). While the film reaches far beyond the story of the find, it is also the case, as Curtis Russell argues, that this is Morrison's "clearest statement yet that nothing is outside the realm of cinema" (2018, 15). In Morrison's practice of archival interventionist historiography, the cinema is the global village.

Dawson City: Frozen Time has been widely praised as an innovative use of documentary remediation. Including clips from Chaplin's work, and those of Edison, the film speaks directly to the Arctic's significance for global film and media history, and not only in terms of the kinds of films that were made in—or, most often, about—the Arctic. As a compilation film, *Dawson City* remediates not only the films found, but one's understanding of Dawson City, not as a far-off outpost, but as a global space, where commerce and entertainment, Indigenous peoples and settlers, meet. Morrison notes both the global interconnectivity and the remoteness of Dawson City:

> You have a military explosive that's holding all of our moving images. Of course, it takes the ephemeral nature of cinema to a different level when you basically have all of your films printed on that explosive medium. It also links the military nature of colonialism to cinema, which was in fact a colonizing agent, not just for bringing the first world into the third world but also for exoticizing the third world and bringing it into the first world. For the people in the Yukon, who were extremely isolated, they didn't know anything about the outside unless they left and came back. (Crafton and Morrison 2018, 94)

The Arctic, then, was not isolated but rather was integrated into a global system of networks and circulation.

Many inhabitants of Dawson City, though, had not moved there to find their riches. Morrison's film also addresses the erasure of Indigenous stories and lives, such as that of Chief Isaac (first name unknown, born around 1847)

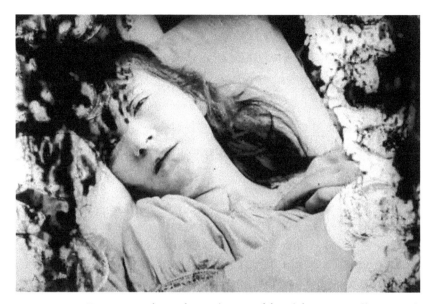

FIGURE II.I. Environmental remediation by way of found footage in Bill Morrison's *Dawson City: Frozen Time* (USA, 2016).

in the Yukon during the gold rush. This can be seen as a form of reverse decentering, highlighting the centrality of Dawson as a globally intercon-nected city at the beginning of the twentieth century.

In this way, Morrison's work complements the emerging trend in Arctic Indigenous documentary cinema discussed in earlier chapters, such as *Sumé: The Sound of a Revolution* (2014) and *Kaisa's Enchanted Forest* (2016), in which archival footage is remediated as a means of discovering various buried pasts. Like *Sumé, Dawson City: Frozen Time* further mediates Dawson's his-tory by including found footage from later works that tell a far more homo-geneous story, from the poetic *City of Gold* (Colin Low and Wolf Koenig, Canada, NFB/ONF, 1957), the inspiration for much of Ken Burns's work, to *Klondike Holiday* (Earl Clark, Canada, Associated Screen News, 1950), an educational film that engages in voice-of-God pastoralism. In a further jux-taposition, the audience sees images of environmental destruction, from aerial shots of the mining slags to shots of the river beneath which reels of film lie decaying. These images of decay and erosion are made visible in a metrical fashion, and one sees how the environment itself—with the effects of decaying in the ground and exposure to the elements visible on the film stock—practices a form of environmental remediation (figure II.I).

As interventionist historiography, then, the archival remediation in *Dawson City: Frozen Time* not only documents the past, but also retrieves it, using pieces of film extracted from the tundra to tell new stories long buried and ignored.

THE FUTURES OF ARCTIC MEDIA PRODUCTION AND CIRCULATION

As we conclude *New Arctic Cinemas,* it is time to affirm, once again, that film-making in and about the Arctic—especially when it engages and features Indigenous voices, perspectives, and priorities—continues to be a formidable and ever-expanding cultural, political, and social force. This book has focused on a curated set of works selected to address a wide range of contemporary Arctic cinemas—and the range of works that could be included in these discussions continues to expand, from realist documentaries and fiction features to experimental media and animation. This diverse body of works increasingly circulates in two interrelated ways outside of traditional modes of distribution: through festivals (both Indigenous and non-Indigenous) and online, whether on platforms such as Vimeo and YouTube or via streaming. These innovations have increased since the outbreak of COVID-19, as new, at first temporary, means of exhibition emerged in the face of the pandemic. Some of these once temporary modes may continue. The Nuuk International Film Festival, Iceland's Northern Wave International Film Festival, the Indigenous Peoples' Film Festival Skábmagovat, the Tromsø International Film Festival, and others have mobilized online platforms, and hybrid festivals may become a new norm. Increasing accessibility is a net positive, with the reduction in travel, but anecdotal evidence from festival viewership suggests that limiting in-person, on-site exhibition is less attractive to existing audiences and does little to draw new ones. This may be especially significant for Indigenous festivals, where a sense of a shared space, community, and *terroir* is as important as the films on the screens. Film festivals such as Skábmagovat have, for years, served as one of the winter's most important social and cultural events in Sápmi, where artists, politicians, activists, and journalists meet to advance public dialogue about Indigenous priorities. At the same time, hybrid festival models could emerge to be politically more significant as, along with screening Arctic Indigenous films, these festivals buttress the creation and perpetuation of globalized counter-public spheres of both artistic practice and political power. New developments in global Arctic Indigenous film circulation outside of festivals and single-channel screenings include the recently announced partnership between the ISFI and Netflix. According to the announcement, "the Netflix partnership . . . will support the Sámi Film Institute's education and training programmes, including more behind-the-camera opportunities and helping Sami creatives break through in the industry" (Mitchell 2022), pointing to collaborations not only for distribution but also for training the next generation of Indigenous practitioners.

Increasingly, festivals are serving as catalysts to generate funding, and coproductions have increased, which helps ameliorate the fact that the structures and infrastructures for funding differ vastly across the Arctic, especially with regard to Indigenous production. There are many reasons for these funding opportunity differentials, including the differing structures of (self-) government in Greenland, Nunavut, and Sápmi; tax-funded support from states such as Norway—and the general absence thereof in Alaska; the presence of organizations like the National Film Board of Canada/Office national du film (NFB/ONF) and the International Sámi Film Institute (ISFI); and the lack of funding for infrastructure in Russia, despite the emergence of new festivals such as the Arctic Open Festival in Arkhangelsk. Working through festival communities allows filmmakers to draw on structures—funding and otherwise—across the Arctic, leading both to coproductions and to works that are situated in the local but that interface with the global Arctic. An example of this is the Arctic Indigenous Film Fund (AIFF), launched by the ISFI "to support the development of Indigenous filmmakers from the circumpolar Arctic and to support the production of their films, TV series, and other screen-based works" (ISFI 2021). Programs such as these allow for reimaginings of the global Arctic from Indigenous perspectives. A related ISFI public relations initiative is the formal appointment of "Ambassadors," or *Verddet* (partners), to advance Arctic Indigenous film priorities: "Indigenous voices must be heard and those of the circumpolar region have urgent, vital, and powerful stories of the kind the world has not yet heard or seen" (ISFI 2021).

The exhibitions *Moratorium Office* and *Toolkit for Revolution: The Poster Art of Suohpanterror & Jay Soule,* mounted in collaboration with imagineNATIVE in Toronto in 2017, provide a key example of the political power premise underlying Arctic Indigenous film. These exhibitions connect to the Ellos Deatnu (Long Live the [River] Deatnu) occupation movement in eastern Sápmi during the summer of 2017, which Kuokkanen describes as originating in "the desire to start a new sovereignty discourse vis-à-vis the Sámi in Scandinavia" (2021, 310). The multimedia exhibitions contain video work, installations, workshops, and framed posters. The latter remediate Soviet, Leninist, leftist, and popular Scandinavian advertising-poster aesthetics in the name of Sámi self-determination. Exploring various understandings of revolution, this collection of activist art forms is used to foreground the political activism of Sámi occupants and Arctic artistic activism (see Jokela et al. 2021). The imagineNATIVE catalog describes the exhibitions as constituting an act of Indigenous self-determination and a way to "enact a revolution based on our core value of love: love for our land, our people, and our cultures" (imagineNATIVE 2017, 24). The exhibitions are likely the first international event to present one of the most significant political interventions in Sápmi in decades. By connecting Indigenous art to global audiences through circulation of films and cinephilia, imagineNATIVE brings Sámi political resistance and Indigenous sovereignty debates to a broader public. The increasing political power of art, media, and moving images to further self-determination is evidenced by the fact that the "Nordic Pavilion" for the 2022 Venice Biennale was renamed the "Sámi Pavilion" and includes work by Sámi filmmakers and video artists such as Pauliina Feodoroff and the collective of directors represented in the "ÁRRAN 360" installation. In the sprit of global Arctic activism, they partnered with Canadian Inuit filmmakers and also screened "ARCTIC XR" with works by Nyla Innuksuk, Melaw Nakehk'o, and Tanya Tagaq, among others. This change of name also decenters the mapping of the European nation-state, remapping it in a new Sámi and global Arctic cartography as an act of interventionist historiography.

Two recent feature documentaries from Sápmi further illustrate this significant trend of explicitly political interventionist historiography in a global media context. A focus on sociopolitical issues, self-determination, and the unveiling of conflicts between the colonial nation-state and the Sámi community is at the heart of Suvi West's documentary *Our Silent Struggle* (*Eatnameamet,* Finland, 2021), which relates to the official launch, in October 2021, of Finland's Truth and Reconciliation Commission, mandated to "address the historical treatment of the indigenous Sami population and promote the attainment of the Sami people's rights" (Finland 2021). Liselotte Wajstedt's *The Silence in Sápmi* (*Tystnaden i Sápmi,* Sweden, 2021) addresses a pervasive culture of violence and sexual abuse of women within the higher echelons of Sámi political life, in which some of the perpetrators are well-known political, business, and community figures. The film integrates several interviews, done by Wajstedt over more than five years, with Sámi victims of this abuse. The film moves between different time periods—filmed using a range of technologies, including a smartphone—and integrates Wajstedt's own artwork through superimposition. The taboo against addressing sexual abuse and violence is examined alongside a stigma surrounding mental health issues and suicide. Focusing on women's testimony, the film is about the courage and pain involved in reliving traumatic experiences by retelling them on screen, to offer evidence, reveal suppressed truths, and share those experiences to advance political and social change, while respecting victims.

As the first Sámi film to engage with the topic, and as part of the #metoo movement, it is significant that it took nearly a decade for *The Silence in Sápmi* to be completed. An experimental precursor of the film is Wajstedt's short fiction film *Jorinda's Journey* (*Jorindas resa,* Sweden, 2014), a loose interpretation of Ann-Marie Ljungberg's novel *Resan till Kautokeino* ("The Journey to Kautokeino," 1999). In *Jorinda's Journey,* the artistic traditions of Sámi yoik and Japanese butoh dance are integrated to begin the breaking of taboos surrounding sexual abuse, underage prostitution, and gender discrimination in Sámi communities. *Jorinda's Journey* was intensely criticized by some Sámi who expressed that it betrayed their community, especially when Wajstedt pinpointed figures in the Kautokeino establishment as the perpetrators of the abuse (see Lindstrand 2014). Indeed, negative media coverage delayed the completion of *The Silence of Sápmi,* with ISFI stepping in to provide production support in 2019. Wajstedt's own reflections on the making of the film are significant. She notes that she started with a fictional, experimental art film because she felt like "there was so much to learn," that she did not want her perspective silenced, that she knew the subject to be taboo, and yet that her ultimate goal was to make a "feminist film for all the young women who have been abused" (Lindstrand 2014; see also Wajstedt 2021). *The Silence in Sápmi*'s tagline is #sapmitoo, in solidarity with the global #metoo movement. Opening at the Tromsø International Film Festival to great acclaim, the film had cinema release in Sweden and Norway and also screened at many dozen museums, libraries, festivals, and community centers all over Sápmi in fall 2022. Robust media coverage and debate ensued, spanning television to daily press and culture magazines (e.g., NRK Sápmi 2022; Haglund 2022).

The works discussed in this concluding chapter, from Disney animation to feminist Indigenous media, point to the diversity of processes and imaginaries used by Indigenous filmmakers and settlers in engaging with Arctic works in the twenty-first century. We juxtapose these quite different cinematic forms as a continuation of the underlying intellectual and ethical principles that underpin the book. Throughout *New Arctic Cinemas,* we have attempted to invert the frame of Arctic cinemas, placing explorer, outsider, or settler cinema at the margins, thereby drawing into focus the diversity of global Indigenous Arctic cinema. The concepts of interventionist historiography, Indigenous media sovereignty, and environmental and climate crisis have informed the examination. This approach is a decentering, self-reflexive process, in terms of both the works themselves and our position as settler allies. The goal of *New Arctic Cinemas,* then, is to engage in a process of radically reimagining, re-presenting, and challenging how the Arctic is understood globally, how it is mediated, and how it bridges boundaries between "here" and "there," to allow for new understandings of inclusion and exclusion, of which stories are told and which ones are not. *New Arctic Cinemas* thereby places the Arctic at the center of twenty-first-century media about climate, Indigenous rights, media sovereignty, geopolitics, cultural hybridity, and social justice.

REFERENCES

Acland, Charles R. 1997. "IMAX in Canadian Cinema: Geographic Transformation and Discourses of Nationhood." *Studies in Cultures, Organizations and Societies* 3: 289–305.

———. 1998. "IMAX Technology and the Tourist Gaze." *Cultural Studies* 12 (3): 429–45.

Agebro, Olle. 2016. "Sameblod." In *Göteborg International Film Festival Program 2016*, 6–8. Gothenburg, Sweden: GIFF.

"Agreement." 2019. Walt Disney Animation Studios and the Sámi Parliament of Finland, the Sámi Parliament of Norway, the Sámi Parliament Sweden, and the Saami Council. https://www.samediggi.fi/wp-content/uploads/2019/09/Agreement_WDAS_SAMI.pdf.

Alia, Valerie. 1999. *Un/Covering the North.* Vancouver: University of British Columbia Press.

Alioff, Maurie. 2001. "From the Edge of the Earth: Zacharias Kunuk's *Atanarjuat.*" *Take One* 34 (10): 17–21.

Allen, Chadwick. 2012. *Trans-Indigenous: Methodologies for Global Native Literary Studies.* Minneapolis: University of Minnesota Press.

Altamirano-Jiménez, Isabel. 2008. "Nunavut: Whose Homeland, Whose Voices?" *Canadian Woman Studies* 26 (3–4): 128–34.

Alter, Charlotte. 2015. "Official Wants *Frozen* to Teach Kids about Climate Change." *Time,* January 23. https://time.com/3680495/frozen-climate-change/.

Amur Pravda. 2014. "Review: *White Moss.*" October 7. http://www.ampravda.ru/2014/10/07/052144.html.

Andersen, Astrid Nonbo. 2019. "The Greenland Reconciliation Commission: Moving Away from a Legal Framework." *Yearbook of Polar Law* 11: 214–44.

Anderson, Benedict. 1991. *Imagined Communities: Reflections on the Origin and Spread of Nationalism.* London: Verso.

Andreassen, Elin, Hein Bjartmann Bjerck, and Bjørnar Olsen. 2010. *Persistent Memories: Pyramiden, A Soviet Mining Town in the High Arctic.* Bergen, Norway: Tapir Academic Press.

Antrim, Caitlyn L. 2011. "The Russian Arctic in the Twenty-First Century." In *Arctic Security in an Age of Climate Change,* edited by James Kraska, 107–28. Cambridge: Cambridge University Press.

Arctic Days. 2012. Press release. http://www.arctic-days.ru/en/arctic-days-2012.html [link now defunct].

———. 2014. Press release. http://www.arctic-days.ru/en/arctic-days-2014.html [link now defunct].

———. 2021. *Arctic Days in St. Petersburg 2021: International Scientific Cooperation in the Arctic in the Era of Climate Change.* http://www.rshu.ru/university/science /konf_arctic2021/info_letter_eng_2.pdf.

———. n.d. ["Days of the Arctic in Moscow."] http://arctic-days.ru/ru/.

Arke, Pia. 2012. "Ethno-Aesthetics." In *Tupilakosaurus: An Incomplete(able) Survey of Pia Arke's Artistic Work and Research,* edited by Kuratorisk Aktion, translated by Erik Gant and John Kendal, 335–43. Copenhagen: Kuratorisk Aktion.

Arnait Video Productions. n.d.[a] "About Us." https://arnaitvideo.ca/about-us/.

———. n.d.[b] "Arnait Ikajurtigiit/Women Helping Each Other: Exhibition of Arnait's Work from 1991 to 2018 at AGYU April 17th to June 23th 2019." http:// arnaitvideo.ca/explorations.

Arnaquq-Baril, Alethea Aggiuq. 2018. "Filmmaking and Media." In *Indigenous Peoples Atlas of Canada.* Ottawa: Canadian Geographic. https://indigenouspeoplesatlas ofcanada.ca/article/inuit-film-and-broadcasting/.

Arthur, Paul. 1996. "In the Realm of the Senses: IMAX 3-D and the Myth of Total Cinema." *Film Comment* 32 (1): 78–81.

Ash, John. 2020. "Svalbard and Conflict Management in a Changing Climate: A Risk-Based Approach." *NordLit* 45: 56–85.

Ashe, Teresa. 2018. "The Importance of the Soviet Experience." In *Climate Change Discourse in Russia: Past and Present,* edited by Marianna Poberezhskaya and Teresa Ashe, 1–16. London: Taylor & Francis.

Asinnajaq. 2019. "Isuma Is a Cumulative Effort." *Isuma TV,* January 12, 2019. http:// www.isuma.tv/isuma-book/essays/isuma-is-a-cumulative-effort.

Athens, Allison K. 2014. "Saviors, 'Sealfies,' and Seals: Strategies for Self-Representation in Contemporary Inuit Films." *Ecozona* 5 (2): 41–56.

Åtland, Kristian, and Torbjørn Pedersen. 2008. "The Svalbard Archipelago in Russian Security Policy: Overcoming the Legacy of Fear—or Reproducing It?" *European Security* 17 (2–3): 227–51.

———. 2014. "Cold War Legacies in Russia's Svalbard Policy." In *Environmental and Human Security in the Arctic,* edited by Andrew Tanentzap, Dawn Bazely, Gunhild Hoogensen Gjørv, and Marina Goloviznina, 17–36. Abingdon, UK: Routledge.

Augé, Marc. 1995. *Non-places: An Introduction to Supermodernity,* translated by John Howe. London: Verso.

Balog, James. 2007. "The Big Thaw." *National Geographic* 211 (6): 56.

———. 2010. "Melt Zone." *National Geographic* 217 (6): 34.

Barclay, Barry. 2003. "Celebrating Fourth Cinema." *Illusions* 35: 7–11.

Barker, Joanne. 2017. "Introduction: Critically Sovereign." In *Critically Sovereign: Indigenous Gender, Sexuality, and Feminist Studies,* edited by Joanne Barker, 1–45. Durham, NC: Duke University Press.

Barnard, Linda. 2014. "Inuit Actors Draw on Real Life for Uvanga Performances." *Toronto Star,* June 19. https://www.thestar.com/entertainment/movies/2014/06/19/inuit_actors_draw_on_real_life_for_uvanga_performances.html.

Bazin, André. 1967. "Cinema and Exploration." In *What Is Cinema?, vol. 1,* 154–63. Berkeley: University of California Press.

Belanger, Noelle, and Anna Westerstahl Stenport. 2017. "The Global Politics of Color in the Arctic Landscape: Blackness at the Center of Frederic Edwin Church's *Aurora Borealis* (1865) and Nineteenth-Century Limits of Representation." *ARTMargins* 6 (2): 6–26.

Belokurova, Anastasia. 2014. Review: *White Moss. Zavtra,* July 31. https://zavtra.ru/blogs/vechnoe-siyanie-chistoj-strasti.

Benjamin, Walter. 1969. *Illuminations: Essays and Reflections.* New York: Schocken Books.

———. 1999. *The Arcades Project,* translated by Howard Eiland and Kevin McLaughlin. London: Belknap Press.

Berckvens, Marie. 2019. "Le Nord de Gabrielle Roy toujours pertinent en film." *La Liberté,* April 24. https://www.la-liberte.ca/2019/04/24/le-nord-de-gabrielle-roy-toujours-pertinent-en-film/.

Bertrand, Karine. 2015. "La Présence autochtone et la figure du médiateur blanc dans le cinéma des Premières Nations." *Recherches amérindiennes au Québec* 45 (1): 51–58.

———. 2017. "Le Collectif Arnait Video Productions et le cinéma engagé des femmes inuits: Guérison communautaire et mémoire Culturelle." *Canadian Review of Comparative Literature/Revue Canadienne de Littérature Comparée* 44 (1): 36–53.

———. 2019. "*Martha of the North* and Nunavik Narratives of Survivance." In *Arctic Cinemas and the Documentary Ethos,* edited by Lilya Kaganovsky, Scott MacKenzie, and Anna Westerstahl Stenport, 289–301. Bloomington: Indiana University Press.

Beumers, Birget, Stephen Hutchings, and Natalia Rulyova, eds. 2009. *The Post-Soviet Russian Media: Conflicting Signals.* New York: Routledge.

Bjørst, Lill Rastad. 2019. "The Right to 'Sustainable Development' and Greenland's Lack of a Climate Policy." In *The Politics of Sustainability in the Arctic: Reconfiguring Identity, Space, and Time,* edited by Ulrik Pram Gad and Jeppe Strandsbjerg, 121–35. London: Routledge.

Bloom, Lisa E. 2022. *Climate Change and the New Polar Aesthetics: Artists Reimagine the Arctic and Antarctic.* Durham, NC: Duke University Press.

Bohr, Marco. 2015. "Cinema of Emancipation and Zacharias Kunuk's *Atanarjuat: The Fast Runner.*" In *Films on Ice: Cinemas of the Arctic,* edited by Scott MacKenzie and Anna Westerstahl Stenport, 84–96. Edinburgh: Edinburgh University Press.

Borisov, Petr Mikhailovich. 1958. "Mozhno li ispravit' klimat?" *Vokrug sveta* 4: 2–6.

———. 1969. "Can We Control the Arctic Climate?" *Bulletin of the Atomic Scientists* 29 (3): 43–48.

———. 1973. *Can Man Change the Climate (Mozhet li chelovek izmenit' klimat)?* Moscow: Progress.

Braun, Marta. 1992. *Picturing Time: The Work of Étienne-Jules Marey (1830–1904)*. Chicago: University of Chicago Press.

Breum, Martin. 2022. "Canada, Denmark Agree on a Landmark Deal over Disputed Hans Island." *Arctic Today,* June 12. https://www.arctictoday.com /canada-denmark-agree-on-a-landmark-deal-over-disputed-hans-island/?wallit _nosession=1.

Broberg, Gunnar. 1975. *Homo sapiens L: studier i Carl von Linnés naturuppfattning och människolära*. Stockholm: Almqvist & Wiksell i distribution.

Broberg, Gunnar, and Nils Roll-Hansen. 1996. *Eugenics and the Welfare State: Sterilization Policy in Denmark, Sweden, Norway, and Finland*. East Lansing: Michigan State University Press.

Brundtland Commission. 1987. "Foreword." In *Our Common Future: The World Commission On Environment And Development,* edited by Volker Hauff, x–xv. Oxford: Oxford University Press

Bruno, Andy. 2016. "Nature and Power in the Soviet North." In *The Nature of Soviet Power: An Arctic Environmental History,* edited by J. R. McNeil and Edmund P. Russell, 1–28. Cambridge: Cambridge University Press.

Burelle, Julie. 2020. "Inuit Visual and Sensate Sovereignty in Alethea Arnaquq-Baril's *Angry Inuk*." *Canadian Journal of Film Studies* 29 (1): 145–62.

Burnett, Ron. 1977. "Film/Technology/Ideology." *Ciné-tracts* (1) 1: 6–14.

Burtynsky, Edward, Jennifer Baichwal, and Nicholas de Pencier. 2018. *Anthropocene*. London: Steidl.

BusinessOulu. n.d. "Who We Are: North Finland Film Commission." https:// www.businessoulu.com/en/frontpage-old/en/businessoulu-2/north-finland-film-commission.html.

Cache Collective. 2008. "Cache: Provision and Productions in Contemporary Igloolik Video." In *Global Indigenous Media: Cultures, Poetic, and Politics,* edited by Pamela Wilson and Michelle Stewart, 74–88. Durham, NC: Duke University Press.

Cahill, James Leo, and Luca Caminati, eds. 2020. *Cinema of Exploration: Essays on an Adventurous Film Practice*. New York: Routledge.

Callison, Candis. 2014. *How Climate Change Comes to Matter: The Communal Life of Facts*. Durham, NC: Duke University Press.

Campion-Smith, Bruce. 2010. "Ottawa Apologizes to Inuit for Using Them as 'Human Flagpoles.'" *Toronto Star,* August 18. https://www.thestar.com/news /canada/2010/08/18/ottawa_apologizes_to_inuit_for_using_them_as_human _flagpoles.html.

Carbon 14: Climate Is Culture. Toronto: Royal Ontario Museum/Cape Farewell.

Carey, Mark M., Alessandro Antonello Jackson, and Jaclyn Rushing. 2016. "Glaciers, Gender, and Science: A Feminist Glaciology Framework for Global Environmental Change Research." *Progress in Human Geography* 40 (6): 770–93.

Carpenter, Edmund. 1955. *Aivilik Eskimo: Time-Space Concepts.* Washington, DC: Smithsonian Libraries.

Carpenter, Edmund, and Robert Flaherty. 1959. *Eskimo.* Toronto: University of Toronto Press.

Carpenter, Edmund, and Marshall McLuhan. 1960. "Acoustic Space." In *Explorations in Communications,* edited by Edmund Carpenter and Marshall McLuhan, 65–70. Boston: Beacon Press.

Carrasco, Carlos Mínguez. 2020. "Kiruna Forever: Relocating a City in Territories of Extraction." In *Kiruna Forever,* edited by Daniel Golling and Carlos Mínguez Carrasco, 31–38. Stockholm: ArkDes.

Catford, Ken. 2002. "The Industrial Archaeology of Spitsbergen." *Industrial Archaeology Review* 24 (1): 23–36.

CBC Arts. 2009. "Film Festival Brings Inuit Perspective to Climate Conference." *CBC News,* December 9. https://www.cbc.ca/news/entertainment/film-festival-brings-inuit-perspective-to-climate-conference-1.836224.

CBC News. 2011. "Inuit Films Move Online and into Northern Communities." *CBC News,* November 2. https://www.cbc.ca/news/entertainment/inuit-films-move-online-and-into-northern-communities-1.1044506.

———. 2014. "Suicide Numbers in Nunavut in 2013 a Record High: Nunavut Youth Decry Lack of Help for Those Thinking about Suicide." *CBC News,* January 10. https://www.cbc.ca/news/canada/north/suicide-numbers-in-nunavut-in-2013-a-record-high-1.2491117.

Chang, Chris. 2012. "*Chasing Ice.*" *Film Comment* 48 (6): 72.

Chartier, Daniel. 2018. *What Is the Imagined North? Ethical Principles.* Montréal: Université du Québec à Montréal.

Chaturvedi, Sanjay. 2000. "Arctic Geopolitics Then and Now." In *The Arctic: Environment, People, Policy,* edited by Mark Nuttall and Terry V. Callaghan, 441–58. Amsterdam: Harwood Academic.

Chew, May, Susan Lord, and Janine Marchessault. 2018. "Introduction." *Public* 57: 5–10.

Chisholm, Dianne. 2016. "The Enduring Afterlife of *Before Tomorrow:* Inuit Survivance and the Spectral Cinema of Arnait Video Productions." *Étudees/Inuit/Studies* 40 (1): 211–27.

Christensen, Cato. 2012. "Reclaiming the Past: On the History-Making Significance of the Sámi Film *The Kautokeino Rebellion.*" *Nordic Journal of Circumpolar Societies* 29 (1): 56–76.

Christensen, Miyase. 2013. "Arctic Climate Change and the Media: The News Story That Was." In *Media and the Politics of Arctic Climate Change: When the Ice Breaks,* edited by Annika E. Nilsson, Miyase Christensen, and Nina Wormbs, 26–31. New York: Palgrave Macmillan.

Cobb, Shelley, and Linda Ruth Williams. 2020. "Histories of Now: Listening to Women in British Film." *Women's History Review* 29 (5): 890–902.

Cochran, Patricia, and Norway Inuit Circumpolar Council (ICC) Members. 2009. "Circumpolar Inuit Launch Declaration on Arctic Sovereignty." https://www .inuitcircumpolar.com/press-releases/circumpolar-inuit-launch-declaration-on-arctic-sovereignty/#:~:text=The%20Circumpolar%20Inuit%20Declaration%20 on,have%20within%20each%20respective%20state.

Cocq, Coppélie, and Thomas A. Dubois. 2019. *Sámi Media and Indigenous Agency in the Arctic North.* Seattle: University of Washington Press.

Columpar, Corinn. 2010. *Unsettling Sights: The Fourth World on Film.* Carbondale: Southern Illinois University Press.

Condee, Nancy. 2015. "*Leviathan.*" *Slavic Review* 74 (3): 607–08.

"Conversation." 2008. In *Ikuma: Tournet du Carnage,* 31–43. Montréal: Mémoire d'encrier.

Corrigan, Timothy. 2011. *The Essay Film: From Montaigne, after Marker.* Oxford: Oxford University Press.

Cousineau, Marie-Hélène. 2010. "Making Connections." *Isuma TV,* July 9. http:// www.isuma.tv/en/making-connections.

Craciun, Adriana. 2016. *Writing Arctic Disaster: Authorship and Exploration.* Cambridge: Cambridge University Press.

Crafton, Donald, and Bill Morrison. 2018. "A Deal with the Devil: Bill Morrison on *Dawson City: Frozen Time.*" *Moving Image* 18 (1): 92–103.

Crewe, Louise. 2017. *The Geographies of Fashion: Consumption, Space, and Value.* London: Bloomsbury.

Cruikshank, Julie. 2005. *Do Glaciers Listen? Local Knowledge, Colonial Encounters, and Social Imagination.* Vancouver: University of British Columbia Press.

Dal'skaia, Svetlana. 2014. Interview, December 9. http://www.arctic-info.ru /Interview/12–09–2014/-belii-agel_novoe-slovo-v-sovremennom-igrovom-kino [link now defunct].

Dalton, Stephen. 2015. "*Sumé: The Sound of a Revolution ('Sumé—mumisitsinerup nipaa'):* Berlin Review." *Hollywood Reporter,* February 2.

Dancus, Adriana M. 2022. "Sámi Identity across Generations: From Passing for Nordics to Sámi Self-Exposure." *Journal of Mixed Race Critical Studies* 1 (2): 262–76.

De Bromhead, Toni. 1996. *Looking Two Ways: Documentary's Relationship with Cinema and Reality.* Aarhus, Denmark: Intervention Press.

Demos, T. J. 2017. *Against the Anthropocene: Visual Culture and Environment Today.* Berlin: Sternberg Press.

Diatchkova, Galina. 2008. "Indigenous Media as an Important Resource for Russia's Indigenous Peoples." In *Global Indigenous Media: Cultures, Poetic, and Politics,* edited by Pamela Wilson and Michelle Stewart, 214–31. Durham, NC: Duke University Press.

Dick, Lyle. 1995. "'Pibloktoq' (Arctic Hysteria): A Construction of European-Inuit Relations?" *Arctic Anthropology* 32 (2): 1–42.

Diesen, Jan Anders. 2015. "The Changing Polar Films: Silent Films from Arctic Exploration 1900–1930." In *Films on Ice: Cinemas of the Arctic,* edited by Scott

MacKenzie and Anna Westerstahl Stenport, 267–80. Edinburgh: Edinburgh University Press.

Dillon, Kimberley A. 2013. "The Emerging Russian Identity as Expressed through Russian Art House Cinema." *Verges: Germanic & Slavic Studies in Review* 2 (1): 1–9.

"Director's Note." n.d. *Uvanga.* Website [link now defunct].

Dixon, Guy. 2010. "New Documentary Recounts Bizarre Climate Changes Seen by Inuit Elders." *Globe and Mail,* October 19. https://www.theglobeandmail.com /arts/film/new-documentary-recounts-bizarre-climate-changes-seen-by-inuit-elders/article1215305/.

Dodds, Klaus, and Rikke Bjerg Jensen. 2019. "Screening Greenland: Popular Geopolitics, Film and the Assemblage of Place." In *Arctic Cinemas and the Documentary Ethos,* edited by Lilya Kaganovsky, Scott MacKenzie, and Anna Westerstahl Stenport, 155–74. Bloomington: Indiana University Press.

Dodds, Klaus, and Mark Nuttall. 2015. *The Scramble for the Poles: The Geopolitics of the Arctic and Antarctic.* London: Polity.

Doose, Katja, and Jonathan Oldfield. 2018. "Natural and Anthropogenic Climate Change Understanding in the Soviet Union, 1960s–1980s." In *Climate Change Discourse in Russia: Past and Present,* edited by Marianna Poberezhskaya and Teresa Ashe, 17–28. London: Routledge.

Dowell, Kristin L. 2013. *Sovereign Screens: Aboriginal Media on the Canadian West Coast.* Lincoln: University of Nebraska Press.

Duarte, José. 2018. "On the 'Ghosts' of Piramida: Ruins, Memory and Music." In *New Approaches to Cinematic Space,* edited by Filipa Rosário and Iván Villarmea Álvarez, 129–39. London: Routledge.

Dunaway, Finis. 2015. *Seeing Green: The Use and Abuse of American Environmental Images.* Chicago: University of Chicago Press.

Duus, Søren Duran. 2014. "Danske medier hylder Sume-film." *Sermitsiaq,* October 16. http://sermitsiaq.ag/danske-medier-hylder-sume-film.

Edwards, Paul N. 2010. *A Vast Machine: Computer Models, Climate Data, and the Politics of Global Warming.* Cambridge, MA: MIT Press.

Eisenstein, Sergei. [1924] 1988. "The Montage of Film Attractions." In *Selected Works vol. 1: 1922–1934,* edited and translated by Richard Taylor, 39–48. London: British Film Institute.

EMA (Electronic Memory of the Arctic: Cultural Communications of the Circumpolar World). 2012. *Arctic Portal.* https://arcticportal.org/ap-library /news/909-electronic-memories-in-st-petersburg.

Ertel, David. 2010. "The Art of Unuit Story-Telling." *Isuma TV,* April 22. http:// www.isuma.tv/isuma-productions/art-inuit-story-telling.

Evans, Michael R. 2008. *Isuma: Inuit Video Art.* Montréal: McGill-Queen's University Press.

Everett, Wendy. 2009. "Lost in Transition? The European Road Movie, or a Genre 'Adrift in the Cosmos.'" *Literature-Film Quarterly* 37 (3): 165–75.

Fakhri, Michael. 2017. "Gauging US and EU Seal Regimes in the Arctic against Inuit Sovereignty." In *The European Union and the Arctic,* edited by

Nengye Liu, Elizabeth A. Kirk, and Tore Henriksen, 200–35. Leiden, The Netherlands: Brill.

Fay, Jennifer. 2018. *Inhospitable World: Cinema in the Time of the Anthropocene.* Oxford: Oxford University Press.

Feldman, Seth. 2014. *"Tiger Child:* IMAX and Donald Brittain Times Nine." In *Cinephemera: Archives, Ephemeral Cinema, and New Screen Histories in Canada,* edited by Zoë Druick and Gerda Cammaer, 159–83. Montréal: McGill-Queen's University Press.

Ferguson, Graeme. 2014. "A Recollection of Polar Life at Expo '67." In *Reimagining Cinema: Film at Expo 67,* edited by Monika Kin Gagnon and Janine Marchessault, 139–45. Montréal: McGill-Queen's University Press.

Fiala, Anthony. 1907. "Polar Photography." *National Geographic* 18 (1–2): 140–42.

Fienup-Riordan, Ann. 1995. *Freeze Frame: Alaska Eskimos in the Movies.* Seattle: University of Washington Press.

———. 2015. "Frozen in Film: Alaska Eskimos in the Movies." In *Films on Ice: Cinemas of the Arctic,* edited by Scott MacKenzie and Anna Westerstahl Stenport, 59–71. Edinburgh: Edinburgh University Press.

FILM.GL. n.d. "FILM.GL Is the Industry Organization of the Small but Vibrant Film Community." https://film.gl/about-2/#top.

Finland. 2021. Sami Truth and Reconciliation Commission Established. Web page. https://www.loc.gov/item/global-legal-monitor/2021–11–18/finland-sami-truth-and-reconciliation-commission-established/.

Fischer, Kurt W., and June Price Tangney. 1995. "Self-Conscious Emotions and the Affect Revolution: Framework and Overview." In *Self-Conscious Emotions: The Psychology of Shame, Guilt, Embarrassment and Pride,* edited by June Price Tangney and Kurt W. Fischer, 3–24. New York: Guilford Press.

Fish, Cheryl J. 2018. "Liselotte Wajstedt's *Kiruna Space Road:* Experimental Ecocinema as Elegiac Memoir in 'Extractivist' Sápmi." *Journal of Scandinavian Cinema* 8 (2): 111–22.

———. 2019. "'Extractivism' in Sápmi: Elegiac Ecojustice in Liselotte Wajstedt's Film *Kiruna Space Road* and Marja Helander's *Silence* Photographs." In *Nordic Narratives of Nature and the Environment: Ecocritical Approaches to Northern European Literatures and Cultures,* edited by Reinhard Hennig, Anna-Karin Jonasson, and Peter Degerman, 209–28. London: Lexington Books.

Fleming, Kathleen. 1996. "Marie-Hélène Cousineau: Videomaker." *Inuit Art Quarterly* 11 (2): 12–20.

Flew, Terry. 2018. *Understanding Global Media,* 2nd ed. London: Macmillan.

Flinn, Andrew. 2011. "Archival Activism: Independent and Community-Led Archives, Radical Public History and the Heritage Professions." *InterActions: UCLA Journal of Education and Information Studies* 7 (2). https://escholarship.org/uc/item/9pt2490x.

Flora, Janne. 2019. *Wandering Spirits: Loneliness and Longing in Greenland.* Chicago: University of Chicago Press.

Forsell, Håkan. 2015. "Modernizing the Economic Landscapes of the North Resource Extraction, Town Building and Educational Reform in the Process of Internal Colonization in Swedish Norrbotten." In *International Journal of History, Culture and Modernity* 3 (2): 195–211.

Forsoningskommissionen [Greenlandic Reconciliation Commission/Saammaatta]. 2017. *Vi forstår fortiden. Vi tager ansvar for nutiden. Vi arbejder for en bedre fremtid: Betænkning udgivet af Grønlands Forsoningskommission.* https://naalakkersuisut .gl/~/media/Nanoq/Files/Attached%20Files/Forsoningskommission/Endelig%20 bet%C3%A6nkning%20DK.pdf. Greenlandic version at https://saammaatta.gl/ [link now defunct].

Fossato, Floriana. 2006. "Vladimir Putin and The Russian Television 'Family.'" *Les Cahiers Russie/The Russia Papers* 1. Paris: Sciences Po.

Fotovystavka "Arktika." 2014. http://arctic-days.ru/ru/fotoexpo.html [link now defunct].

Frank, Ivalo. 2019. "Reconciling the Past: Greenlandic Documentary in the Twenty-First Century." In *Arctic Cinemas and the Documentary Ethos,* edited by Lilya Kaganovsky, Scott MacKenzie, and Anna Westerstahl Stenport, 335–45. Bloomington: Indiana University Press.

Frank, Susi K. 2010. "*City of Sun* on Ice: The Soviet (Counter-) Discourse of the Arctic in the 1930s." In *Arctic Discourses,* edited by Anka Ryall, Johan Schimanski, and Henning Howlid Wærp, 106–31. Newcastle upon Tyne, UK: Cambridge Scholars.

Frank, Susi K., and Kjetil A. Jakobsen. 2019. "Introduction: The Arctic as an Archive." In *Arctic Archives: Ice, Memory, and Entropy,* edited by Susi K. Frank and Kjetil A. Jakobsen, 9–21. Bielefeld, Germany: Transcript Verlag.

Friedman, Ingrid Burke. 2022. "After Ukraine, Can the Arctic Peace Hold?" *Foreign Policy,* April 4. https://foreignpolicy.com/2022/04/04/arctic-council-members-russia-boycott-ukraine-war/.

Gabrielsen, Geir Wing. 2005. "The Arctic Paradox." *Nature* 436: 177–78.

Gad, Ulrik Pram. 2009. "Post-colonial Identity in Greenland? When the Empire Dichotomizes Back—Bring Politics Back In." *Journal of Language and Politics* 8: 136–58.

———. 2017. *National Identity Politics and Postcolonial Sovereignty Games: Greenland, Denmark, and the European Union.* Copenhagen: Museum Tusculanum Press.

Gagnon, Monika Kin. 2019. "The Two Lives of Expo 67's *Polar Life* (*La Vie Polaire*)." In *Arctic Cinemas and the Documentary Ethos,* edited by Lilya Kaganovsky, Scott MacKenzie, and Anna Westerstahl Stenport, 191–206. Bloomington: Indiana University Press.

Gates, Michael. 2022. *Hollywood in the Klondike: Dawson City's Great Film Find.* Madeira Park, BC: Lost Moose/Harbour Publishing.

Gerlach, Julia, and Nadir Kinossian. 2016. "Cultural Landscape of the Arctic: 'Recycling' of Soviet Imagery in the Russian Settlement of Barentsburg, Svalbard (Norway)." *Polar Geography* 39 (1): 1–19.

Gesink, Dionne, Gert Mulvad, and Anders Koch. 2010. "Developing a Culturally Competent and Socially Relevant Sexual Health Survey with an Urban Arctic Community." *International Journal of Circumpolar Health* 69 (1): 25–37.

Ghosh, Amitav. 2016. *The Great Derangement: Climate Change and the Unthinkable.* Chicago: University of Chicago Press.

Ginsburg, Faye. 1991. "Indigenous Media: Faustian Contract or Global Village?" *Cultural Anthropology* 6 (1): 92–112.

———. 1995. "The Parallax Effect: The Impact of Aboriginal Media on Ethnographic Film." *Visual Anthropology Review* 11 (2): 64–76.

———. 2016. "Indigenous Media from U-Matic to Youtube: Media Sovereignty in the Digital Age." *Sociologia & Antropologia* 6 (3): 581–99.

———. 2019. "Isuma TV, Visual Sovereignty, and the Arctic Media World." In *Arctic Cinemas and the Documentary Ethos,* edited by Lilya Kaganovsky, Scott MacKenzie, and Anna Westerstahl Stenport, 254–74. Bloomington: Indiana University Press.

Goldberg, Susan. 2018. "For Decades, Our Coverage Was Racist. To Rise above Our Past, We Must Acknowledge It." *National Geographic* 233 (5): 4–6.

"Golden Raven." 2019. http://en.goldenravenfilmfest.ru/competition_program -feature-length_films.

Government of Norway. 2019. *Report no. 22: Svalbard.* https://www.regjeringen.no /en/dokumenter/meld.-st.-22–2008–2009/id554877/.

Graff, James. 2007. "Fight for the Top of the World." *Time,* October 1. http:// content.time.com/time/subscriber/article/0,33009,1663848,00.html.

Graffy, Julian. 2019. *"Leviathan."* In *The Contemporary Russian Cinema Reader 2005–2016,* edited by Rimgalia Salys, 305–28. Boston: Academic Studies Press.

Green, Joyce. 2007. "Taking Account of Aboriginal Feminism." In *Making Space for Indigenous Feminism,* edited by Joyce Green, 20–32. London: Zed Books.

Greimas, Algirdas Julien. [1985] 1992. *Of Gods and Men: Studies in Lithuanian Mythology.* Bloomington: Indiana University Press.

Griffiths, Alison. 2006. "Time Travel IMAX Style." In *Virtual Voyages: Cinema and Travel,* edited by Jeffrey Ruoff, 238–59. Durham, NC: Duke University Press.

Gritsenko, Daria. 2016. "Vodka on Ice? Unveiling Russian Media Perceptions of the Arctic." *Energy Research & Social Science* 16: 8–12.

Grønlund, Anders. 2021a. "Fra film om Grønland til grønlandsk film." *Kosmorama* 280. https://www.kosmorama.org/280/groenlandsk-film.

———. 2021b. "I hele mit liv har jeg forsøgt at fortælle mine venner om det liv, jeg kender i Grønland." *Kosmorama* 280. https://www.kosmorama.org/280 /otto-rosing.

———. 2021c. "Jeg tror, at vi står lige før 'The Big Bang' i grønlandsk films historie." *Kosmorama* 280. https://www.kosmorama.org/280/nina-kristiansen.

———. 2021d. "Man laver film med den ene hånd, og så laver man branchen med den anden." *Kosmorama* 280. https://www.kosmorama.org/280/emile-peronard.

Groo, Katherine. 2019. *Bad Film Histories: Ethnography and the Early Archive.* Minneapolis: University of Minnesota Press.

Hackett, Sophie. 2018. "Far and Near: New Views of Anthropocene." In *Anthropocene: Burtynsky, Baichwal, DePencier,* edited by Stephan Jost, Marc Mayer, and Isabella Seràgnoli, 13–33. Toronto: Art Gallery of Ontario.

Hafsteinsson, Sigurjón Baldur, and Marian Bredin, eds. 2010. *Indigenous Screen Cultures in Canada.* Winnipeg, Canada: University of Manitoba Press.

Hagen, Rune Blix. 2006. "Female Witches and Sami Sorcerers in the Witch-Trials of Arctic Norway (1593–1695)." *ARV: Nordic Yearbook of Folklore* 62: 123–42.

Haglund, Birgitta. 2022. "Samtalet: Liselotte Wajstedt. Jag släpper en feministisk minoritetsfilm i en blåbrun tid." *ETC,* September 28.

Hall, Shannon. 2016. "Earth Is Tipping Because of Climate Change." *Scientific American,* April 16. https://www.scientificamerican.com/article/earth-is-tipping-because-of-climate-change1/.

Hamilton, Su. 2011. "Journal of the Inuit Sled Dog International: Unikkausivut—Sharing Our Stories." *Fan Hitch* 13 (4). https://thefanhitch.org/V14N1/V14,N1Unikkausivut.html.

Hansson, Staffan. 1998. "Malm, räls och elektricitet—skapandet av ett teknologiskt megasystem i Norrbotten 1880–1920." In *Den konstruerade världen,* edited by Pär Blomkvist and Arne Kaijser, 45–76. Lund, Sweden: Symposion.

Haraway, Donna J. 2016. *Staying with the Trouble: Making Kin in the Chthulucene.* Durham, NC: Duke University Press.

Harding, Luke. 2009. "Yamal Peninsula: The World's Biggest Gas Reserves." *The Guardian,* October 20. https://www.theguardian.com/environment/2009/oct/20/yamal-gas-reserves.

Haugdal, Elin. 2020. "Photographs of the Soviet Settlements on Svalbard." *Nordlit* 45: 104–38.

Haynes, Suyin, and Madeline Roache. 2020. "Why the Film Industry Is Thriving in the Russian Wilderness." *Time,* January 31. https://time.com/longform/film-industry-russia-yakutia/.

Hearne, Joanna. 2017. "'Who We Are Now': Iñupiaq Youth *On the Ice.*" *MediaTropes eJournal* 7 (1): 185–202.

Heinrich, Jens. n.d. "Forsoningskommissionen og fortiden som koloni." *Baggrund.* http://baggrund.com/une-belle-robe-demoiselle-dhon neur/.

Henderson, Emily. 2019. "Aka Hansen: Circumpolar Cinema." *Inuit Art Quarterly,* Summer. https://www.inuitartfoundation.org/iaq-online/circumpolar-cinema-aka-hansen.

Hennig, Martin, and Richard Caddell. 2017. "On Thin Ice? Arctic Indigenous Communities, the European Union and the Sustainable Use of Marine Mammals." In *The European Union and the Arctic,* edited by Nengye Liu, Elizabeth A. Kirk, and Tore Henriksen, 296–341. Leiden, The Netherlands: Brill.

Heuer, Thomas, and Patrick Rupert-Kruse. 2015. "Virtuelle Realität als Forschungsfeld. Die Oculus Rift in Forschung und Lehre des Instituts für Immersive Medien." *Jahrbuch immersiver Medien* 7: 77–88.

Hicks, Jack, and Graham White. 2015. *Made in Nunavut: An Experiment in Decentralized Government.* Vancouver: University of British Columbia Press.

Hill, Jen. 2008. *The White Horizon: The Arctic in the Nineteenth-Century British Imagination.* Albany, NY: SUNY Press.

Hjort, Mette. 2005. *Small Nation, Global Cinema: The New Danish Cinema.* Minneapolis: University of Minnesota Press.

Holden, Stephen. 2009. "Ancient Inuit Wisdom as Sustenance in Dire Times." *New York Times,* December 1. https://www.nytimes.com/2009/12/02/movies/02before.html.

Hønneland, Geir. 2020. *Russia and the Arctic Environment, Identity and Foreign Policy,* 2nd ed. London: Bloomsbury.

Huang, Heidi Yu. 2015. "The Hong Jong Dilemma and a Constellation Solution." *Modern Language Quarterly: A Journal of Literary History* 76 (3): 369–91.

Huhndorf, Shari. 2008. "Atanarjuat, The Fast Runner: Culture, History, and Politics in Inuit Media." *American Anthropologist* 105 (4): 822–26.

Hulme, Mike. 2009. *Why We Disagree about Climate Change: Understanding Controversy, Inaction and Opportunity.* Cambridge: Cambridge University Press.

Ihle, Johanne Haaber. 2015. "The Tour: A Film about Longyearbyen, Svalbard. An Interview with Eva la Cour." In *Films on Ice: Cinemas of the Arctic,* edited by Scott MacKenzie and Anna Westerstahl Stenport, 255–65. Edinburgh: Edinburgh University Press.

Ikuma: Carnet de Tournage. 2008. Montréal: Mémoire d'encrier.

imagineNATIVE. 2014. imagineNATIVE 2014 catalogue. https://issuu.com/imaginenative/docs/2158_in_catalog__2014_.

———. 2017. imagineNATIVE 2017 catalogue. https://static1.squarespace.com/static/5711573b044262398e3acb85/t/5bcb19108165f5678d1aa41d/1540036905132/iN_catalogue%282018%29.pdf.

Inuiaat Isaat ("Eyes of the People"). 2015. http://www.inuiaatisaat.com.

Inuit Circumpolar Council. 2009. *A Circumpolar Inuit Declaration on Sovereignty in the Arctic.* https://iccalaska.org/wp-icc/wp-content/uploads/2016/01/Signed-Inuit-Sovereignty-Declaration-11x17.pdf.

ISFI (International Sámi Film Institute). 2021. "Arctic Indigenous Film Fund Announces Ambassadors." https://isfi.no/article/arctic-indigenous-film-fund-announces-ambassadors/.

———. 2022. "Ofelaš—The Pathfinder: Guidelines for Responsible Filmmaking with Sámi Culture and People," revised edition. https://isfi.no/wp-content/uploads/2021/04/OFELAS%CC%8C-FINAL-2022.pdf.

Isuma TV. 2009. "Arnait History: Arnait Video Productions: Voicing a Unique Canadian View." December 8. http://www.isuma.tv/arnaitvideo/arnait-history.

———. n.d. "IsumaTV: World: 7800+ Indigenous Videos in 70+ Languages." http://www.isuma.tv/isumatvworld.

Iversen, Gunnar. 2005. "Learning from Genre: Genre Cycles in Modern Norwegian Cinema." In *Transnational Cinema in a Global North,* edited by Andrew Nestigen and Trevor G. Elkington, 261–78. Detroit, MI: Wayne State University Press.

———. 2019. "A Sámi in Hollywood: Nils Gaup's Transnational and Generic Negotiations." In *Nordic Film Cultures and Cinemas of Elsewhere,* edited by Anna Westerstahl Stenport and Arne Lunde, 157–68. Edinburgh: Edinburgh University Press.

Iversen, Gunnar, and Scott MacKenzie. 2021. "Introduction: Images of Sound and Fury." In *Mapping the Rockumentary: Images of Sound and Fury,* edited by Gunnar Iversen and Scott MacKenzie, 1–17. Edinburgh: Edinburgh University Press.

Jaafar, Ali. 2008. "Greenland Is Getting into the Film Biz." *Variety,* May 16. https://variety.com/2008/film/news/greenland-is-getting-into-the-film-biz-1117985860/.

Jackson, Jerilynn M. 2015. "Glaciers and Climate Change: Narratives of Ruined Futures." *WIREs Climate Change* 6: 479–92.

Johnson, Martin. 2014. "Rolling Programme: Archigram's Cities Walked—This One's Creeping East, Away from the Mine That Feeds It." *RIBA Journal.* https://www.ribaj.com/culture/rolling-programme.

Johnson, Vida. 2014. "Vladimir Tumaev: *White Moss (Belyi iagel',* 2014)." *KinoKultura* 46. http://www.kinokultura.com/2014/46r-bely-yagel.shtml.

Jokela, Timo, Maria Huhmarniemi, Ruth Beer, and Anna Soloviova. 2021. "Mapping New Genre Arctic Art." In *Defining and Mapping the Arctic: Sovereignties, Policies and Perceptions,* edited by Lasse Heininen, Heather Exner-Pirot, and Justin Barnes, 1–16. Akureyri, Iceland: Arctic Portal.

Jones, Nicola. 2012. "Chasing Ice." *Nature Climate Change* 2: 142.

Jonsson, Grimar, Emile Hertling Péronard, Mikkjal Niels Helmsdal, Jákup Veyhe, and Prami Larsen. 2016. *NAT North Atlantic Talents: The North Atlantic Film Collaboration.* Copenhagen: Christensen Grafisk.

Judell, Brandon. 2002. "INTERVIEW: The Nimble Duo behind 'The Fast Runner': Zacharias Kunuk and Norman Cohn." *IndieWire,* June 5. https://www.indiewire.com/2002/06/interview-the-nimble-duo-behind-the-fast-runner-zacharias-kunuk-and-norman-cohn-80364/.

Juhasz, Alexandra, ed. 2001. *Women of Vision: Histories in Feminist Film and Video.* Minneapolis: University of Minnesota Press.

Kääpä, Pietari. 2015. "Northern Exposures and Marginal Critiques: The Politics of Sovereignty in Sámi Cinema." In *Films on Ice: Cinemas of the Arctic,* edited by Scott MacKenzie and Anna Westerstahl Stenport, 45–58. Edinburgh: Edinburgh University Press.

Kaganovsky, Lilya. 2017. "The Negative Space in the National Imagination: Russia and the Arctic." In *Arctic Environmental Modernities: From the Age of Polar Exploration to the Era of the Anthropocene,* edited by Lill-Ann Körber, Scott MacKenzie, and Anna Westerstahl Stenport, 169–82. London: Palgrave-Macmillan.

———. 2019. "'The Threshold of the Visible World': Dziga Vertov's *A Sixth Part of the World* (1926)." In *Arctic Cinemas and the Documentary Ethos,* edited by Lilya Kaganovsky, Scott MacKenzie, and Anna Westerstahl Stenport, 46–67. Bloomington: Indiana University Press.

Kaganovsky, Lilya, Scott MacKenzie, and Anna Westerstahl Stenport. 2019. "What Is the Documentary Ethos and Why Does It Matter for the Arctic?" In *Arctic Cinemas and the Documentary Ethos,* edited by Lilya Kaganovsky, Scott MacKenzie, and Anna Westerstahl Stenport, 1–28. Bloomington: Indiana University Press.

Kalmus, Peter. 2021. "I'm a Climate Scientist. *Don't Look Up* Captures the Madness I See Every Day." *The Guardian,* December 29. https://www.theguardian.com /commentisfree/2021/dec/29/climate-scientist-dont-look-up-madness.

Kara, Selmin. 2016. "Anthropocenema: Cinema in the Age of Mass Extinctions." In *Post-Cinema: Theorizing 21st-Century Film,* edited by Shane Denson and Julia Leyda, 750–85. Brighton, UK: REFRAME: Open-Access and Multimedia Publishing in Media, Film and Music Studies.

Kichin, Valery. 2014. Review: *White Moss. Rossiyskaya Gazeta,* June 24. http:// www.rg.ru/2014/06/24/festival-site.html.

Kinney, D.J. 2013. "Selling Greenland: The *Big Picture* Television Series and the Army's Bid for Relevance during the Early Cold War." *Centaurus* 55: 344–57.

Kinossian, Nadir. 2020. "Svalbard's Haunted Landscapes." *Nordlit* 45: 86–103.

Klein, Naomi. 2015. "'The Right to Be Cold': A Revelatory Memoir That Looks at What Climate Change Means for the North." *Globe and Mail,* March 13. https:// www.theglobeandmail.com/arts/books-and-media/book-reviews/the-right-to-be-cold-a-courageous-and-revelatory-memoir/article23449642/.

Klimenko, Ekaterina, Annika E. Nilsson, and Miyase Christensen. 2019. "Narratives in the Russian Media of Conflict and Cooperation in the Arctic." *SIPRI Insights on Peace and Security* 2019 (5).

Knauff, Markus. 2013. *Space to Reason: A Spatial Theory of Human Thought.* Cambridge, MA: MIT Press.

Körber, Lill-Ann. 2017. "Toxic Blubber and Seal Skin Bikinis, or: How Green Is Greenland? Ecology in Contemporary Film and Art." In *Arctic Environmental Modernities: From the Age of Polar Exploration to the Era of the Anthropocene,* edited by Lill-Ann Körber, Scott MacKenzie, and Anna Westerstahl Stenport, 145–68. London: Palgrave Macmillan.

Kral, Michael J. 2019. *The Return of the Sun: Suicide and Reclamation among Inuit of Arctic Canada.* Oxford: Oxford University Press.

Kramvig, Britt, and Rachel Andersen Gomez. 2019. "From Dreamland to Homeland: A Journey toward Futures Different Than Pasts." In *Arctic Cinemas and the Documentary Ethos,* edited by Lilya Kaganovsky, Scott MacKenzie, and Anna Westerstahl Stenport, 322–34. Bloomington: Indiana University Press.

Kran Film Collective. n.d. "Kran Film Productions." http://kranfilm.net /productions/.

Kremlin. 2018. "Vladimir Putin: Greetings to Arctic Days in Moscow Federal Arctic Forum." November 21. http://en.kremlin.ru/events/president/news/59173http:// en.kremlin.ru/events/president/news/59173.

Krupnik, Igor. 2016. "From Boas to Burch: Transitions in Eskimology." In *Early Inuit Studies: Themes and Transitions, 1850s–1980s,* edited by Igor Krupnik, 1–30. Washington, DC: Smithsonian Institution Scholarly Press.

Kula, Sam. 1979a. "Rescued from the Permafrost: The Dawson Collection of Motion Pictures." *Archivaria* 8: 141–48.

———. 1979b. "There's Film in Them Thar Hills." *American Film* 4 (9): 14–18.

Kuokkanen, Rauna. 2007. "Myths and Realities of Sami Women: A Post-colonial Feminist Analysis for the Decolonization and Transformation of Sami Society." In *Making Space for Indigenous Feminism,* edited by Joyce Green, 73–89. London: Zed Books.

———. 2009. "Indigenous Women in Traditional Economies: The Case of Sámiu Reindeer Herding." *Journal of Women in Culture and Society* 34 (3): 499–504.

———. 2011. "Self-Determination and Indigenous Women: 'Whose Voice Is It We Hear in the Sámi Parliament?'" *International Journal on Minority and Group Rights* 18 (1): 39–62.

———. 2019. *Restructuring Relations: Indigenous Self-Determination, Governance, and Gender.* Oxford: Oxford University Press.

———. 2021. "Ellos Deatnu and Post-state Indigenous Feminist Sovereignty." In *Routledge Handbook on Critical Indigenous Studies,* edited by Brendan Hokowhitu, Aileen Moreton-Robinson, Linda Tuhiwai-Smith, Chris Andersen, and Steve Larkin, 310–23. London: Routledge.

Kuptana, Rosemarie. 1993. "Ilira, or Why It Is Unthinkable for Inuit to Challenge Qallunaat Authority." Paper presented to the Royal Commission on Aboriginal Peoples, April 5, 1993, Ottawa, Ontario, Canada.

Kvidal-Røvik, Trine, and Ashley Cordes. 2022. "Into the Unknown [Amas Mu Vuordá]? Listening to Indigenous Voices on the Meanings of *Disney's Frozen 2* [*Jikŋon 2*]." *Journal of International and Intercultural Communication* 15 (1): 17–35.

KYUK. n.d. "Inside KYUK." https://www.kyuk.org/inside-kyuk.

La Biennale di Venezia. 2019. "Canada: Isuma." https://www.labiennale.org/en/art/2019/national-participations/canada.

Lagutina, Maria L. 2019. *Russia's Arctic Policy in the Twenty-First Century: National and International Dimensions.* Lanham, MD: Lexington Books.

Lajus, Julia. 2103. "In Search of Instructive Models: The Russian State at a Crossroads to Conquering the North." In *Northscapes: History, Technology, and the Making of Northern Environments,* edited by Dolly Jørgensen and Sverker Sörlin, 110–36. Vancouver: University of British Columbia Press.

Lantto, Patrik. 2012. *Lappväsendet: tillämpningen av svensk samepolitik 1885–1971.* Umeå, Sweden: Centre for Sámi Research, Umeå University.

Laruelle, Marlene. 2014. *Russia's Arctic Strategies and the Future of the Far North.* Armonk, NY: M. E. Sharp.

Last, John. 2019. "Hollywood Gets Indigenous Consultation Right in *Frozen 2*, Sami Experts Say." *CBC Canada,* November 23. https://www.cbc.ca/news/canada/north/frozen-2-consultation-sami-1.5370801?fbclid=IwAR3xjqoNn7UqR3W6Y-WEbeC3LaTepn5cnP22tK1l3cYNWahnIuDAh0ENnYA.

Lebow, Alisa. 2012. "Introduction." In *The Cinema of Me: The Self and Subjectivity in First Person Documentary,* edited by Alisa Lebow, 1–15. London: Wallflower Press.

Légaré, André. 1998. "An Assessment of Recent Political Development in Nunavut: The Challenges and Dilemmas of Inuit Self-Government." *Canadian Journal of Native Studies* 16 (2): 271–99.

Lehtola, Jorma. 2000. *Lailasta Lailan.* Inari: Kustannus-Puntsi, 2000.

———. 2014. *Skábmagovat Festival Program 2014.* Inari: Skábmagovat Indigenous Film Festival.

Lehtola, Veli-Pekka. 2000. "The Idea of the Skolt Sámi Territory in the 1930s." In *Den komplexa kontinenten: staterna på Nordkalotten och samerna i ett historiskt perspektiv,* edited by Peter Sköld and Patrik Lantto, 209–15. Umeå, Sweden: Umeå universitet, Institutionen för historiska studier.

———. 2005. *The Sámi People: Traditions in Transitions,* 2nd ed. Fairbanks: University of Alaska Press.

———. 2017. "Vanishing Lapps, Progress in Action. Finnish Lappology and Representations of the Sámi in Publicity in the Early 20th Century." *Arctic and North* 27: 83–102.

———. 2018. "'The Soul Should Have Been Brought Along': The Settlement of Skolt Sami to Inari in 1945–1949." *Journal of Northern Studies* 12 (1): 53–72.

Lewis, Sue, and A. J. Russell. 2011. "Being Embedded: A Way Forward for Ethnographic Research." *Ethnography* 12 (3): 398–416.

Lien, Sigrid. 2018. "Performing Academic Masculinity in the Arctic: Sophus Tromholt and Roland Bonaparte's Photographic Accounts of Sámi Peoples and Northern Landscapes." *Journal of Aesthetics & Culture* 10 (4): 1–17.

Lindroth, Marjo. 2019. "Greenland and the Elusive Better Future." In *Resources, Social and Cultural Sustainabilities in the Arctic,* edited by Monica Tennberg, Hanna Lempinen, and Susanna Pirnes, 15–26. London: Routledge.

Lindstedt, Krister. 2016. "How to Move a City? The Story of Kiruna, Sweden." *reSITE.* https://www.resite.org/talks/how-we-moved-a-city-kiruna-story-by-krister-lindstedt.

Lindstrand, Åsa. 2014. "En Film För De Utsatta Tjejerna." *Samefolket,* February 24. https://samefolket.se/en-film-for-de-utsatta-tjejerna/.

Lipovetsky, Mark. 2019. "Introduction: Fedorchenko's *Angels of Revolution.*" *Studies in Russian and Soviet Cinema* 13 (3): 244–46.

Liu, Nengye, Elizabeth A. Kirk, and Tore Henriksen. 2017. *The European Union and the Arctic.* Leiden, The Netherlands: Brill Nijhoff.

Longfellow, Brenda. 2019. "Political Modernism, Policy Environments, and Digital Daring: The Changing Politics and Practice of Cine-feminism in Quebec." In *The Oxford Handbook of Canadian Cinema,* edited by Janine Marchessault and Will Straw, 307–33. Oxford: Oxford University Press.

Lundmark, Lennart. 1998. *Så länge vi har marker: Samerna och staten under sexhundra år.* Stockholm: Prisma.

Lúthersdóttir, Helga Hlaðgerður. 2015. "Transcending the Sublime: Arctic Creolisation in the Works of Isaac Julien and John Akomfrah." In *Films on Ice: Cinemas of the Arctic,* edited by Scott MacKenzie and Anna Westerstahl Stenport, 325–44. Edinburgh: Edinburgh University Press.

Lyall, Sutherland. 1994. "Obituary: Professor Ron Heron." *The Independent,* October 4. https://www.independent.co.uk/news/people/obituary-professor-ron-herron-1440981.html.

Lyubimov, Boris. 1960. "Bering Strait Dam." Offices of Technical Services, U.S. Department of Commerce. JPRS (Joint Publications Research Service) 2390. [Contains translation of Lyubimov's treatise published in *Literaturnaya Gazeta*, October 24, 1959: 1–2.]

MacDonald, Scott. 2016. "The Filmmaker as Miner: An Interview with Bill Morrison." *Cineaste* (42) 1: 40–43.

MacKenzie, Scott. 2004. *Screening Québec: Québécois Moving Images, National Identity and the Public Sphere.* Manchester, UK: Manchester University Press.

———. 2008. "Life Drawings: Reflection on the Animated Documentary." *POV Magazine* 70: 22–25.

———. 2014. Interview with Graeme Ferguson on *Polar Life. La Vie polaire/Polar Life CINEMAexpo67 Exposition,* La Cinémathèque québéçoise, Montréal. November 1. https://vimeo.com/user35006702.

———. 2015. "The Creative Treatment of Alterity: Nanook as the North." In *Films on Ice: Cinemas of the Arctic,* edited by Scott MacKenzie and Anna Westerstahl Stenport, 201–14. Edinburgh: Edinburgh University Press.

———. 2018. "The Cadaver's Pulse: Cinema and the Modern Prometheus." In *Global Frankenstein,* edited by Carol Margaret Davison and Marie Mulvey-Roberts, 149–66. London: Palgrave.

MacKenzie, Scott, and Anna Westerstahl Stenport. 2013. "All That's Frozen Melts into Air: Arctic Cinemas at the End of the World." *Public* 48: 81–91.

———. 2015. Research interview with Inuk Silis Høegh. Nuuk, Greenland, October 2.

———. 2016. "Arnait Video Productions: Inuit Women's Collective Filmmaking, Coalitional Politics, and Globalized Arctic." *Camera Obscura* 31 (3): 153–63.

———. 2017. "Sumé, Grønland og den sosialhistoriske rockumentaren." *Z filmtidsskrift* 138: 28–39.

———. 2019a. "The Polarities and Hybridities of Arctic Cinemas." In *The Oxford Handbook of Canadian Cinema,* edited by Janine Marchessault and Will Straw, 125–46. Oxford: Oxford University Press.

———. 2019b. "Tearing Up the Screen: Pia Arke's Post-colonial Processes." In *Process Cinema: Handmade Film in the Digital Age,* edited by Scott MacKenzie and Janine Marchessault, 390–406. Montréal: McGill-Queen's University Press.

———. 2019c. "Visualizing Climate Change in the Arctic and Beyond: Participatory Media and the United Nations Conference of the Parties (COP), and Interactive Indigenous Arctic Media." *Journal of Environmental Media* 1 (1): 79–99.

———. 2020. "An Alternative History of the Arctic: The Origins of Ethnographic Filmmaking, the Fifth Thule Expedition, and Indigenous Cinema." *Visual Anthropology Review* 36 (1): 137–62.

———. 2021. "Tracing the Lost Films of the Fifth Thule Expedition in Alaska." *Alaska Journal of Anthropology* 19 (1–2): 160–75.

———. 2023. "The Seal in Arctic Documentary and Global Media since the 1950s: Perspectives on Indigenous Sovereignty, Gendered Geopolitics, and the EU's Seal

Regime." In *Mediated Arctic Geographies,* edited by Johannes Riquet. Manchester, UK: Manchester University Press, forthcoming.

Magomedov, Arbakhan K. 2019a. "Oil Derricks or Reindeer? A Clash of Economics and Traditional Lifeway in Russia's Far North." *The Russia File* blog (Wilson Center, Kennan Institute), February 22. https://www.wilsoncenter.org/blog-post/oil-derricks-or-reindeer-clash-economics-and-traditional-lifeway-russias-far-north.

———. 2019b. "'Where Is Our Land?': Challenges for Indigenous Groups in the Russian Arctic." *The Russia File* blog (Wilson Center, Kennan Institute), November 14. https://www.wilsoncenter.org/blog-post/where-our-land-challenges-for-indigenous-groups-the-russian-arctic.

Manzo, Kate. 2010. "Beyond Polar Bears? Re-envisioning Climate Change." *Meteorological Applications* 17: 196–208.

Marchessault, Janine. 2005. *Marshall McLuhan: Cosmic Media.* London: Sage.

Marcus, Alan Rudolph. 1995. *Relocating Eden: The Image and Politics of Inuit Exile in the Canadian Arctic.* Lebanon, NH: University Press of New England.

Marcus, Daniel, and Selmin Kara. 2015. *Contemporary Documentary.* London: Routledge.

Marcus, George E. 1995. "Ethnography in/of the World System: The Emergence of Multi-sited Ethnography." *Annual Review of Anthropology* 24: 95–117.

Marino, Elizabeth. 2015. *Fierce Climate, Sacred Ground: An Ethnography of Climate Change in Shishmaref, Alaska.* Fairbanks: University of Alaska Press.

Marks, Laura U. 2000. *The Skin of the Film: Intercultural Cinema, Embodiment, and the Senses.* Durham, NC: Duke University Press.

Marmysz, John. 2003. *Laughing at Nothing: Humor as a Response to Nihilism.* Albany, NY: SUNY Press.

Martin-Nielsen, Janet. 2013. "'The Deepest and Most Rewarding Hole Ever Drilled': Ice Cores and the Cold War in Greenland." *Annals of Science* 70 (1): 47–70.

Maxwell, Richard, and Toby Miller. 2012. *Greening the Media.* Oxford: Oxford University Press.

Mazzullo, Nuccio. 2017. "And People Asked: 'We Want to Have Lakes to Fish!' and Lakes Were Given. Skolt Sámi Relocation after WWII in Finland." *Arctic Anthropology* 54 (1): 46–60.

McElroy Ann. 2007. *Nunavut Generations: Change and Continuity in Canadian Inuit Communities.* Long Grove, IL: Waveland Press.

McGinity-Peebles, Adelaide. 2021. "Sakha Film: The History of a Post-Soviet Cultural Phenomenon." *Klassiki,* October 20. https://klassiki.online/sakha-film-the-history-of-a-post-soviet-cultural-phenomenon/.

McGough, Laura. 2019. "Arnait Video Productions: A Conversation with Marie-Hélène Cousineau." *Incite!,* November 5. http://incite-online.net/arnait.html.

McGwin, Kevin. 2020. "How a Collaboration with Disney Shaped the Way Sámi Cultural Details Were Portrayed in *Frozen 2*" *Arctic Today,* February 3. https://www.arctictoday.com/how-a-collaboration-with-disney-shaped-the-way-sami-cultural-details-were-portrayed-in-frozen-2/?wallit_nosession=1.

McLaughlin, Peter. 2010. "Interview: Sealskin Clothing Designer and Lawyer Aaju Peter." *This,* February 17. https://this.org/2010/02/17/aaju-peter-interview/.

McLuhan, Marshall. 1962. *The Gutenberg Galaxy: The Making of Typographic Man.* Toronto: University of Toronto Press.

———. 1964. *Understanding Media: The Extensions of Man.* New York: McGraw-Hill.

———. 1970. *Culture Is Our Business.* Toronto: McGraw-Hill.

McLuhan, Marshall, Eric Larrabee, Erik Barnouw, Robert King Merton, and Richard Schickel. 1966. *McLuhan on McLuhanism.* New York: Educational Broadcasting Corporation. [Television episode, first broadcast May 15, 1966.]

McNeill, J. R. 2016. *The Great Acceleration: An Environmental History of the Anthropocene since 1945.* Cambridge, MA: Belknap Press of Harvard University Press.

McNeill, J. R., and Corinna M. Unger. 2010. "Introduction: The Big Picture." In *Environmental Histories of the Cold War,* edited by J. R. McNeill and Corinna M. Unger, 1–18. Cambridge: Cambridge University Press.

Mecsei, Monica. 2015. "Cultural Stereotypes and Negotiations in Sámi Cinema." In *Films on Ice: Cinemas of the Arctic,* edited by Scott MacKenzie and Anna Westerstahl Stenport, 72–81. Edinburgh: Edinburgh University Press.

———. 2019. "Hybrid First-Person Sámi Documentaries: Identity Construction and Contact Zones in the Twenty-First Century." In *Arctic Cinemas and the Documentary Ethos,* edited by Lilya Kaganovsky, Scott MacKenzie, and Anna Westerstahl Stenport, 302–21. Bloomington: Indiana University Press.

Mikhailova, Tatiana. 2019. "The Myth of Two Goddesses." *Studies in Russian and Soviet Cinema* 13 (3): 255–59.

Miller, Toby, and Marwan M. Kraidy. 2016. *Global Media Studies.* Cambridge, UK: Polity Press.

Millman, Lawrence. 1990. *Last Places: A Journey in the North.* New York: Houghton Mifflin.

Milne, Charles C., ed. 1967. *Expo 67: Official Guide.* Toronto: Maclean-Hunter and the Canadian Corporation for the 1967 World Exhibition.

Minde, Henry. 2003. "The Challenge of Indigenism: The Struggle for Sami Land Rights and Self-Government in Norway 1960–1990." In *Indigenous Peoples: Resource Management and Global Rights,* edited by Svein Jentoft and Ragnar Nilsen, 75–104. Utrecht, The Netherlands: Eburon.

Mitchell, Wendy. 2022. "Netflix Signs Partnership with International Sámi Film Institute." *Screen Daily,* August 31. https://www.screendaily.com/news/netflix-signs-partnership-with-international-sami-film-institute-exclusive/5173946.article?fbclid=IwAR0DbFOjoRXlWaudeNOIVIvTBGrxxCv5uigroOr3H3C3bzvGu1_2Ox_WiZ8.

Montgomery-Andersen, Ruth. 2021. "From Palo to Sumé: The Development of the Modern Greenlandic Film Industry and Nation-Building." In *Arctic Cinemas: Essays on Polar Spaces and the Popular Imagination,* edited by Kylo-Patrick R. Hart, 84–94. Jefferson, NC: McFarland.

Montpetit, Caroline. 2019. "Dans les pas de Gabrielle Roy au Nunavik." *Le Devoir,* October 22. https://www.ledevoir.com/culture/cinema/565297/ entrevue-dans-les-pas-de-gabrielle-roy-au-nunavik.

Mortensen, Bent Ole Gram, and Ulrike Barten. 2016. "The Greenland Self-Government Act: The Pitfall for the Inuit in Greenland to Remain an Indigenous People?" *Yearbook of Polar Law* 8: 103–28.

Morton, Timothy. 2013. *Hyperobjects: Philosophy and Ecology after the End of the World.* Minneapolis: University of Minnesota Press.

Mowat, Farley. 1976. *Canada North Now: The Great Betrayal.* Toronto: McClelland & Stewart.

National Film Board of Canada. 2014. "What Is the National Film Board of Canada (NFB)?" December 3. https://help.nfb.ca/knowledge-base/what-is-the-national-film-board-of-canada-nfb/.

———. 2017a. "Redefining the NFB's Relationship with Indigenous Peoples." http://films.nfb.ca/pdfs/NFB_Indigenous_Peoples_Action_Plan.pdf?_gl=1*jpirno*_ga*MTA3ODcxNjYoMC4xNjMzMjczOTI5*_ga_EP6WV87GN V*MTYzMzI3NDAxNC4xLjAuMTYzMzI3NDAxNC4w.

———. 2017b. "Sharing Our Stories." August 15. http://onf-nfb.gc.ca/en/unikkausivut-sharing-our-stories/the-project/.

Nichols, Bill. 1991. *Representing Reality: Issues and Concepts in Documentary.* Bloomington: Indiana University Press.

———. 1994. *Blurred Boundaries: Questions of Meaning in Contemporary Culture.* Bloomington: Indiana University Press.

———. 2017. *Introduction to Documentary,* 3rd ed. Bloomington: Indiana University Press.

Nilsson, Mats. 1991. *Rebell i verkligheten: Stefan Jarl och hans filmer.* Göteborg: Filmkonst.

NIMMIWG (National Inquiry into Missing and Murdered Indigenous Women and Girls). 2019. *Reclaiming Power and Place: The Final Report of the National Inquiry into Missing and Murdered Indigenous Women and Girls.* https://www.mmiwg-ffada.ca/final-report/#:~:text=Women%20and%20Girls-,Reclaiming%20Power%20and%20Place,women%2C%20girls%20and%202SLGBTQQIA%20people.

NoiseCat, Julian Brave. 2019. "Can Film Save Indigenous Languages?" *New Yorker,* November 14.

North Finland Film Commission. n.d. "Who We Are." https://www.businessoulu.com/en/frontpage-old/en/businessoulu-2/north-finland-film-commission.html.

Norum, Roger. 2016. "Barentsburg and Beyond: Coal, Science, Tourism, and the Geopolitical Imaginaries of Svalbard's 'New North.'" In *Postcolonial Perspectives on the European High North: Unscrambling the Arctic,* edited by Graham Huggan and Lars Jensen, 31–65. London: Palgrave Macmillan.

Norwegian Ministry of Justice and the Police. 2009. *Svalbard: Report no. 22 to the Storting (2008–2009).* Norway: author.

NRK Sápmi. 2022. "Tungt å komme tilbake." October 16. https://www.nrk.no/sapmi/tystnaden-i-sapmi_-tungt-a-komme-tilbake-til-bygda-1.16140951.

Nunatsiaq News. 2013. "Russian Indigenous Activist Arrested, Then Released after Indigenous Conference in Norway." *Nunatsiaq News,* June 14. http://nunatsiaq .com/stories/article/65674russian_activist_arrested_after_indigenous_conference _in_norway/.

Nuttall, Mark. 2017. *Climate, Society and Subsurface Politics in Greenland: Under the Great Ice.* London: Routledge.

Nyyssönen, Jukka. 2009. "The Control of the Cultural Borders between the Finns and the Skolt Sami: From Ambivalence to Respect for the Border." *Journal of Northern Studies* 1: 43–54.

Oppenheimer, Rebecca. 2021. "Hollywood Can Take On Science Denial: *Don't Look Up* Is a Great Example." *Scientific American,* December 30. https://www .scientificamerican.com/article/hollywood-can-take-on-science-denial-dont-look-up-is-a-great-example/.

Ostermann, Hother, and Erik Holtved. 1952. *The Alaskan Eskimos, as Described in the Posthumous Notes of Dr. Knud Rasmussen.* Report of the Fifth Thule Expedition 1921–24, vol. 10, no. 3. Copenhagen: Gyldendal.

Palosaari, Teemu. 2019. "The Arctic Paradox (and How to Solve It): Oil, Gas and Climate Ethics in the Arctic." In *The GlobalArctic Handbook,* edited by Mattias Finger and Lassi Heininen, 141–52. New York: Springer.

Parfitt, Tom. 2009. "Profile: Artur Chilingarov. Russia's Polar Hero." *Science* 324: 1382–84.

Payer, Julius von. 1895. "An Artistic Expedition to the North Pole." *Geographical Journal* 5 (2): 106–11.

Pearce, Warren, Brian Brown, Brigitte Nerlich, and Nelya Koteyko. 2015. "Communicating Climate Change: Conduits, Content, and Consensus." *Wiley Interdisciplinary Reviews: Climate Change* 6 (6): 613–26.

Pearson, Wendy G. 2016. "Memories of Cultural Dismemberment." In *A Companion to Nordic Cinema,* edited by Mette Hjort and Ursula Lindqvist, 377–95. London: Wiley-Blackwell.

Pietikäinen, Sari. 2008. "'To Breathe Two Airs': Empowering Indigenous Sámi Media." In *Global Indigenous Media: Cultures, Poetics, and Politics,* edited by Michelle Stewart and Pamela Wilson, 197–213. Durham, NC: Duke University Press.

Pinney, Christopher. 2011. *Photography and Anthropology.* London: Reaktion.

PlanetFriendly. 2009. "Countdown to Copenhagen—Live from the Floe Edge." May 29. http://www.planetfriendly.net/calendar/events.php?id=10867.

Pollack, Henry. 2012. Letter to the Editor. *New York Times,* April 3, A26.

Poole, Robert M. 2004. *Explorers House: National Geographic and the World It Made.* New York: Penguin.

Porter, John. 1965. *The Vertical Mosaic: An Analysis of Social Class and Power in Canada.* Toronto: University of Toronto Press.

Potter, Russell A. 2007. *Arctic Spectacle: The Frozen North in Visual Culture, 1818–1875.* Seattle: University of Washington Press.

Prokhorov, Alexander. 2019. "Images of 'Posthumous Subjectivity' in Fedorchenko's *Angels of Revolution.*" *Studies in Russian and Soviet Cinema* 13 (3): 260–67.

Propp, Vladimir. [1928] 1971. *Morphology of the Folktale*, 2nd ed. Austin: University of Texas Press.

Putin, Vladimir. 2010. "V. Putin vystupil na forume pro 'Arktika—territoriia dialoga.'" http://www.priroda.ru/news/detail.php?ID=10277.

———. 2013. "Presidential Executive Order Establishing Polar Explorers' Day." Kremlin. May 21. http://en.kremlin.ru/events/president/news/18145.

Qikiqtani Inuit Association. 2013. *Qikiqtani Truth Commission: Thematic Reports and Special Studies 1950–1975*. Iqaluit: Inhabit Media.

Raheja, Michelle H. 2007. "Reading Nanook's Smile: Visual Sovereignty, Indigenous Revisions of Ethnography, and Atanarjuat (The Fast Runner)." *American Quarterly* 59 (4): 1159–85.

Rasmussen, Laura Renée Lucht. 2015. "LYKKELÆNDER: Interview & Portræt." *Magasinet Kunst,* May 2. http://www.magasinetkunst.dk/Home/Detail/NTEzMg%3D%3D.

Renninger, Bryce J. 2011. "The Kickstarting Never Stops: Three Sundance Films Are Looking for Distribution Dough." *IndieWire,* August 12. https://www.indiewire.com/2011/08/the-kickstarting-never-stops-three-sundance-films-are-looking-for-distribution-dough-52837/.

Renov, Michael. 2004. *The Subject of Documentary*. Minneapolis: University of Minnesota Press.

———. 2013. "Art, Documentary as Art." In *The Documentary Film Book,* edited by Brian Winstron, 345–52. London: British Film Institute.

Rickard, Jolene, George Longfish, Zig Jackson, Pamela Shields Carroll, Ron Carraher, and Hulleah Tsinhnahjinnie. 1995. "Sovereignty: A Line in the Sand." *Aperture* 139: 50–59.

Rippl, Gabriele. 1996. "E. A. Poe and the Anthropological Turn in Literary Studies." In *Real: Yearbook of Research in English and American Literature: The Anthropological Turn in Literary Studies, vol. 12,* edited by Jürgen Schlaeger, 223–42. Tübingen: Gunter Narr Verlag Tübingen.

Riquet, Johannes, and Anna Westerstahl Stenport. 2023. "Migrants, Refugees, and Arctic Homelands Mediating Displacement in Cinematic Geographies of the Circumpolar World." In *Mediated Arctic Geographies.* Manchester, UK: Manchester University Press, forthcoming.

Roberts, Tom. 2019. "The Violence of Antiquated Forms: Aleksei Fedorchenko's *Angels of Revolution*." *Studies in Russian and Soviet Cinema* 13 (3): 247–54.

Robin, Libby, Paul Warde, and Sverker Sörlin. 2013. "Introduction: Documenting Global Change." In *The Future of Nature: Documents of Global Change,* edited by Libby Robin, Paul Warde, and Sverker Sörlin, 1–14. New Haven, CT: Yale University Press.

Robinson, Dylan, and Keavy Martin, eds. 2016. *Arts of Engagement: Taking Aesthetic Action In and Beyond the Truth and Reconciliation Commission of Canada.* Waterloo, Canada: Wilfrid Laurier University Press.

Rockström, Johan, Will Steffen, Kevin Noone, Åsa Persson, F. Stuart Chapin III, Eric F. Lambin, Timothy M. Lenton, et al. 2009. "A Safe Operating Space for Humanity." *Nature* 461: 472–75.

Roe, Annabelle Honess. 2013. *Animated Documentary*. London: Palgrave Macmillan.

Roesch, Stefan. 2020. "The Northuldra of Scandinavia." *FilmQuest,* April 3. https://www.filmquest.co/stories/the-northuldra-of-scandinavia/.

Rogatchevski, Andrei. 2020a. "Avant-Garde Artists vs. Reindeer Herders: The Kazym Rebellion in Aleksei Fedorchenko's *Angels of the Revolution* (2014)." In *The Russian Revolutions of 1917: The Northern Impact and Beyond,* edited by Kari Aga Myklebost, Jens Petter Nielsen, and Andrei Rogatchevski, 133–52. Brookline, MA: Academic Studies Press.

———. 2020b. "Svalbard on the (Post-) Soviet Screen." *Nordlit* 45: 150–74.

Rony, Fatimah Tobing. 1996. *The Third Eye: Race, Cinema, and Ethnographic Spectacle*. Durham, NC: Duke University Press.

Ross, Sara. 2012. "Invitation to the Voyage: The Flight Sequence in Contemporary 3D Cinema." *Film History* 24 (2): 210–20.

Roth, Lorna. 2005. *Something New in the Air: The Story of First Peoples Television Broadcasting in Canada*. Montréal: McGill-Queen's University Press.

Rothrock, Drew, Yangling Yu, and Gary Maykut. 1999. "Thinning of Arctic Sea Ice Cover." *Geophysical Research Letters* 26 (23): 3469–72.

Rowe, Elana Wilson. 2013. *Russian Climate Politics: When Science Meets Policy*. London: Palgrave Pivot.

Roy, Gabrielle. 1970. *La rivière sans repos*. Montréal: Librairie Beauchemin.

Rud, Søren. 2017. *Colonialism in Greenland: Tradition, Governance and Legacy*. London: Palgrave Macmillan.

Russell, Catherine. 1999. *Experimental Ethnography: The Work of Film in the Age of Video*. Durham, NC: Duke University Press.

Russell, Curtis. 2018. "*Dawson City: Frozen Time* Review." *Film & History* 48 (2): 14–16.

Ryall, Anka. 2017. "Svalbard in and beyond European Modernity." In *Arctic Modernities,* edited by Heidi Hansson and Anka Ryall, 232–58. Newcastle upon Tyne, UK: Cambridge Scholars.

Saami Council. 2019. "The Sámi Arctic Strategy: Securing Enduring Influence for the Sámi People in the Arctic through Partnerships, Education, and Advocacy." https://static1.squarespace.com/static/5dfb35a66f00d54ab0729b75/t/5e81b3211173962ccof18c71/1585558362311/Samer%C3%A5det+arctic+strategies+Web.pdf.

Sakakibara, Chie. 2018. "Challenging Anti-Sealing Campaigns in the Arctic: *Angry Inuk*." *Visual Anthropology Review* 34 (1): 105.

Sammaattaa.gl (Greenlandic Reconciliation Commission main site). 2017. https://saammaatta.gl/da/Aktiviteter/Projekter/Dokumentarfilm/Inooqataanermi-unammillernartut [link now defunct].

Sand, Stine. 2022. "Dealing with Racism: Colonial History and Colonization of the Mind in the Autoethnographic and Indigenous Film *Sami Blood*." *Journal of International and Intercultural Communication*. https://doi.org/10.1080/17513057.2022.2052156.

Sandemose, Aksel. 2003 [1933]. *En flyktning krysser sitt spor: Fortelling om en morders barndom*. Oslo: Tiden Norsk Forlag.

Sapmi, Yle. 2017. "'Like Terrorists at the Bottom of an Outhouse': Being Sámi in the 100-Year-Old Finland." *Barents Observer,* August 28. https://thebarentsobserver.com/en/life-and-public/2017/08/terrorists-bottom-outhouse-being-sami-100-year-old-finland.

Sarkisova, Oksana. 2019. "Moving through the Century: The Far North in Soviet and Contemporary Russian Nonfiction." In *Arctic Cinemas and the Documentary Ethos,* edited by Lilya Kaganovsky, Scott MacKenzie, and Anna Westerstahl Stenport, 231–53. Bloomington: Indiana University Press.

Schneider, Birgit, and Thomas Nocke. 2014. "Image Politics of Climate Change: Introduction." In *Image Politics of Climate Change: Visualizations, Imaginations, Documentations,* edited by Birgit Schneider and Thomas Nocke, 9–27. Bielefeld, Germany: Transcript Verlag.

Schuppli, Susan. 2020. *Material Witness: Media, Forensics, Evidence.* Cambridge, MA: MIT Press.

Scobie, Willow, and Kathleen Rodgers. 2019. "Diversions, Distractions, and Privileges: Consultation and the Governance of Mining in Nunavut." *Socialist Review* 100 (3): 232–51.

Sejersen, Frank. 2015. *Rethinking Greenland and the Arctic in the Era of Climate Change: New Northern Horizons.* London: Routledge.

Shadian, Jessica M. 2015. *The Politics of Arctic Sovereignty: Oil, Ice, and Inuit Governance.* London: Routledge.

———. 2018. "Indigeneity, Sovereignty, and Arctic Indigenous Internationalism." In *The Routledge Handbook of the Polar Regions,* edited by Mark Nuttall, Torben R. Christensen, and Martin J. Siegert, 331–47. London: Routledge.

Silis Høegh, Inuk. 2015. Berlinale interview. http://www.thesoundofarevolution.com/about/.

Silvén, Eva. 2014. "Constructing a Sami Cultural Heritage: Essentialism and Emancipation." *Ethnologia Scandinavica* 44: 59–74.

Simonpillai, Radheyan. 2019. "Disney Signed a Contract with Indigenous People before Making *Frozen II.*" *Now Toronto,* November 19. https://nowtoronto.com/movies/news-features/disney-frozen-2-indigenous-culture-sami.

Simpson, Audra. 2007. "On Ethnographic Refusal: Indigeneity, 'Voice' and Colonial Citizenship." *Junctures: The Journal for Thematic Dialogue* 9: 67–80.

Skarðhammar, Anne-Kari. 2008. "Changes in Film Representations of Sami Culture and Identity." In *Arktiske Diskurser,* edited by Anka Ryall, Johan Schimanski, and Henning Howlid Wærp. *Nordlit* 23: 293–303.

Slezkine, Yuri. 1994. *Arctic Mirrors: Russia and the Small Peoples of the North.* Ithaca, NY: Cornell University Press.

Smith, Linda Tuhiwai. 1999. *Decolonizing Methodologies: Research and Indigenous Peoples.* London: Zed Books.

Smolka, Harry P. 1937. *Forty Thousand against the Arctic: Russia's Polar Empire.* London: Hutchinson.

Sol. n.d. "The Facts." http://soldocumentary.com/microsite/the-facts/.

Sörlin, Sverker. 1988. *Framtidslandet. Debatten om Norrland och naturresurserna under det industriella genombrottet.* Stockholm: Carlssons.

———. 2015. "Cryo-history: Narratives of Ice and the Emerging Arctic Humanities." In *The New Arctic,* edited by Birgitta Evengard, Joan Nyman Larsen, and Øyvind Paasche, 327–39. New York: Springer.

Spivak, Gayatri Chakravorty. 1988. "Subaltern Studies: Deconstructing Historiography." In *Selected Subaltern Studies,* edited by Ranajit Guha and Gayatri Chakravorty Spivak, 3–34. New York: Oxford University Press.

Steffen, Will, Wendy Broadgate, Lisa Deutsch, Owen Gaffney, and Cornelia Ludwig. 2015. "The Trajectory of the Anthropocene: The Great Acceleration." *Anthropocene Review* 2 (1): 1–18.

Steinberg, Philip E., Jeremy Tasch, and Hannes Gerhardt, eds. 2015. *Contesting the Arctic: Politics and Imaginaries in the Circumpolar North.* London: I. B. Tauris.

Stenport, Anna Westerstahl. 2013. Research Interview with Liselotte Wajstedt. Kiruna, June 4.

———. 2015. "The Threat of the Thaw: The Cold War on the Screen." In *Films on Ice: Cinemas of the Arctic,* edited by Scott MacKenzie and Anna Westerstahl Stenport, 163–77. Edinburgh: Edinburgh University Press.

———. 2019. Research interview with Liselotte Wajstedt. Stockholm, August 26.

———. 2021. "Madonna on Film: Geopolitics, Globalization, and Gender Politics." In *Mapping the Rockumentary: Images of Sound and Fury,* edited by Gunnar Iversen and Scott MacKenzie, 97–112. Edinburgh: Edinburgh University Press.

Stenport, Anna Westerstahl, and Richard S. Vachula. 2017. "Polar Bears and Ice: Cultural Connotations of Arctic Environments that Contradict the Science of Climate Change." *Media, Culture & Society* 39 (2): 282–95.

Storfjell, Troy. 2019. "Elsewheres of Healing: Trans-Indigenous Spaces in Elle-Máijá Apiniskim Tailfeathers' *Bihttoš*." In *Nordic Film Cultures and Cinemas of Elsewhere,* edited by Anna Westerstahl Stenport and Arne Lunde, 278–86. Edinburgh: Edinburgh University Press.

Strukov, Vlad. 2015. "How I Ended This Summer." In *The Contemporary Russian Cinema Reader 2005–2016,* edited by Rimgalia Salys, 330–34. Boston: Academic Studies Press.

———. 2019. "Introduction: Russian Cinema in the Era of Globalization." In *The Contemporary Russian Cinema Reader 2005–2016,* edited by Rimgalia Salys, 9–35. Boston: Academic Studies Press.

Stuhl, Andrew. 2016. *Unfreezing the Arctic: Science, Colonialism, and the Transformation of Inuit Lands.* Chicago: University of Chicago Press.

Sumé Electronic Press Kit. 2015. Courtesy Emile Péronard.

Sumé: The Sound of a Revolution. 2015. Official website. http://www.thesoundofarevolution.com/about/.

Tarasov, Aleksei. 2012. "Narod tol'ko meshaet: vpervye ob etom zaiavleno otkryto." *Novaia Gazeta* 130.

Taveras, Hanna M. 2021. "A Prelude to Postcolonial Cultural Histories of Education: 'Reading'Amanda Kernell's *Sami Blood.*" In *Paedagogica Historica: International Journal of Education* 57: 1–22.

Theroux, Paul. 2000. "Introduction." In *Last Places: A Journey in the North,* edited by Lawrence Millman, vii–ix. New York: Houghton Mifflin.

Thisted, Kirsten. 2011. "Nationbuilding-Nationbranding: Identitetspositioner og tilhørsforhold under det selvstyrede Grønland." In *Fra Vild til Verdensborger: Grønlandsk identitet fra kolonitiden til nutidens globalitet,* edited by Ole Høiris and Ole Marquardt, 597–637. Aarhus, Denmark: Aarhus University.

———. 2013. "Discourses of Indigeneity: Branding Greenland in the Age of Self-Government and Climate Change." In *Science, Geopolitics and Culture in the Polar Region,* edited by Sverker Sörlin, 228–58. Farnham, UK: Ashgate.

———. 2015. "Cosmopolitan Inuit: New Perspectives on Greenlandic Film." In *Films on Ice: Cinemas of the Arctic,* edited by Scott MacKenzie and Anna Westerstahl Stenport, 97–104. Edinburgh: University of Edinburgh Press.

———. 2017a. "Blubber Poetics: Emotional Economies and Post-postcolonial Identities in Contemporary Greenlandic Literature and Art." In *Sámi Art and Aesthetics: Contemporary Perspectives,* edited by Ulla Angkjær Jorgensen, Elin Kristine Haugdal, and Svein Aamold, 267–96. Aarhus, Denmark: Aarhus University Press.

———. 2017b. "The Greenlandic Reconciliation Commission: Ethnonationalism, Arctic Resources, and Post-colonial Identity." In *Arctic Environmental Modernities: From the Age of Polar Exploration to the Era of the Anthropocene,* edited by Lill-Ann Körber, Scott MacKenzie, and Anna Westerstahl Stenport, 231–46. London: Palgrave.

Thomson, Patricia. 2012. "An Award-Winning Document of Climate Change." *American Cinematographer* 93 (12): 20–24.

Thorsen, Isak, and Emile Hertling Péronard. 2021. "'Finally, We're Beginning to Tell Our Own Stories': Filmmaking in Greenland." In *A History of Danish Cinema,* edited by C. Claire Thomson, Isak Thorsen, and Pei-Sze Chow, 263–76. Edinburgh: Edinburgh University Press.

Thunberg, Greta. 2019. *No One Is Too Small to Make a Difference,* 2nd ed. London: Penguin.

To the Arctic Featurette no. 1: Meryl Streep. 2012. Greg MacGillivray, Warner Bros., USA. [DVD extra.]

To the Arctic Featurette no. 4: Challenges of Filming in the Arctic. 2012. Greg MacGillivray, Warner Bros., USA. [DVD extra.]

Toulouze, Eva, Laur Vallikivi, and Art Leete. 2017. "The Cultural Bases in the North: Sovietisation and Indigenous Resistance." In *Ethnic and Religious Minorities in Stalin's Soviet Union: New Dimensions of Research* (Södertörn Academic Studies 72), edited by Andrej Kotljarchuk and Olle Sundström, 199–223. Huddinge, Sweden: Södertörn University.

Truth and Reconciliation Commission of Canada (TRC). 2015. *Canada's Residential Schools: The Inuit and Northern Experience, vol. 2.* Montréal: McGill-Queen's University Press.

Tuck, Eve, and K. Wayne Yang. 2013. "R-words: Refusing Research." In *Humanizing Research: Decolonizing Qualitative Inquiry with Youth and Communities,* edited by Django Paris and Maisha T. Winn, 223–46. New York: Sage.

Turner, Mark David, ed. 2022. *Inuit TakugatsaliuKatiget: On Inuit Cinema.* St. John's, NL: Memorial University Press.

Tynkkynen, Veli-Pekka. 2018. "The Environment of an Energy Giant: Climate Discourse Framed by 'Hydrocarbon Culture.'" In *Climate Change Discourse in Russia: Past and Present,* edited by Marianna Poberezhskaya and Teresa Ashe, 49–52. London: Routledge.

Tysiachniouk, Maria, Andrey N. Petrov, and Violetta Gassiy. 2020. *Benefit Sharing in the Arctic: Extractive Industries and Arctic People.* Basel, Switzerland: MDPI.

United Nations. 2007. *United Nations Declaration on the Rights of Indigenous Peoples.* Adopted by the General Assembly on September 13. https://www.un.org /development/desa/indigenouspeoples/wp-content/uploads/sites/19/2018/11 /UNDRIP_E_web.pdf.

Valkonen, Jarno, Sanna Valkonen, and Tim Ingold. 2019. "On Knowing from the North: Introduction." In *Knowing from the Indigenous North: Sámi Approaches to History, Politics and Belonging,* edited by Thomas Hylland Eriksen, Sanna Valoknen, and Jarno Valkonen, 3–11. New York: Routledge.

Vanstone, Gail, and Brian Winston. 2019. "'This Would Be Scary to Any Other Culture . . . but to Us It's So Cute!' The Radicalism of Fourth Cinema from Tangata Whenua to *Angry Inuk*." *Studies in Documentary Film* 13 (3): 233–49.

Vidal, John. 2013. "Russian Military Storm Greenpeace Arctic Oil Protest Ship." *The Guardian,* September 19. https://www.theguardian.com/environment/2013 /sep/19/greenpeace-protesters-arrested-arctic.

Vizenor, Gerald. 2008. *Survivance: Narrative of Native Presence.* Lincoln: University of Nebraska Press.

Wachowich, Nancy, and Willow Scobie. 2010. "Uploading Selves: Inuit Digital Storytelling on Youtube." *Études/Inuit/Studies* 34 (2): 81–105.

Wajstedt, Liselotte. 2021. "Frispel 2020." *Tidskrift för genusvetenskap* 42 (4): 123–35.

Walsh, Katie. 2015. "LAFF Review: 'White Moss' an Exploration into the Siberian Nenets Tribe, with Drama and Heart."*IndieWire,* June 19. https://www .indiewire.com/2015/06/laff-review-white-moss-an-exploration-into-the-siberian-nenets-tribe-with-drama-and-heart-262759/.

Walsh, Lynda. 2015. "The Visual Rhetoric of Climate Change." *Wiley Interdisciplinary Reviews: Climate Change* 6 (4): 361–68.

Warde, Paul, Libby Robin, and Sverker Sörlin. 2018. *The Environment: A History of the Idea.* Baltimore: Johns Hopkins University Press.

Watt-Cloutier, Sheila. 2005. Petition to the Inter-American Commission on Human Rights Seeking Relief from Violations Resulting from Global Warming Caused by Acts and Omissions of the United States. http://blogs2.law.columbia .edu/climate-change-litigation/wp-content/uploads/sites/16/non-us-case-documents/2005/20051208_na_petition.pdf.

———. 2015. *The Right to Be Cold: One Woman's Story of Protecting Her Culture, the Arctic, and the Whole Planet.* Toronto: Random House.

Waugh, Thomas, Ezra Winton, and Michael Brendan Baker, eds. 2010. *Challenge for Change: Activist Documentary at the National Film Board of Canada.* Montréal: McGill-Queen's University Press.

Weiner, Douglas. 1999. *A Little Corner of Freedom: Russian Nature Protection from Stalin to Gorbachev.* Berkeley: University of California Press.

White Arkitekter. 2014. Press Release. https://whitearkitekter.com/project/kiruna-masterplan/.

———. 2016. "Kiruna 4-ever." *88 Designbox,* May 7. http://88designbox.com/architecture/kiruna-4-ever-by-white-arkitekter-1085.html.

Wijermars, Marielle. 2016. "Memory Politics beyond the Political Domain: Historical Legitimation of the Power Vertical in Contemporary Russian Television." *Problems of Post-Communism* 63 (2): 84–93.

Williams, Linda. 1981. "Film Body: An Implantation of Perversions." *Ciné-tracts* 3 (4): 19–35.

Wilson, Pamela. 2015. "Indienous Documentary Media." In *Contemporary Documentary,* edited by Daniel Marcus and Selmin Kara, 87–104. London: Routledge.

Wilson, Pamela, and Michelle Stewart. 2008. "Introduction: Indigeneity and Indigenous Media on the Global Stage." In *Global Indigenous Media: Cultures, Poetics, and Politics,* edited by Pamela Wilson and Michelle Stewart, 1–38. Durham, NC: Duke University Press.

Windsor, H. H. 1956. "Ocean Dams Would Thaw the North." *Popular Mechanics,* June 6, 135.

Wolfe, Judy. 2019. "Making Films for Her Community: Alethea Arnaquq-Baril." In *Arctic Cinemas and the Documentary Ethos,* edited by Lilya Kaganovsky, Scott MacKenzie, and Anna Westerstahl Stenport, 275–88. Bloomington: Indiana University Press.

World Commission on Environment and Development. 1987. *Our Common Future.* Oxford: Oxford University Press.

Wormbs, Nina. 2013. "Eyes on the Ice: Satellite Remote Sensing and the Narratives of Visualized Data." In *Media and the Politics of Arctic Climate Change: When the Ice Breaks,* edited by Miyase Christensen, Annika E. Nilsson, and Nina Wormbs, 26–31. New York: Palgrave Macmillan.

Wright, Shelley. 2014. *Our Ice Is Vanishing/Sikuvut Nunguliqtuq: A History of Inuit, Newcomers, and Climate Change.* Montréal: McGill-Queen's University Press.

Yamal Region TV. 2013. "The First Feature Film in Nenets Will Be Released at the End of This Year." February 26. https://yamal-region.tv/news/6845/?black=on.

Youngblood, Gene. 1970. *Expanded Cinema.* New York: E. P. Dutton.

Yunes, Erin. 2016. "Arctic Cultural (Mis)Representation: Advocacy, Activism, and Artistic Expression on Social Media." *Public* 57: 98–103.

Yusoff, Kathryn. 2019. *A Billion Black Anthropocenes or None.* Minneapolis: University of Minnesota Press.

INDEX

Hammond, Aleqa, 221
Hans Island, 260–1
Hansen, Aka, 198, 199, 204, 217–18, 221
Hansen, Leo, 79–80
Hawks, Howard, 284
Hearne, Joanna, 22
Heart of Light (*Lysets hjerte,* Denmark,
    1998), 237–38
Helander, Marja, 163
Henriksen, Christina, 306
Henson, Matthew, 211
heritage ruin, 298–99
Herron, Ron, 190
Hertling Péronard, Emile, 198, 222, 225,
    231, 233
High School (USA, 1968), 196
Hinrik's Dream (*Hinnarik Sinnattunilu,*
    Greenland, 2009), 204
Høegh, Inuk Silis, 198–9, 218–9, 222,
    227–28, 231
Høegh, Malik, 224–25, 232
Holmberg, Liisa Elisabet, 158
Hønneland, Geir, 247, 259–60
How I Ended This Summer (*Kak ia provel
    etim letom,* Russia, 2010), 31, 280–86,
    289, 292
Hulme, Mike, 214
Hydro Québec, 46

I Am Greta (*Greta,* Sweden/UK/USA/
    Germany, 2020), 57
ILO-169, 131, 168–9, 201
imagineNATIVE Film + Media Arts
    Festival, 20, 27–28, 105, 109, 159,
    171, 175, 209, 224, 249, 251,
    313–14
IMAX, 29, 37, 41–47, 57
Ingemann, Ulannaq, 198
Ingold, Tim, 188
Innuksuk, Nyla, 314
intercultural cinema, 144
Intergovernmental Panel on Climate
    Change (IPCC), 8, 213, 274
International Humane Society, 113
International Sámi Film Institute (ISFI),
    17, 20–21, 25–26, 117, 140, 142–43, 153,
    155, 157–61, 164, 170–71, 203, 210, 305,
    313, 315

Interstellar (USA, 2014), 2, 301–302
interventionist historiography, 13–15, 19,
    23, 29–31, 98, 115–55, 163, 166–67,
    169–70, 173, 180–81, 183, 185, 191,
    196, 219, 224, 225–30, 241, 306–307,
    309–310, 316
Inuiaat Isaat ("Eyes of the people"), 231
Inuit Broadcasting Corporation (IBC), 21,
    63–65, 67, 93, 108–109, 199
Inuit Circumpolar Council (ICC), 14, 72
Inuit Nunaat (album, 1974), 225
Issaittuq (Canada, 2011), 106
Isuma (Igloolik Isuma Productions), 20,
    26–27, 29, 59–88, 91, 93, 99, 105–106,
    108–109, 113–14, 156, 203
Ivalu, Madeline Piujuq, 29, 90, 92, 94, 96,
    100–102

Ja de boðii dulvi ("And Then Came the
    Flood," Finland/Norway/Sweden,
    1976), 118
Jante law (*Jantelagen*), 237
Jarl, Stefan, 30, 120, 122–27, 151
Jåvna: Reindeer Herdsman in the Year 2000
    (*Jåvna—Renskötare år 2000,* Sweden,
    1991), 123
Joli-Coeur, Claude, 66
Jørgensen, Pipaluk K., 199, 209, 211, 233
Jorinda's Journey (*Jorindas resa,* Sweden,
    2014), 315
Journals of Knud Rasmussen, The (Canada/
    Denmark, 2006), 69, 79
Julien, Isaac, 211
Juvenile Court (USA, 1973), 196

Kaganovsky, Lilya, 243
Kaisa's Enchanted Forest (*Kuun metsän
    Kaisa,* Finland, 2016), 130, 156, 164,
    175–84, 311
Kalvemo, Johs., 118
Kautokeino Rebellion, The (*Kautokeino-
    opprøret,* Norway, 2008), 13, 129, 132–33,
    154, 162
Kautokeino uprising, 128–29, 131, 132, 135
Kennedy, John F., 283–84
Kernell, Amanda, 30, 130, 139–54, 306
Khrushchev, Nikita, 276, 284
Kill Buljo (Norway, 2007), 162

Founded in 1893,
UNIVERSITY OF CALIFORNIA PRESS
publishes bold, progressive books and journals
on topics in the arts, humanities, social sciences,
and natural sciences—with a focus on social
justice issues—that inspire thought and action
among readers worldwide.

The UC PRESS FOUNDATION
raises funds to uphold the press's vital role
as an independent, nonprofit publisher, and
receives philanthropic support from a wide
range of individuals and institutions—and from
committed readers like you. To learn more, visit
ucpress.edu/supportus.